# Interactions between Herbicides and the Soil

# Interactions between Herbicides and the Soil

*Edited by* R. J. Hance

Agricultural Research Council
Weed Research Organization
Yarnton
Oxford

1980

EWRS

Published for the
European Weed Research Society

ACADEMIC PRESS

*A Subsidiary of Harcourt Brace Jovanovich. Publishers*

London New York Toronto Sydney San Francisco

ACADEMIC PRESS INC. (LONDON) LTD.
24/28 Oval Road
London NW1

*United States Edition published by*
ACADEMIC PRESS INC.
111 Fifth Avenue
New York, New York 10003



British Library Cataloguing in Publication Data

Interactions between herbicides and the soil.
1. Herbicides 2. Soil chemistry
I. Hance, R. J.
II. European Weed Research Society
631.4′1 SB951.4 80-41233

ISBN 0-12-323840-4



Printed in Great Britain by
Page Bros (Norwich) Ltd

# Contributors


C. F. VAN DE BUND

*Plant Protection Service, Geertjesweg 15, 6706 EA Wageningen, The Netherlands*

R. CALVET

*Institut National de la Recherche Agronomique, Station de Science du Sol, Route de St-Cyr, 78000 Versailles, France*

H. EIJSACKERS

*Research Institute for Nature Management, Kemperbergerweg 67, 6816 RM Arnhem, The Netherlands*

M. P. GREAVES

*Agricultural Research Council, Weed Research Organization, Begbroke Hill, Yarnton, Oxford OX5 1PF, UK*

J. A. GUTH

*CIBA-GEIGY Ltd, Agricultural Division, PO Box CH-4002, Basle, Switzerland*

R. J. HANCE

*Agricultural Research Council, Weed Research Organization, Begbroke Hill, Yarnton, Oxford OX5 1PF, UK*

K. HURLE

*Institut für Phytomedizin, Universität Hohenheim, Postfach 106, 7000 Stuttgart 70, F. R. Germany*

M. LEISTRA

*Institute for Pesticide Research, Marijkeweg 22, 6709 PG Wageningen, The Netherlands*

H. P. MALKOMES

*Biologische Bundesanstalt für Land- und Forstwirtschaft, Institut für Unkrautforschung, 3300 Braunschweig, F.R. Germany*

W. PESTEMER

*Biologische Bundesanstalt für Land- und Forstwirtschaft, Institut für Unkrautforschung, 3300 Braunschweig, F.R. Germany*

R. R. SCHMIDT

*Bayer AG, Sparte Pflanzenschutz, Anwendungstechnik, Biologische Forschung, 5090 Leverkusen-Bayerwerke, F.R. Germany*

L. TORSTENSSON
*Swedish University of Agricultural Sciences, Department of Microbiology, 5-750 07 Uppsala 7, Sweden*

A. WALKER
*National Vegetable Research Station, Wellesbourne, Warwick CV35 9EF, UK*

# Preface

The behaviour of herbicides in the soil has been studied now for several decades. This book is an attempt by members of the Herbicide-Soil Working Group of the European Weed Research Society to summarize the available information and then suggest where the most important gaps lie with respect to agricultural practice, legislation and academic science.

At first, work on this subject was designed to provide background information with the object of allowing compounds that act through the soil to be used more reliably and efficiently. Later, a second aim was added which can perhaps be described as "safety" in the general rather than in the specifically agricultural context. The need for this became apparent as we learned more about the fate of agricultural chemicals in the environment and became aware of the possibilities for both parent compounds and their metabolites to affect non-target organisms and ecological systems. These developments quite properly were drawn to the attention of politicians and government officials and demands for further information followed. In turn these political pressures have led to laws to control pesticide use in many countries and in most others legislation is pending or is being seriously considered.

Herbicides are essential to Western European agriculture as it is currently practised. Other systems are technically possible but would require substantial economic and social change so that in the medium term they are unlikely to be politically acceptable. The desirability or otherwise of changes of this nature are outside the scope of this book but whether or not they occur there will still be a need for many years to use herbicides.

Scientists are often accused of failing to contribute adequately to policy discussions. This book is not primarily intended as a contribution in this respect but the Group hopes that by setting down our assessment of current knowledge of what happens to herbicides in the soil and sometimes speculating on the possibilities for future developments we may improve the basis for rational thought.

This may be an embarrassing exercise for some of us who have been working in the field for many years. For example, are we able to use soil-applied herbicides more efficiently now than we were in 1960, when the European Weed Research Council, the forerunner of EWRS, first sponsored a book on the subject (Woodford and Sagar, 1960). If so, is this because of a better understanding of the physics, chemistry and biology involved, or is it simply a consequence of empirical experience?

We measure a wide range of herbicide characteristics such as solubility, volatility and adsorption, we construct laboratory systems to measure leaching, degradation rates and metabolic pathways but how reliable are these measurements and how valuable are they in predicting what will happen in the farmer's field? The registration authorities in some countries require data derived from the same sort of tests but can this information be interpreted validly to assess safety and efficiency or are the requirements merely based on available experimental techniques regardless of their relevence? The chapters of this book attempt to review our knowledge in relation to these questions. This naturally leads to consideration of the major deficiencies in our methodology and the possibilities for improvement both from the agricultural and legislative viewpoints.

Problems of interpretation and doubts concerning the validity of information also occur in the academic context. Often it seems that in order to make a measurement we use a system that is so artificial that the results cannot be extrapolated to natural conditions, sometimes not even to other experimental systems. Another difficulty is that in some cases we do not know what is the normal range of a "natural" soil property so that we cannot decide whether or not a herbicide-induced perturbation is of any significance. This uncertainty has clear implications for the farmer and the lawyer as well. We hope that this book will draw attention to these limitations.

However, all is not ignorance and doubt. There are a number of areas in which our information is based on sound observation and can be interpreted with confidence. We will try to identify these topics.

This book deals first with what could be termed the processes, physical, chemical and biological, which act upon a herbicide in the soil, and then goes on to consider the biological consequences of the presence of herbicides in the soil with respect both to the crop and the soil.

Adsorption and desorption influence almost all the other processes occurring in the soil and, for this reason, in the first chapter Dr Calvet reviews the factors which control sorption, essentially from a theoretical viewpoint, after which he goes on to discuss the experimental aspects of the subject. The next two chapters are concerned with the movement of herbicides in the soil and thus continue the physico-chemical theme. Dr Leistra discusses transport in solution and lays particular emphasis on the role of mathematical models while in Chapter 3 I deal with transport in the vapour phase.

The subjects then become more biological. In Chapter 4 Drs Hurle and Walker consider the kinetics of the decomposition of herbicides in the soil and the factors involved before assessing the prediction of persistence in the field. Dr Guth in the next chapter then discusses the nature of the transformations that occur and the methods for studying them, while Dr Torstensson in Chapter 6 outlines the part played by micro-organisms in the processes of conversion.

Up to this point the chapters describe the factors which determine the distribution, persistence and availability of a herbicide. Chapter 7 by Drs Schmidt and Pestemer introduces the plant into the discussion with a survey of the routes and mechanisms by which herbicides enter plants from the soil. This is followed by Dr Walker's chapter on activity and selectivity to crops and weeds, a subject central to the successful agricultural use of herbicides. The next two chapters are concerned with effects of herbicides on soil inhabitants other than plants, Mr Greaves and Dr Malkomes deal with micro-organisms and Drs Eijsackers and van de Bund with soil fauna. They concentrate on interpreting such effects in the context of agriculture, forestry and range management. The final chapter is a synthesis of the conclusions and recommendations which emerge from the earlier sections of the book. Since this summary is the most important section, French and German versions are also included.

The aim of this book is to be useful to the active scientist, the research administrator and to those concerned with regulation. For the scientist there will be an evaluation of techniques and pointers to areas of high potential research interest. These things should interest the administrator too but we will also try to put the various subjects in a perspective that will be helpful in assessing research priorities. To the legislator we hope to show what is scientifically valid and hence legally useful but also to indicate clearly what is uncertain or irrelevant.

Though each chapter is attributed to only one or two authors they have all been reviewed by the Group as a whole in order to reduce to a minimum personal prejudices without, we hope, inhibiting the ventilation of controversial issues. Errors, of course, remain the responsibility of the editor.

## Reference

Woodford E. K. & Sagar G. R. (1960) *Herbicides and the Soil.* Blackwells Scientific Publications, Oxford.

*R. J. Hance*
*November 1980*

# Contents

# 1 Adsorption–Desorption Phenomena

*R. Calvet*
Institut National de la
Recherche Agronomique
Station de Science du Sol
Versailles France



Adsorption of herbicides by soils and their constituents has been reviewed by several authors (Hamaker and Thompson, 1972; Guenzi, 1974; Kearney and Kaufman, 1975; Calvet *et al.*, 1980). The aim of this chapter is to recall the main features of adsorption–desorption phenomena and to draw attention to some points which are not well understood. At the same time aspects which may be interesting to pursue in the future will be emphasized.

Most herbicides have low vapour pressures and even in water unsaturated soils there are generally several layers of water molecules on the solid surfaces. Therefore, the dominant phenomenon is adsorption at the solid–water interface. Adsorption at the solid–gas interface will not be described in this chapter, details can be found in de Boer (1953) or Brunauer (1945). However, this type of adsorption may play an important part in the movement of herbicides in the gaseous phase (see Chapter 3).

## I. Description of Adsorption and Desorption

### A. Amount Adsorbed at Equilibrium

#### 1. *Neutral molecules*

(*a*) *Adsorption isotherms.* "A complete description of adsorption processes requires consideration of both solute and solvent behaviour" (Kipling, 1965). However, for herbicides the mole fraction in saturated solution rarely exceeds $10^{-4}$ so knowledge of solute concentrations only is sufficient for the process to be described, usually as an adsorption isotherm (the relation between amount adsorbed and solution concentration). This does not mean that water molecules do not play a part in adsorption because they may be concerned in solute-adsorbent interactions and in solute-water exchanges.

Adsorption of a single gas on a solid has been extensively described (de Boer, 1953; Brunauer, 1945; Fripiat *et al.*, 1971) and adsorption isotherms have been derived from physical theories. Unfortunately, this is not the case for adsorption from aqueous solutions and we have only numerical descriptions of isotherms. It is thus difficult to infer physical or chemical mechanisms from these descriptions. The most general description is that proposed by Giles *et al.* (1960) and is of the form of the Langmuir equation

$$Q = \frac{k\,K_L\,C}{1 + kC} \qquad (1)$$

where $k = \omega C^{\beta}$, $Q$ = amount adsorbed/unit mass of adsorbent, $C$ = equilibrium concentration, $K_L$, $\omega$, $\beta$ = constants.

This relation allows a description of the various kinds of adsorption isotherms according to the value of $\frac{\mathrm{d}Q}{\mathrm{d}C}$ and of $\beta$. When $k$ is constant, (1) is the classical Langmuir equation: however, this has been used very little in herbicide adsorption studies so far.

For dilute solutions, that is for low values of $C$, we have

$$Q \simeq K_L \omega\, C^{(\beta + 1)} \quad \text{or} \quad Q = K_f C n_f \qquad (2)$$

which is the Freundlich relation. However this relation only has a physical significance for the adsorption of gases on solids, in which case, the constant $n_f$ is less than one. Here it can take any values less or greater than one.

Relation (1) may be expressed by a polynomial the first term of which is (2). This has been done by Lambert (1967), but the more frequently used expression is (2).

It is interesting to consider further the constant, $n_f$. Its value is sometimes unity or nearly unity so that (2) describes a linear isotherm. This is a very simple situation which greatly facilitates the description of adsorption and its introduction in transport analysis. It is only in this case that a distribution coefficient $K_d$ has a physical significance. Table I gives the mean values of $n_f$ taken from several publications. It shows that isotherms are often not linear, except to a first approximation.

(*b*) *Desorption isotherms*. Hysteresis is frequently observed and the use of desorption isotherms has been proposed (Savage and Wauchope, 1974;

Table I. Some average values of $n_f$ coefficients determined from published results.

| Herbicide | $n_f \pm 2\,\sigma$ | Number of observations |
|---|---|---|
| Atrazine | 0·91 ± 0·04 | 29 |
| Prometryn | 0·81 ± 0·02 | 9 |
| Diuron | 0·74 ± 0·02 | 27 |
| Linuron | 0·85 ± 0·08 | 17 |
| Monuron | 0·78 ± 0·10 | 17 |
| Picloram | 0·87 ± 0·06 | 19 |
| Terbacil | 1·05 ± 0·16 | 12 |
| 2,4-D | 0·97 ± 0·06 | 7 |

$\sigma$ = standard error.

Hornsby and Davidson, 1973; Swanson and Dutt, 1973; Farmer and Aochi, 1974). Figure 1 gives a typical example. It could be due either to modification of the adsorbents occurring during the shaking period or to relatively low rates of desorption. It may also be the result of errors caused by the decomposition of the solute during the experiment. Van Genuchten *et al.* (1974) have tentatively proposed that desorption isotherms could be described with a Freundlich type equation. However, before developing mathematical descriptions, research is needed to define the true nature of

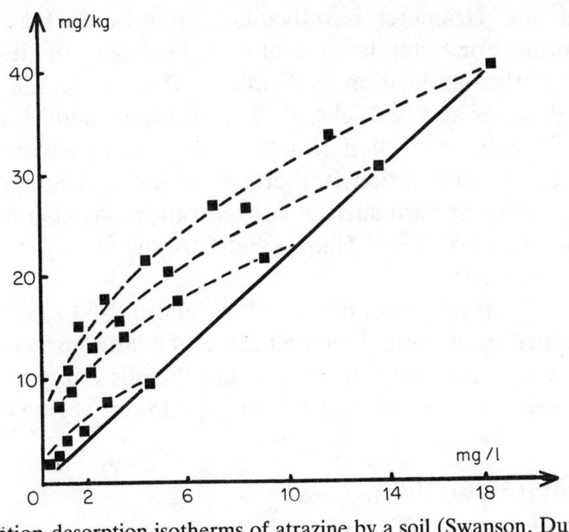


Fig. 1. Adsorption-desorption isotherms of atrazine by a soil (Swanson, Dutt, 1973). Reproduced from Soil Sci. Soc. Am. Proc 37, (6) p. 874 by permission of the Soil Science Society of America.

what is called the “hysteresis of desorption”. This is important since transport phenomena and biological activity of herbicides may be closely controlled by desorption.

### 2. *Charged molecules*

Some herbicide molecules carry electrical charges. These are either permanent, as in the case of quaternary ammonium compounds, or result from ionization processes such as protonation of weak bases (e.g.: triazines) or by release of a proton from weak acids (e.g.: 2,4-D, picloram).

Charged molecules may be adsorbed by ion exchange with ions adsorbed on the soil constituents. Adsorption isotherms of electrolytes have been extensively studied and may be analysed in several ways (Helferich, 1962; Babcock, 1963; Fripiat *et al.*, 1971). However, these theories have not been applied to the adsorption of charged herbicide molecules. This is probably difficult to do for weak bases and weak acids but it presents no *a priori* difficulties for permanently charged molecules.

## B. Thermodynamic Characteristics

In the general case the exchange of solute water molecules has to be taken into account in the description of the thermodynamics of adsorption (Calvet *et al.*, 1980). Derivation of expressions for the enthalpy of adsorption requires the introduction of the solute surface concentration (Fu *et al.*, 1948; Biggar and Cheung, 1973; Moreale and Van Bladel, 1979). Since the evaluation of this parameter is difficult, it may be useful to deal with simpler situations. For example, an estimate of enthalpy of adsorption may be obtained by the application of Van't Hoff's law to the distribution coefficient (Leenheer and Ahlrichs, 1971; Van Bladel and Moreale, 1974). However, this is only valid when no solute–water molecule exchanges take place and when the adsorption isotherm is linear. Extrapolation of heat of adsorption values at zero surface concentration has also been used by Haque and Sexton (1968) and Haque and Coshow (1971) to characterize adsorption phenomena.

Values of the heat of adsorption are low, about 8–33 kJ/mol. It is very difficult to deduce from these data the nature of solute surface interactions and any conclusion must be tentative because the possibility of a molecular exchange between the solvent and the solute can not yet be excluded.

## C. Kinetic Characteristics

Since most published work deals with equilibrium situations, little information exists about the kinetics of adsorption–desorption phenomena.

Kinetic studies on soil–herbicide systems are generally made by measuring the composition of the solution phase. The observed variations of composition with time are controlled by two phenomena: molecular diffusion into the adsorbent (soil aggregates); rate of adsorption at the surface. We do not know which of these is rate limiting.

A general expression for the rates of adsorption and desorption may be written in the following way (Fripiat *et al.*, 1971).

$$\frac{dn_s}{dt} = \begin{bmatrix}\text{number of}\\ \text{encounters}\\ \text{between the}\\ \text{molecules}\\ \text{and the}\\ \text{surface}\end{bmatrix} \times \begin{bmatrix}\text{efficacy of}\\ \text{encounters}\end{bmatrix} \times \begin{bmatrix}\text{sticking}\\ \text{probability}\end{bmatrix} \times \begin{bmatrix}\text{temperature}\\ \text{factor}\end{bmatrix}$$

$$\frac{dn_d}{dt} = E' \times \begin{bmatrix}\text{desorption}\\ \text{probability}\end{bmatrix} \times \begin{bmatrix}\text{temperature}\\ \text{factor}\end{bmatrix}, E' = \text{constant}$$

where $n_s$ is the number of adsorbed molecules and $n_d$ the number of desorbed molecules. This approach has been used by Lindstrom *et al.*, (1970).

Usually, more concise relations are used which state that the rate of adsorption is proportional to either the concentration in solution ($C$)

$$\frac{dQ_a}{dt} = k_a C \tag{3}$$

or to the amount which remains available for adsorption

$$\frac{dQ_a}{dt} = k'_a (Q - Q_\infty) \tag{4}$$

where $Q_a$ = adsorbed quantity, $Q_\infty$ = maximum adsorbed quantity, $k_a$ and $k'_a$ = rate constants for adsorption.

Expression (4) was used by Swearingen and Dickinson (cited in Kipling, 1965) and also by Lindstrom and Boersma (1970).

In the same way the rate of desorption is written as proportional to the quantity in the adsorbing phase

$$\frac{dQ_d}{dt} = k_{\text{des}} Q_a \tag{5}$$

where $Q_d$ = desorbed quantity, $k_{\text{des}}$ = rate constant for desorption.

These expressions of the rate of adsorption and desorption are generally used in the modelling of non-equilibrium transport (see Chapter 2).

Adsorption equilibrium is attained more or less rapidly, 4 to 24 h in general (Hance, 1967a). This holds for suspensions but may not be valid for systems with low water contents.

The curve representing the variations of quantities adsorbed with time, often shows two parts (Haque *et al.*, 1968; Haque and Sexton, 1968; Leehneer and Ahlrichs, 1971; Khan, 1973a). This is illustrated in Fig. 2. The first part could be due to adsorption of molecules on accessible surfaces and the second part could correspond with molecular diffusion into the adsorbent particles.

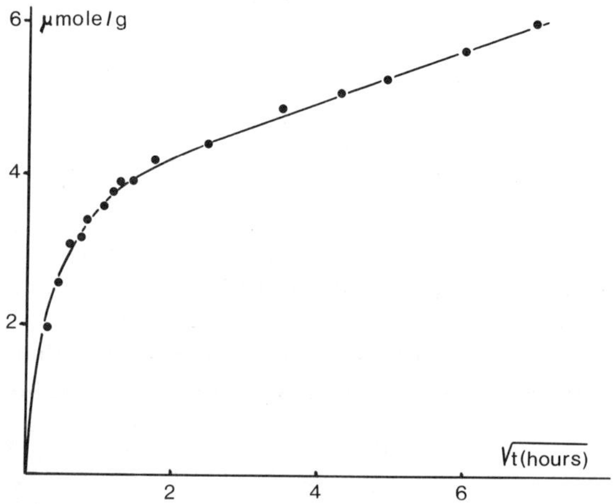


Fig. 2. Adsorption kinetics of 2,4-D on humic acids (Khan, 1973).

Measurements of temperature effects on the rate constant allow activation energy values to be determined. These are used to predict the variation of adsorption rate with temperature, but are not a reliable means of describing the mechanisms of adsorption. Values of activation energies of the adsorption rate have been reported for picloram on humic acids (Khan, 1973a), carbaryl on organic matter (Leehneer and Ahlrichs, 1971) and 2,4-D on clay minerals (Haque *et al.*, 1968). They range from 6·3 to 45·8 kJ/mole.

These values are of the same order of magnitude as values of enthalpy of adsorption. This indicates that solute–surface interactions may be similar to those undergone by molecules during their transport from the solution to the adsorbed phase.

Desorption is generally slower than adsorption, but its kinetic characteristics are less well known. The low rate of desorption leads to difficulties in the interpretation of herbicide–soil interactions since long time periods may be necessary to reach equilibrium.

## D. Conclusions from the Descriptive Aspects

Some theoretical treatments for describing adsorption exist but they are of limited value for describing herbicide behaviour in soil, mainly because of

(i) the relatively low solubility of most herbicides,
(ii) the lack of knowledge about adsorption mechanisms,
(iii) the complexity of the soil–water system.

It is possible to describe adsorption–desorption phenomena with a Freundlich type equation. However, it is misleading to give physical meaning to the constants so determined. Thermodynamic and kinetic studies are useful to predict the effect of temperature on either equilibrium or rate characteristics but have only provided physico-chemical data that can be interpreted when model adsorbents are used.

# II. Interactions Involved in Adsorption

Only a short description of the interactions involved in adsorption is given here. More details on the possible bonds between organic molecules and organic and mineral adsorbents can be found in several reviews such as those of Mortland (1970), Calvet and Chassin (1973), Theng (1974), Hayes *et al.*, (1975).

Adsorbate-adsorbent bonds belong to two categories:

(i) high energy bonds (>80 kJ/mole): ionic bonds, ligand exchange,
(ii) low energy bonds (<80 kJ/mole): ion–dipole and dipole–dipole interactions, hydrogen bonds and London–Van der Waals bonds.

## A. High Energy Bonds

### 1. *Ionic bonds*

These bonds occur between organic anions or cations and positive or negative electric charges located at the adsorbent surface. Some herbicide molecules are always in the cationic form (paraquat, diquat) while others can be ionized, the ionization depending on the acidity of the medium. These compounds are either weak acids (e.g. phenoxyacetic acids) or weak bases (e.g. triazines). Adsorption by ionic bonds is due to ion exchange and has been studied essentially only for permanently charged molecules (Weber *et al.*, 1965; Knight and Tomlinson, 1967; Knight and Denny, 1970).

### 2. *Ligand exchange*

Such interactions are possible with the water molecule usually being the exchange ligand. Infrared spectroscopy, suggests that this can be the case for the adsorption of aminotriazole on $Cu^{2+}$ and $Ni^{2+}$ montmorillonites (Russell *et al.*, 1968a), for the adsorption of linuron on montmorillonite (Khan, 1974a) and for the adsorption of EPTC on montmorillonite (Mortland and Meggitt, 1966).

Ligand exchange was also postulated for the binding of 1,3,5-triazines onto the residual transition metals of humic acids (Hamaker and Thomson, 1972) for the adsorption of linuron on clays (Hance, 1971) and for the adsorption of terbutryne on aluminum saturated clays (Terce and Calvet, 1975).

## B. Low Energy Bonds

### 1. *Charge–dipole bonds and dipole–dipole bonds*

Electric charges on adsorbents and compensating ions can attract polar molecules, which may compete for adsorption with water molecules. This type of bonding has not been studied in depth for herbicides.

### 2. *Hydrogen bonds*

These are mainly associated with N–H and O–H groups and N and O atoms so that all herbicide molecules are able to establish hydrogen bonds. No very convincing experimental proof of the part they play has yet been furnished and these bonds are always postulated from the properties of adsorbed molecules. In spite of this limitation, two kinds of hydrogen bonds have been described,

(i) between adsorbed water molecules and adsorbed organic molecules,
(ii) between surface groups and organic molecules.

The first type has been suggested as a mechanism for the adsorption on montmorillonite of carbamates (Mortland and Meggitt, 1966; Bailey *et al.*, 1968) and atrazine (Calvet and Terce, 1975a). The second has been proposed for a variety of systems by Dieguez Carbonell and Rodriguez Pascal (1975), Van Bladel and Moreale (1974), Biggar and Cheung (1973), Li and Felbeck (1972), Sullivan and Felbeck (1968).

### 3. *Charge transfer bonds*

These are due to the transfer of electrons between the surface of the adsorbent and the organic molecule. Structures containing either $\pi$ bonds or lone pairs of electrons can give rise to such interactions. They were proposed for bipyridilium cations adsorbed on clays (Haque *et al.*, 1970; Khan, 1973b, 1973c) on spectroscopic evidence. There is an increase in the wavelength of maximum u.v. absorption and a decrease in the wavelength of the out-of-plane C–H vibration in the i.r. spectrum. Charge transfer bonds were also postulated for the adsorption of methylthiotriazines on organic matter (Hayes, 1970).

#### 4. *London–Van der Waals bonds*

These are due to dispersion forces and they probably exist with all herbicides. They have not been specifically studied for pesticides.

#### 5. *Entropy generation*

Water molecules may be displaced when organic molecules interact with the surface of adsorbents. The stability of the system with adsorbed molecules is thus increased by the resulting change in entropy (Hamaker and Thompson, 1972). The expression "hydrophobic bond" has sometimes been used for this phenomenon but this may be misleading since it is a consequence not a cause of adsorption.

### C. Conclusions from Studies of Interactions

Solute–surface interactions are complex and involve a variety of types of bond. For example, paraquat molecules are adsorbed through ionic bonds and charge transfer bonds, whereas adsorption of substituted ureas is probably through charge–dipole bonds and/or ligand exchange and hydrogen bonds.

Such a complicated situation makes the comprehensive study of herbicide adsorption difficult. Beside ionic bonds, other bonds can only be identified easily when u.v. or i.r. spectroscopy gives unambiguous results. The numerous descriptions in the literature are usually best considered only as hypotheses at the moment.

## III. Factors Controlling Adsorption

Adsorption depends on three factors which often interact in a complicated way: the molecular properties of the solute, soil constitution and the experimental conditions under which the adsorption is studied.

### A. Molecular Properties

The molecular properties of particular importance in solute–surface interactions are: electronic structure, molecular volume and water solubility.

### 1. *Electronic structure*

The presence of charges is an important characteristic as it permits adsorption by ion exchange. As previously noted, some molecules bear permanent charges (paraquat, diquat), some may be ionized according to the acidity of the medium (triazines) and others are neutral (ureas). Ionization depends on molecular structure and Weber (1967) has discussed this in detail for 1,3,5-triazines. The possibility for a molecule to be ionized or neutral explains why herbicide–soil interactions may vary greatly depending on the pH of the soil.

Electron distribution and electron mobility can be expressed respectively by polarity and by polarizability which greatly condition the various interactions described in Section II.

Several properties have been used to characterize electronic structure, particularly, the $\sigma$ Hammett function and the Taft function (Hamaker and Thompson, 1972). Briggs (1969) studied several substituted ureas and a homologous series of alkyl-*N*-phenyl carbamates on four soils, and showed the existence of relationships between the quantity adsorbed and values of the Hammett and Taft functions.

### 2. *Molecular volume*

On the assumption that molecular volume is a factor controlling adsorption, Lambert (1967) studied the relationship between the parachor and the equilibrium adsorption constant. He found a linear relationship for several 2,6-dinitro-4-alkyl-sulphonyl-*N*,*N*-dialkylamines and phenylureas. Hance (1969a) proposed a modification in order to take into account the influence of hydrogen bonding. However Briggs (1969) did not consider such relationships to be of general validity.

### 3. *Water solubility*

Since adsorption in saturated soils takes place in an aqueous medium, interactions between water and solutes are of particular importance. The complexity of adsorption mechanisms, however, leads to a situation which is confusing to study and where it is difficult to find a general rule. Several workers have tried to establish a relationship between adsorption and water solubility, but the results obtained are contradictory. For example, with substituted ureas, either no correlation has been found (Briggs, 1969; Hance, 1965) or a decrease of adsorption with increasing solubility (Hilton and Yuen, 1963; Bailey *et al.*, 1968; Grover, 1975). Other contradictions occur with the triazines when comparing the observations of Bailey *et al.*, (1968) Talbert and Fletchall (1965) and Ward and Holly (1966).

Beside water solubility, some other properties have been used in order to relate adsorption to the hydrophylic character of molecules. This has been done by measuring partition coefficients between water and an organic solvent. Briggs (1969) used the Hansch $\pi$ function (determined from partition between 1-octanol and water) and found a relationship with the adsorption of alkyl-*N*-phenylcarbamates. Ward and Holly (1966) observed a linear variation of log $K_d$ for triazines against the log of the hexane–water partition coefficient. Similarly Hance (1967b) using reverse-phase thin layer chromatography found a good linearity between chromatographic movement and log $K_f$.

### 4. *General comments*

Theoretically, a complete knowledge of electronic structure would enable prediction of the adsorption behaviour of herbicides. This is not the case since several types of interactions act simultaneously with a given surface. It is always possible to obtain more or less empirical relationships between adsorption and molecular properties but these must be considered rather as predictive tools than complete explanations.

## B. Soil Constitution

### 1. *Adsorption on soil constituents*

(*a*) *Soil minerals.* Clays are the most important soil minerals when considering adsorption because of their abundance and their surface properties. These characteristics vary according to mineralogical nature, so different types of clays possess different adsorptive properties. Table II gives examples for montmorillonite, kaolinite and illite clays.

Adsorption on clays is also greatly conditioned by the ionic composition of the surface. Compensating cations can act in three ways.

(1) by competition for adsorption sites with positively charged organic molecules. This has been observed for paraquat (Weber and Weed, 1968; Hayes *et al.*, 1975) and for atrazine (Hance, 1969b; Vallet *et al.*, 1973).

(2) by acting directly as adsorption sites through the formation of coordination bonds. Table III gives some examples together with the effects of the nature of the cation. Cations can act indirectly via their hydration water molecules as it is well known that these water molecules are responsible for the surface acidity (Mortland *et al.*, 1963; Calvet *et al.*, 1964). Organic molecules may thus establish a variety of bonds with the surface; for example

| surface | $(M^{n+})$—O(—H)—H . . . . . B; | surface | $(M^{n+})$—$O^-$(—H) $HB^+$ |
|---|---|---|---|
| | hydrogen bond | | ionic bond |

Table II. Characteristics of the adsorption of some herbicides by clay minerals

| Herbicide | Clay | Freundlich constants $n_f$ | Freundlich constants $K_f$ | Authors |
|---|---|---|---|---|
| Atrazine | M-Ca; pH = 2·8-3·0 | 0·78 | 65 | Terce and Calvet (1978) |
| M > I | I-Ca; pH = 2·8-3·0 | 0·35 | 7·2 | Terce and Calvet (1978) |
| Terbutryne | M-Ca; pH = 2·8-3·0 | 1·60 | 3065·0 | Terce and Calvet (1978) |
| M > K > I | K-Ca; pH = 2·8-3·0 | 1·60 | 447·0 | Terce and Calvet (1978) |
| | I-Ca; pH = 2·8-3·0 | 1·20 | 199·0 | Terce and Calvet (1978) |
| Tri-allate | M-Na; pH = 9·9 | 0·99 | 10·0 | Grover (1974) |
| K > M | K-Na; pH = 7·1 | 0·97 | 16 | Grover (1974) |
| 2, 4-D | M[a]; pH = 6·5 | 1·0 | 9 | Haque *et al.* (1968) |
| I > M > K | K[a] | 1·0 | 0·43 | Haque *et al.* (1968) |
| | 1[a] | 1·0 | 13·8 | Haque *et al.* (1968) |
| Trifluralin | M-Na; pH = 9·5 | 1 | 35 | Grover (1974) |
| K > M | K-Na; pH = 6·7 | 0·96 | 65 | Grover (1974) |

Letters M, K, I stand respectively for montmorillonite, kaolinite, illite.
Classification is indicated under the herbicide name and corresponds to a decreasing adsorption.
[a] Authors have not stated the nature of compensating cations.

Under these conditions, the electronegativity of the cation is an important characteristic as has been shown by Saltzman and Yariv (1975).

(3) cations such as $Al^{3+}$ and $Fe^{3+}$ can form hydroxides on the clay surface. This results in a great increase of the adsorption capacity of the mineral (Terce and Calvet, 1977). Thus, any weathering process will be able to modify adsorption by clays.

Crystallized oxides and hydroxides are in general poor adsorbents (Terce and Calvet, 1975) except for compounds which are weak acids (Cheung, 1973). Amorphous oxides and hydroxides are efficient adsorbents when associated with clays, and they are probably responsible for some adsorption properties of the clay fraction but in the form of free compounds they are also able to adsorb some herbicides. This has been shown for some triazines and substituted ureas by Terce and Calvet (1975, 1977) and for glyphosate by Hance (1976).

(*b*) *Soil organic constituents*. Soil organic matter is not yet sufficiently well characterized to allow the relationships between its structure and its properties to be defined exactly. Adsorption of herbicides varies greatly according to the nature of organic matter as is illustrated by the results obtained for trifluralin (Grover, 1974) and for atrazine (Dunigan and MacIntosh, 1971).

Table III. Effect of compensating cations on the adsorption of several pesticides

| Herbicides | Classification of cations according to decreasing adsorption | Authors |
|---|---|---|
| Linuron | $Fe^{3+} > Ce^{4+} > Cu^{2+} > Ni^{2+} > Ca^{2+}$ | Hance (1971) |
| Linuron | $Al^{3+} > Cu^{2+} > Ni^{2+} > Mg^{2+}$ | Khan and Mazurkewich, (1974) |
| Picloram | $Cu^{2+} > Al^{3+} > Zn^{2+} > Fe^{3+}$ | |
| Atrazine | $Fe^{3+} > Al^{3+} > Ca^{2+}$ | Terce and Calvet (1977) |
| Terbutryn | $Fe^{3+} > Al^{3+} > Ca^{2+}$ | Terce and Calvet (1977) |
| Methabenzthiazuron | $Al^{3+} > Fe^{3+} > Ca^{2+}$ | Terce and Calvet (1977) |

It appears that both humic substances and non-humified organic materials may play various parts so that it is difficult to establish general rules.

The large surface area and the chemical nature of humic substances can explain their high adsorptive properties. Carboxylic and hydroxylic groups are important (Li and Felbeck, 1972) as they can form ionic bonds with ionized molecules (Burns *et al.*, 1973a; Khan, 1974b) and participate in cation–dipole bonds (Khan and Mazurkevitch, 1974). The aromatic nature of soil organic matter is probably also important in the formation of charge transfer complexes (Hayes, 1970). Finally hydrogen bonds are certainly possible either directly between solute and adsorbent or indirectly through hydration water molecules.

A last, but very important point, is that the properties of organic materials with respect to adsorption are not constant and vary as a result of the action of soil chemical and biological processes. Table IV shows two examples of such a variation during humification.

(*c*) *Organo-mineral association.* Humified compounds are always associated with minerals. The exact nature of these associations and their properties is not well understood, but it has been observed that $Fe^{3+}$ and $Al^{3+}$ ions and their hydrous-oxides could be very important for the bonding of humic substances to soil minerals (Greenland, 1971; Schnitzer and Khan, 1972; Orlov *et al.*, 1973). The main conclusion is that the properties of these associations are not simply related to those of the constituents (Table V).

(*d*) *Other remarks*. The properties of soil constituents and molecular properties of the solute are complementary in controlling the adsorption of herbicides. We have to note that methods of extraction of these constituents from the soil do not permit us to be sure that the observed properties are always those which exist in the soil. Nevertheless, studies on simple materials can give interesting information about herbicide-adsorbent interactions.

Mineral and organic soil constituents are not stable and undergo various transformations as they age. As a consequence, when describing the properties of soil materials, it is necessary to take into account this evolution especially in the case of organic matter. This situation, plus the fact that mineral and organic constituents are frequently associated, explains why it is difficult to predict soil adsorption behaviour simply from the gross soil composition.

### 2. *Adsorption by soils*

In spite of these considerations it is useful from a practical point of view, to try to relate the adsorption of herbicides to the soil composition. This

Table IV. Variations of adsorption properties with the nature of organic matter

| State of evolution | Peat (Morita, 1976) Freundlich constant values | Vegetal material incorporated in soil (Gaillardon *et al.*, 1978) Distribution coefficient values | | | |
|---|---|---|---|---|---|
| | | Wheat straw | | lucerne | |
| | Linuron | Diuron | Terbutryne | Diuron | Terbutryne |
| Initial state | $K_f = 24$; $n_f = 0{\cdot}922$ | 9·3 | 20 | 8·3 | 36 |
| Humified product | $K_f = 297$; $n_f = 0{\cdot}671$ | 16·3 | 28 | 6 | 8 |

Table V. Adsorption by calcium humate-montmorillonite associations

| Herbicide | Adsorbent: Montmorillonite-Ca % | Adsorbent: Humate Ca % | pH | Adsorption characteristics | Authors |
|---|---|---|---|---|---|
| Atrazine | 100 | 0 | 7·8 | $K_d = 11{\cdot}7$ | Hance (1969b) |
| | 0 | 100 | 7;4 | $K_d = 59$ | |
| | 85 | 15 | 7·6 | $K_d = 14$ | |
| Terbutryn | 100 | 0 | 3-3·2 | $K_f = 0{\cdot}011$; $n_f = 2{\cdot}74$ | Gaillardon *et al.* (1977) |
| | 0 | 100 | 3-3·2 | $K_f = 0{\cdot}71$; $n_f = 0{\cdot}90$ | |
| | 75 | 25 | 3-3·2 | $K_f = 0{\cdot}38$; $n_f = 1{\cdot}51$ | |

may be done by observing the effect of the elimination of a constituent or by correlating adsorption characteristics with soil composition.

(*a*) *Elimination of constituents*. Walker and Crawford (1968) noted both increases and decreases in the adsorption of triazine herbicides following substantial extraction of the soil organic matter. When several treatments are used (to extract both organic matter and inorganic compounds) results are more complicated. The interpretation of such observations must take into account several factors. For example extraction is never complete and residual material can still be adsorbent. Also, treatments can act more by their effects on organo–mineral or mineral–mineral associations than by the elimination of a constituent and they can affect the properties of remaining materials.

(*b*) *Correlation between adsorption and soil composition*. Clay and organic matter content have been the variables most emphasized in this type of study whereas oxide content, specific surface area, exchange capacity and pH have been less frequently considered. In general, adsorption characteristics ($K_d$, $K_f$, $n_f$) are not correlated with clay content except for cationic molecules. This is due (at least partly) to the definition of the clay fraction on the basis of particle size which does not take into account the chemical nature of surfaces. For example, hydroxides associated with clays play an important part but they are not considered in these correlations.

The majority of the published work deals with organic matter content. Its role is very important, but the situation is confused when one looks for a relation between organic matter content and adsorption characteristics. Using a large number of published figures, Calvet *et al.* (1980) studied the correlation between $K_d$, $K_f$, $n_f$ and organic matter content. Their conclusions are

(i) $K_d$ increases as the organic matter content increases;

(ii) $n_f$ values are not correlated to the organic matter content;

(iii) $K_d$ values are influenced more by organic matter for substituted ureas than for triazines, whereas the reverse is true for $K_f$ values;

(iv) the correlation coefficient relating $K_f$ to organic matter is high when the range of organic matter content is large, but it is often small over the range 0–4% which includes most agricultural soils in France and similar temperate regions.

Table VI shows the results for atrazine and diuron.

In order to produce a more universal characteristic, distribution coefficients calculated on the basis of the organic carbon content of the soil have been proposed. As Hamaker and Thompson (1972) have noted, the variability of this new constant is generally less than the constants calculated on a total soil basis, but, nevertheless, the variability remains large enough to render it unsuitable for precise analysis. Similarly, Lambert (1968) has used an empirical correction by introducing an efficiency factor for the organic matter. This factor was calculated relative to a standard soil and implies that organic matter in this soil has the same properties as in the other soils studied.

Table VI. Relations between adsorption constants $K_f$ and the soil organic matter content

| Molecule | Organic matter content < 4% | | | Full range of organic matter content | | |
|---|---|---|---|---|---|---|
| | *n* | regression | *r* | *n* | regression | *r* |
| Atrazine | 23 | (1·7) + (0·6) OM% | 0·11 | 29 | (8) + (7·4) OM% | 0·93 |
| Diuron | 17 | (3·4) + (2·9) OM% | 0·92 | 27 | (0·2) + (4·1) OM% | 0·98 |

$n$ = number of observations
$r$ = correlation coefficient

Thus the situation is far from clear and this is due to our lack of knowledge of soil organic matter. In particular, most published works do not take into account such factors as, the origin of organic matter, the possible modifications of a small active part of the organic matter by biological activities, associations with mineral materials and the possible effects of experimental conditions. Consequently, correlations should be used cautiously for prediction purposes.

## C. Characteristics of the Medium in Which Adsorption Occurs

Besides the nature of adsorbent and that of the herbicide, adsorption depends on several characteristics of the experimental system. These include: temperature, ionic composition of the solution and soil water ratios.

### 1. *Temperature*

As the temperature is raised adsorption may increase, decrease or remain constant. Table VII shows some published results which illustrate this. It is not possible to propose a quantitative explanation, but a qualitative interpretation can be suggested. The variation of temperature has two effects, (i) on surface–solute interactions and (ii) on water–solute interactions. It is the balance between these two effects which determines the

Table VII. Effect of the temperature on adsorption by organic materials, minerals and soils

| Authors | Adsorbent | Herbicide | Effect of temperature |
|---|---|---|---|
| Harris and Warren (1964) | montmorillonite (pH 8,5) | Simazine | – |
| Haque and Sexton (1968) | illite | 2,4-D | – |
| | montmorillonite | 2,4-D | 0 |
| Hayes *et al.* (1975) | montmorillonite Na | Paraquat | 0 |
| | vermiculite Na | Paraquat | + |
| Harris and Warren (1964) | peat | Monuron | 0 |
| | | Simazine | 0 |
| | | Atrazine | 0 |
| | | 2,4-D | 0 |
| MacGlamery Slife (1966) | humic acid | Atrazine | + |
| Dunigan and MacIntosh (1971) | humic acid lignin | Atrazine | + |
| Li and Felbeck (1972) | humic acid | Atrazine | + |
| Weber *et al.* (1965) | charcoal | Prométon | 0 |
| | | 2,4-D | – |
| Talbert and Fletchall (1965) | soils | 1,3,5- triazines | – |
| Sethi and Chopra (1975) | soils | Alachlor | – |
| Farmer and Aochi (1974) | soils | Picloram | – |
| Harvey (1974) | soils | 12 dinitroanilines | – |

0 = no effect
– = the adsorption decreases as the temperature increases.
+ = the adsorption increases as the temperature increases.

observed behaviour and this may result in adsorption increasing, decreasing or remaining unaffected as a consequence of a change in temperature.

## 2. *Composition of the solution*

(*a*) *Organic solvents.* Pure organic solvents have been frequently used in studies of adsorption because they permit the use of herbicide solutions of high concentration. This procedure does not give information about adsorption from aqueous solutions. However, adsorption from mixed organic–water solutions can provide interesting observations. For example, the use of mixtures containing hydroxylic solvents (ethanol, diethylene-glycol) and nonhydroxylic solvents (acetone, dimethylsuloxide) provides information about possible hydrogen bonding (Calvet and Terce, 1975a).

(*b*) *Ionic composition of the aqueous phase.*

(i) Acidity of the solution. The acidity of an adsorbent system is usually measured by the pH of the solution in contact with the solid phase. It is important to note that the colloïd surface will be more acid than this, due to the surface acidity effect (Mortland *et al.*, 1963).

Figure 3 shows the possible variations of the adsorption with acidity. Curves of type A are observed with weak bases adsorbed as electrically neutral molecules (e.g. adsorption of atrazine and simazine on charcoal, Yamane and Green, 1972). Curves B and C are observed in three circumstances.

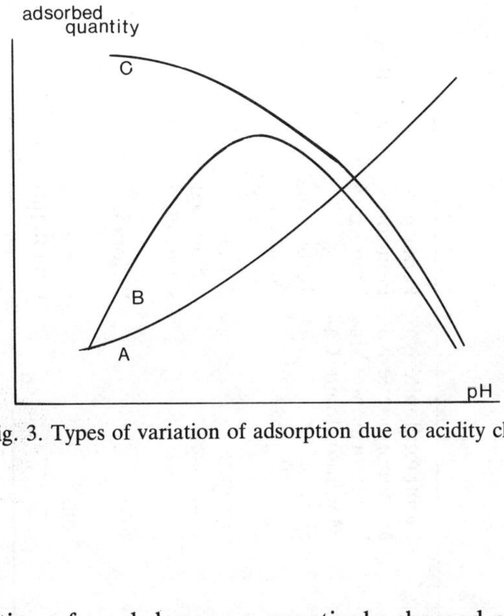


Fig. 3. Types of variation of adsorption due to acidity changes.

(1) adsorption of weak bases on negatively charged adsorbents (clays, humic acids).

(2) adsorption of weak acids on positively charged adsorbents (oxides and hydroxides).

(3) adsorption of neutral molecules on clays.

When adsorption is at a maximum the corresponding pH is sometimes near the $pK_a$ value. However this is not a general rule and is probably coincidental.

Various observed effects may be explained on the basis of the effect of hydrogen ions on solute molecules and on the adsorbents. Protons cause

Table VIII. Effects of salts on adsorption

| Authors | Adsorbent | Herbicide | Salt effect: ionic strength <1 | Salt effect: ionic strength >1 | Experimental conditions | |
|---|---|---|---|---|---|---|
| Van Bladel (1971) | Montmorillonite Na | Fenuron | − | + | NaCl | |
| Van Bladel and Moreale (1974) | Montmorillonite Na | Fenuron | 0 or − | + | NaCl | identical concentrations |
| | Montmorillonite Mg | Monuron | | | $MgCl_2$ | |
| | Montmorillonite Ca | | | | $CaCl_2$ | |
| Gaynor (1973) | Soils | Terbutryne | − | + | NaCl | identical ionic strength |
| | | Atrazine | | | KCl | |
| | | | | | $CaCl_2$ | |
| Watson *et al.* (1973) | Goethite | 2,4-D | − | + | NaCl | |
| Hurle and Freed (1972) | Soils | 1,3,5-triazines | + | | KCl | identical ionic strength |
| | | substituted ureas | | | $NH_4Cl$ | |
| | | | | | $CaCl_2$ | |
| Farmer and Aochi (1974) | Soils | Picloram | + | | KCl | identical concentrations |
| | | | | | $CaCl_2$ | |
| Abernathy and Davidson (1971) | Soils | Prometryne | + | | NaCl | identical concentrations |
| | | Fluometuron | − | | $CaCl_2$ | |

0 = no effect.
− = the adsorption decreases as the ionic strength increases.
+ = the adsorption increases with the ionic strength.

conformational modifications of humic substances and hydrolyse the clay lattices. Hydrolysis of clays brings $Al^{3+}$ and $Fe^{3+}$ ions to the surface which becomes more or less covered by hydroxides. Because such compounds are highly adsorbent, they may frequently be responsible for observed increases in adsorption at low pH.

(ii) Dissolved salts. When cationic organic molecules are present, competition with mineral cations may occur, as we have seen. With neutral molecules the situation is more complicated (Table VIII). For ionic strengths greater than one, adsorption is increased, probably due to a salting-out effect. For ionic strength less than one, a range of behaviour is observed and so far no generally satisfactory explanation seems to have been proposed.

(iii)Formulation additives. Several papers report studies on the effect on adsorption of additives included in formulated products (Bayer, 1967; Huggenberger *et al.*, 1973; Yamane and Green, 1972; Kanellopoulos, 1974; Morita, 1976). The results are contradictory and it is not possible to draw any definite conclusion from them.

### 3. *Soil : water ratio*

The water content of the system can influence adsorption in two ways. It can modify the aggregation of adsorbents and increase or decrease the accessibility of surface to the solute (Grover and Hance, 1970). The water content can also affect the physico-chemical properties of the adsorbent. In acid suspensions, the hydrolysis of clay lattices increases as the adsorbent:solution ratio decreases (Frenkel and Suarez, 1977) and it has been shown by Calvet and Terce (1975b) that both dissolution and adsorption were dependent on the water content of clay pastes. Even in the absence of effects on aggregation, adsorption generally increases as the water content decreases (Savage and Wauchope, 1974; Farmer and Aochi, 1974; Hance, 1977).

### 4. *General comments*

Examination of the effects of the characteristics of the medium on adsorption raises several questions which have not yet received complete answers. Soil constituents are very often fragile materials and the herbicide solutions used can modify their properties so that an adsorption study may be performed on an adsorbent other than the one intended. This is particularly the case for clays in acid conditions.

An important point is that adsorption is usually studied in suspensions while herbicides are used in soils which have relatively low water contents. There is a gap in our knowledge and future research should be oriented towards this problem.

## IV. Methods Used in Adsorption Studies

### A. Preparation of Adsorbents

Some preliminary remarks on the preparation of adsorbents are appropriate before describing the various methods used in adsorption studies. Adsorption is a surface phenomenon and thus depends upon the area of the solid–liquid interface, the accessibility of the surface and the physico-chemical properties of the surface.

Any treatment which modifies these factors can also modify adsorption. This could be the case for the ionic saturation treatments which are frequently used to prepare mineral or organic materials. In particular, it is necessary to be very careful when using acid treatments, as they can hydrolyse clay lattices and can profoundly alter the structure of humic substances. In general one should be aware of possible changes in properties of adsorbents following any modifications of the ionic composition.

Dried soil is often used and the effect of drying on adsorption behaviour needs study because it is known that dehydration increases clay aggregation and decreases the swelling of organic materials.

These considerations could explain some contradictory results found in the literature. It is probable that many of these apparent contradictions are simply a consequence of the fact that the test adsorbents are not strictly comparable because of differences in pretreatments.

### B. Determination of the Quantity Adsorbed

The quantity adsorbed is usually determined from the change in concentration of the solution after contact with the adsorbent. There are several ways to do this and details are given by Calvet *et al.* (1980).

#### *1. Determination from adsorbent–solution mixtures*

Various adsorbent : solution ratios (from 25 to $6 \times 10^{-5}$) have been used. It is not possible to recommend one particular value for this ratio but it is necessary to keep in mind that;

(i) it partly determines the area of the solid liquid interface and thus affects the amount adsorbed (Grover and Hance, 1970).

(ii) its variations may modify the physico-chemical properties of the surface (Vallet *et al.*, 1973; Terce and Calvet, 1979).

(iii) it affects the precision of the measurements since it determines the magnitude of the change of the solute concentration resulting from adsorption.

Methods differ in the way that the liquid phase is separated from the solid. They include centrifugation (the most frequently used) pressure extraction (Burns *et al.*, 1973b; Savage and Wauchope, 1974; Hance and Embling, 1979) and gel filtration (Khan, 1973b; Burns *et al.*, 1973b).

### 2. *Determination from flow experiments*

One method is to equilibrate a column of adsorbent by percolation of a solution and then to extract the pesticide by a suitable organic solvent (Green and Corey, 1971). Another method is to monitor concentrations in the effluent obtained from a column of adsorbent. With these procedures, a numerical analysis is needed in the general case and this is a solute transport problem as discussed by Leistra in Chapter 2. A simple situation can, however, be dealt with if it is assumed that the adsorption isotherm is linear, with instantaneous equilibrium, no hysteresis, and with no stationary component in the liquid phase. Under these conditions the adsorption constant may be calculated from the position of the elution peak (Swoboda and Thomas, 1968). Grice and Hayes (1972) suggested a different type of flow experiment with suspensions, using dialysis cells. Many difficulties may arise including those due to the microporous membrane, which may have adsorptive properties and poor mechanical resistance, and those related to the kinetics of adsorption if equilibrium is not attained instantaneously.

### 3. *General comments*

The current situation is not satisfactory for two reasons. First, there is a need for unambiguous methods for adsorption–desorption studies at soil water contents similar to those existing in the field. Second, it is often difficult to compare experimental results found in the literature because there is no standard procedure for sample preparation and for the measurements. There is thus a need for continued methodological research in order to be able to propose comparable experimental procedures for adsorption studies.

## C. Characterization of Adsorbed Molecules

Characterization of adsorbed molecules can lead to information about adsorbate–surface interactions and permit the detection of chemical transformations. Absorption spectroscopy is one of the most interesting techniques for this type of study but unfortunately it is only in special cases that absorption spectra can be recorded. There are very few examples of u.v. studies but i.r. studies are more numerous. They have usually been limited to adsorption on clay minerals and very infrequently on soils because of the technical difficulties encountered. Infra-red spectra are often used to study clays that have been treated with organic solutions of herbicides, a procedure which greatly limits their interest. Furthermore they are difficult to interpret due to uncertainties in assigning the absorption bands. However, apart from these difficulties such an approach has sometimes provided interesting results; for example, Russell *et al.* (1968b), Brown and White (1969), Mortland and Meggitt (1966) have obtained useful data on herbicide–surface bonds.

X-ray diffraction has been used to provide information about the localization of adsorbed molecules of paraquat, which can occupy interlamellar positions in clays (Knight and Denny, 1970) but otherwise has not been applied to herbicide studies.

Much important information has been obtained from u.v., i.r. and X-ray studies in other fields of soil science. These methods may have some use in increasing the knowledge of herbicide–adsorbent interactions in model systems such as pure clays or pure organic materials. With soils however, the situation is so complex that only in exceptional cases are these methods likely to give any useful practical information.

## V. Conclusions and Suggestions for Future Research

Adsorption–desorption phenomena occupy a central position in controlling the distribution of herbicides in soils. They influence concentrations in the soil solution and thus are closely involved in transport and absorption by living organisms. The distribution of a herbicide between the liquid (or the gas) phase and the adsorbed phase affects the phytotoxicity and selectivity of herbicides, their effects on non-target organisms and the risk of surface and ground water pollution.

Academically much research is necessary to obtain a better comprehension of adsorption–desorption phenomena. Further understanding of the relationship between the electronic structure of molecules and adsorption behaviour would facilitate formulation of general rules which could be useful for predicting solute–surface interactions. More complete information about the role played by soil constituents is necessary. Soil minerals should be characterized by other criteria than the "clay content" given by mechanical analysis. The role played by organic matter, is of great interest and information about the activity of different fractions and the effects of weathering and humification processes would be especially welcome. Careful work is needed to determine whether or not there is hysteresis in desorption since the consequences for the dynamic aspects of herbicide behaviour in soils are very important.

From a practical point of view convenient methods are needed to describe the behaviour of herbicides in the soil. There are two possibilities. The first is to determine adsorption distribution coefficients at low concentrations (say 1 μg/ml or less) so that the isotherms are linear or near linear. These coefficients are easily obtained from measurements in suspension but the results depend on the soil : water ratio and they do not give any information about desorption. Furthermore, the validity of these coefficients for describing field behaviour remains uncertain at the moment. The second is to use organic solvent/water partition coefficients and convert them to distribution coefficients in soils calculated from the correlation between adsorption by soil organic matter and water : octanol partition ratios. This may be useful for general prediction purposes but further studies are needed to check the validity of this approach and to determine its limits.

The relationships of laboratory measurements to field studies require critical study. More information about the influence of soil water content on the distribution of solute between the solution and the solid is of particular importance.

Finally, from both agronomic and environmental considerations we need a better understanding of the quantitative relationships of adsorption–desorption phenomena with transport and biological activity.

## References

Abernathy, J. R. and Davidson, J. M. (1971). Effect of calcium on prometryn and fluometuron adsorption in soil. *Weed Sci.*, **19**, 517–521.

Babcock, K. L. (1963). Theory of chemical properties of soil colloïdal systems at equilibrium. *Hilgardia*, **34**, 417–542.

Bailey, G. W., White, J. L. and Rothbert, T. (1968). Adsorption of organic herbicides by montmorillonite, role of pH and chemical character of adsorbate. *Soil Sci. Soc. Am. Proc.*, **32**, 222–234.

Bayer, D. E. (1967). Effect of surfactants on leaching of substituted urea herbicides in soils. *Weeds*, **15**, 249–252.

Biggar, J. W. and Cheung, M. W. (1973). Adsorption of picloram, 4-Amino-3, 5, 6-trichloropicolinic acid on Panoche, Ephrata and Palouse soils. A thermodynamic approach to the adsorption mechanism. *Soil Sci. Soc. Am. Proc.*, **37**, 863–868.

Briggs, G. G. (1969). Molecular structure of herbicides and their sorption by soils. *Nature*, **223**, 1288–1289.

Brown, C. B. and White, J. L. (1969). Reactions of 12 *s*-triazines with soil clays. *Soil Sci. Soc. Am. Proc.*, **33**, 863–867.

Brunauer, S. (1945). *The Adsorption of Gases and Vapors. Vol I, Physical Adsorption*, Princeton University Press.

Burns, R. G., Hayes, M. H. B. and Stacey, M. (1973a). Spectroscopic studies on the mechanism of adsorption of paraquat by humic acid and model compounds. *Pestic. Sci.*, **4**, 201–209.

Burns, I. G., Hayes, M. H. B. and Stacey, M. (1973b). Studies of the adsorption of paraquat on soluble humic fractions by gel filtration and ultrafiltration techniques. *Pestic. Sci.*, **4**, 629–641.

Calvet, R. and Chassin, P. (1973). Complexes organiques des argiles: Mécanismes de formation. Méthodes d'études et applications. *Bul. Groupe Fr. Argiles*, **XXI**, 87–113.

Calvet, R., Chaussidon, J., Cloos, P., De Kimpe, C., Fripiat, J., Gastuche, M. C., Helsen, J., Jelli, A., Leonard, A., Poncelet, G. and Uytterhoeven, J. (1964). Protons de constitution, protons libres et eau adsorbée. *Bul. Groupe Fr. Argiles*, **XIV**, 59–98.

Calvet, R. and Terce, M. (1975a). Adsorption de l'atrazine par des montmorillonites-Al. *Ann. agron.*, **26**, 693–707.

Calvet, R. and Terce, M. (1975b). Solubilité de l'atrazine dans des gels de montmorillonite calcique. *C. R. Acad. Sci.*, Paris, **281**, 247–250.

Calvet, R., Terce, M. and Arvieu, J. C. (1980). Adsorption des pesticides par les sols et leurs constituants. *Ann. agron.* (to be published).

Cheung, H. W. (1973). *Equilibrium and kinetic process of the interactions of 4-amino 3,5,6-trichloropicolinic acid (Picloram) and 0,0-diethyl-O-p-nitrophenyl phosphorothioate (Parathion) with soils.* Ph. D. Thesis, University of Oxford.

De Boer, J. H. (1953). *The Chemical Character of Adsorption.* Clarendon Press, Oxford.

Dieguez-Carbonell, D. and Rodriguez Pascual, C. (1975). Adsorption of the herbicide 2,4-D by montmorillonite. *Envir. Qual. Safety Suppl. 3 Pesticides*, 237–242.

Dunigan, E. P. and MacIntosh, Th. (1971). Atrazine–soil organic matter interaction. *Weed Sci.*, **19**, 279–282.

Farmer, W. J. and Aochi, Y. (1974). Picloram sorption by soils. *Soil Sci. Soc. Am. Proc.*, **38**, 418–423.

Frenkel, H. and Suarez, D. L. (1977). Hydrolysis and decomposition of calcium montmorillonite. *Soil Sci. Soc. Am. Proc.*, **41**, 887–891.

Fripiat, J., Chaussidon, J. and Jelli, A. (1971). *Chimie-physique des Phénomènes de Surface. Application aux Oxydes et aux Silicates.* Masson et Cie, Paris.

Fu, Y., Hamsen, R. S. and Bartell, F. E. (1948). Thermodynamics of adsorption from solutions. I. Mobility and activity coefficients of adsorbent layers. *J. Phys. Chem.*, **52**, 374–386.

Gaillardon, P., Calvet, R., Rougetet, E. and Gaudry, J. C. (1978). Etudes préliminaires du rôle de la nature des matières organiques dans les phénomènes d'adsorption des herbicides. *Ann. agron.* **29**, 243–256.

Gaillardon, P., Calvet, R. and Terce, M. (1977). Adsorption et désorption de la terbutryne par une montmorillonite Ca et des acides humiques seuls ou en mélanges. *Weed Res.*, **17**, 41–48.

Gaynor, J. D. (1973). Atrazine, terbutryn and S 14254 adsorption, desorption and solubility in salt solutions and movement in soil materials. *Diss. Abst. Int. Ser.* B, **34**, 951.

Giles, H., MacEwan, T. H., Nakhwa, S. N. and Smith, D. (1960). Studies in adsorption. Part XI. A system of classification of solution adsorption isotherms and its use in diagnosis of adsorption mechanisms and in measurement of specific surface areas of solids. *J. Chem. Soc.*, **111**, 3973–3993.

Green, R. E. and Corey, J. C. (1971). Pesticides adsorption measurement by flow equilibrium and subsequent displacement. *Soil Sci. Soc. Am. Proc.*, **35**, 561–565.

Greenland, D. J. (1971). Interaction between humic and fulvic acids and clays. *Soil Sci.*, **111**, 34–41.
Grice, R. E. and Hayes, M. H. B. (1972). A continuous flow method for studying adsorption and desorption of pesticides in soil colloïd systems. *Proc. 11 Br. Weed Control. Conf.*, 784–790.
Grover, R. (1974). Adsorption and desorption of trifluralin, triallate, and diallate by various adsorbents. *Weed Sci.*, **22**, 405–408.
Grover, R. (1975). Adsorption and desorption of urea herbicides on soils. *Canadian J. Soil Sci.*, **55**, 127–136.
Grover, R. and Hance, R. J. (1970). Effect of ratio of soil to water on adsorption of linuron and atrazine. *Soil Sci.*, **109**, 136–138.
Guenzi, W. D. (1974). *Pesticides in Soil and Water.* Soil Sci. Soc. Am., Madison Wisconsin USA.
Hamaker, J. W. and Thompson, J. M. (1972). Adsorption. In, *Organic Chemicals in the Soil Environment.* (C. A. I. Goring and J. W. Hamaker, Eds) pp. 49–143. Dekker, New York.
Hance, R. J. (1965). The adsorption of urea and some of its derivatives by a variety of soils. *Weed Res.*, **5**, 108–114.
Hance, R. J. (1967a). The speed of attainment of sorption equilibria in some systems involving herbicides. *Weed Res.*, **7**, 29–36.
Hance, R. J. (1967b). Relationship between partition data and the adsorption of some herbicides by soils. *Nature*, **214**, 630–631.
Hance, R. J. (1969a). An empirical relationship between chemical structure and the sorption of some herbicides by soil. *J. agric. Fd. Chem.*, **17**, 630–631.
Hance, R. J. (1969b). Influence of pH, exchangeable cations and the presence of organic matter on the adsorption of some herbicides by montmorillonite. *Can. J. Soil Sci.*, **49**, 357–364.
Hance, R. J. (1971). Complex formation as an adsorption mechanism for linuron and atrazine. *Weed Res.*, **II**, 106–110.
Hance, R. J. (1976). Adsorption of glyphosate by soils. *Pestic. Sci.*, **7**, 363–366.
Hance, R. J. (1977). The adsorption of atraton and monuron by soils at different water contents. *Weed Res.*, **17**, 197–201.
Hance, R. J. and Embling, S. J. (1979). Effect of soil water content at the time of application on herbicide content in soil solution extracted in a pressure membrane apparatus. *Weed Res.* **19**, 201–205.
Haque, R., Lindstrom, F. T., Free, V. H. and Sexton, R. (1968). Kinetic study of the sorption of 2,4-D on some clays. *Envir. Sci. Technol.*, **2**, 207–211.
Haque, R. and Sexton, R. (1968). Kinetic equilibrium study of the adsorption of 2,4-dichlorophenoxy acetic acid on some surfaces. *J. Colloïd Interface Sci.*, **27**, 818–287.
Haque, R., Lilley, S. and Coshow, W. R. (1970). Mechanism of adsorption of diquat and paraquat on montmorillonite surface. *J. Colloïd Interface Sci.*, **3**, 185–188.
Haque, R. and Coshow, W. R. (1971). Adsorption of isocil and bromacil from aqueous solutions onto some mineral surfaces. *Envir. Sci. Technol.*, **5**, 139–141.
Harris, C. I. and Warren, C. F. (1964). Adsorption and desorption of herbicides by soil. *Weeds*, **12**, 120–126.
Harvey, R. G. (1974). Soil adsorption and volatility of dinitroaniline herbicides. *Weeds Sci.*, **22**, 120–124.

Hayes, M. H. B. (1970). Adsorption of triazine herbicides on soil organic matter, including a short review on soil organic chemistry. *Residue Rev.*, **32**, 131–168.

Hayes, M. H. B., Pick, M. E. and Toms, B. A. (1975). Interactions between clay minerals and bipyridylium herbicides. *Residue Rev.*, **57**, 1–25.

Helferich, F. (1962). *Ion exchange*, McGraw-Hill, New York, San Francisco, Toronto, London.

Hilton, H. V. and Yuen, Q. H. (1963). Adsorption of several preermergence herbicides by Hawaian sugar cane soils. *J. agric. Fd. Chem.*, **11**, 230–234.

Hornsby, A. G. and Davidson, J. M. (1973). Solution and adsorbed fluometuron concentration distribution in a water saturated soil. Experimental and predicted evaluation. *Soil Sci. Soc. Am. Proc.*, **37**, 823–828.

Huggenberger, F., Letey, J. and Farmer, W. J. (1973). Effect of two nonionic surfactants on adsorption and mobility of selected pesticides in a soil system. *Soil Sci. Soc. Am. Proc.*, **37**, 215–219.

Hurle, K. and Freed, V. H. (1972). Effect of electrolytes on the solubility of some 1, 3, 5-triazines and substituted ureas and their adsorption on soil. *Weed Res.*, **12**, 1–10.

Kanellopoulos, A. G. (1974). Additives in herbicides formulations. *Eng. Chem. Ind.*, **23**, 951–955.

Kearney, P. C. and Kaufman, D. D. (1975). *Herbicides: Chemistry Degradation, and Mode of Action.* Dekker, New York.

Khan, S. U. (1973a). Equilibrium and kinetic studies of the adsorption of 2,4-D and picloram on humic acid. *Can. J. Soil Sci.*, **53**, 429–434.

Khan, S. U. (1973b). Interaction of humic substances with bipyridylium herbicides. *Can. J. Soil. Sci.*, **53**, 199–204.

Khan, S. U. (1973c). Interaction of bipyridylium herbicides with organo-clay complex. *J. Soil Sci.*, **24**, 244–248.

Khan, S. U. (1974a). Adsorption of linuron by montmorillonite. *Can. J. Soil Sci.*, **54**, 235–237.

Khan, S. U. (1974b). Adsorption of bipyridylium herbicides by humic acid. *J. Envir. Qual.*, **3**, 202–206.

Khan, S. U. and Mazurkewich, R. (1974). Adsorption of linuron on humic acid. *Soil Sci. Soc. Am. Proc.*, **118**, 339–344.

Kipling, J. J. (1965). *Adsorption from Solutions of Non-electrolytes*, Academic Press. London and New York.

Knight, B. A. G. and Tomlinson, T. E. (1967). The interaction of paraquat (I,I-dimethyl 4,4 dipyridylium dichloride) with mineral soils. *J. Soil Sci.*, **18**, 233–243.

Knight, B. A. G. and Denny, P. J. (1970). The interaction of paraquat with soil; adsorption by an expanding lattice clay mineral. *Weed Res.*, **10**, 40–48.

Lambert, S. M. (1967). Functional relationships between sorption in soil and chemical structures. *J. agric. Fd. Chem.*, **15**, 572–576.

Lambert, S. M. (1968). Omega, a useful index of soil sorption equilibria. *J. agric. Fd. Chem.*, **16**, 340–343.

Leenheer, J. A. and Ahlrichs, J. L. (1971). A kinetic and equilibrium study of the adsorption of carbaryl and parathion upon soil organic matter surfaces. *Soil Sci. Soc. Am. Proc.*, **35**, 700–704.

Li, G. C. and Felbeck, G. T. (1972). A study of the mechanism of atrazine adsorption by humic acid from muck soils. *Soil Sci.*, **113**, 140–148.

Lindstrom, F. T. and Boersma, L. (1970). Theory of chemical transport with simultaneous sorption in a water saturated porous medium. *Soil Sci.*, **110**, 1–9.
Lindstrom, F. T., Haque, R. and Coshow, W. R. (1970). Adsorption from solution. III-A new model for the kinetic of adsorption desorption processes. *J. Phys. Chem.*, **74**, 495–502.
MacGlamery, M. and Slife, F. W. (1966). The adsorption and desorption of atrazine as affected by pH, temperature and concentration. *Weeds*, **14**, 237–239.
Moreale, A. and Van Bladel, R. (1979). Soil interactions of herbicide derived aniline residues: A thermodynamic approach. *Soil Sci.*, **127**, 1–9.
Morita, H. (1976). Linuron adsorption and the degree of decomposition of peats as measured by rubbed fiber content and pyrophosphate index. *Can. J. Soil Sci.*, **56**, 105–109.
Mortland, M. M., Fripiat, J. J., Chaussidon, J. and Uytterhoeven, J. (1963). Interaction between ammonia and expending lattices of montmorillonite and vermiculite. *J. Phys. Chem.*, **67**, 248–258.
Mortland, M. M. and Meggitt, W. F. (1966). Interaction of ethyl *N*,*N*-di-*n*-propylthiocarbamate (EPTC) with montmorillonite. *J. agric. Fd. Chem.*, **14**, 126–129.
Mortland, M. M. (1970). Clay organic complexes and interactions. *Adv. Agron.*, **22**, 75–114.
Orlov, D. S., Pivovarova, I. A. and Gorbunov, N. I. (1973). Interaction of humic substances with minerals and the nature of their bond; a review. *Sov. Soil Sci.*, **5**, 568–581.
Russell, J. D., Cruz, M. I. and White, J. L. (1968a). The adsorption of 3-aminotriazole by montmorillonite. *J. agric. Fd. Chem.*, **16**, 21–24.
Russell, J. D., Cruz, M. I., White, J. L., Bailey, G. W., Payne, W. R., Pope, J. D. and Treasley, J. I. (1968b). Mode of chemical degradation of *s*-triazine by montmorillonite. *Science*, **160**, 1340–1342.
Saltzman, S. and Yariv, S. (1975). Infrared study of the sorption of phenol and *p*-nitrophenol by montmorillonite. *Soil Sci. Soc. Am. Proc.*, **39**, 474–479.
Savage, K. E. and Wauchope, R. D. (1974). Fluometuron adsorption-desorption equilibria in soil. *Weed Sci.*, **22**, 106–110.
Schnitzer, M. and Khan, S. U. (1972). *Humic Substances in the Environment.* Dekker, New York.
Sethi, R. K. and Chopra, S. L. (1975). Adsorption, degradation and leaching of alachlor in some soils. *J. Indian Soc. Soil Sci.*, **23**, 184–194.
Sullivan, J. D. and Felbeck, G. T. (1968). A study of the interaction of *s*-triazine herbicides with humic acids from three different soils. *Soil Sci.*, **106**, 42–52.
Swanson, R. A. and Dutt, G. R. (1973). Chemistry and physical processes that affect atrazine movement and distribution in soil systems. *Soil Sci. Soc. Am. Proc.*, **37**, 872–876.
Swoboda, A. R. and Thomas, G. W. (1968). Movement of parathion in soil columns. *J. agric. Fd. Chem.*, **16**, 923–927.
Talbert, R. E. and Fletchall, H. (1965). The adsorption of some *s*-triazines in soils. *Weeds*, **13**, 46–52.
Terce, M. and Calvet, R. (1975). Role of aluminum in the adsorption of atrazine by clay minerals. *Proceedings of the Israel–France Symposium "Behaviour of Pesticides in Soil."* (M. Horowitz, Ed.) Bet Dagen, Israel, 33–40.
Terce, M. and Calvet, R. (1977). Some observations on the role of Al and Fe and

their hydroxides in the adsorption of herbicides by montmorillonite. *Z. Pfl-Krankh. Pfl Schutz Sonderheft* **VIII**, 237–243.

Terce, M. and Calvet, R. (1978). Adsorption of several herbicides by montmorillonite, kaolinite and illite clays. *Chemosphere*, **4**, 365–370.

Terce, M. and Calvet, R. (1979). Effet de l'acidité sur l'adsorption de la terbutryne par la montmorillonite. *Ann. Fac. Phar. Montpellier* (in press).

Theng, B. K. G. (1974). *The Chemistry of Clay Organic Reactions*, Hilger, London.

Vallet, M., Calvet, R. and Chaussidon, J. (1973). Remarques complémentaires sur quelques aspects physico-chimiques de l'adsorption de l'atrazine par les montmorillonites. *Proc. Eur. Weed Res. Coun. Symp. Herbicides-Soil*, 41–50.

Van Bladel, R. (1971). Physico-chemical study of the adsorption of organic pesticides by clay minerals. *Ann. de Gemboux*, **77**, 183–194.

Van Bladel, R. and Moreale, A. (1974). Adsorption of fenuron and monuron (substituted ureas) by two montmorillonite clays. *Soil Sci. Soc. Am. Proc.*, **38**, 244–249.

Van Genuchten, M. Th., Davidson, J. M. and Wierenga, P. J. (1974). An evaluation of kinetic and equilibrium equations for the prediction of pesticides movement through porous media. *Soil Sci. Soc. Am. Proc.*, **38**, 29–36.

Ward, T. M. and Holly, K. (1966). The sorption of *s*-triazines by model nucleophiles as related to their partitioning between water and cyclohexane. *J. Colloïd Interface Sci.*, **22**, 221–230.

Walker, A. and Crawford, D. V. (1968). The role of organic matter in adsorption of the triazine herbicides by soils. Isotopes and radiation in soil organic matter studies. *Proc. 2nd Symp. Int. Atomic Energy Agency* (Vienna), 91–108.

Watson, J. R., Posner, A. M. and Quirk, J. P. (1973). Adsorption of the herbicide 2,4-D on Goethite. *J. Soil Sci.*, **24**, 503–511.

Weber, J. B. (1967). Spectrophotometrically determined ionisation constants of 13 alkylamino *s*-triazines and the relationships of molecular structure and basicity. *Spectrochimica Acta*, **23**, 458–461.

Weber, J. B., Perry, P. W. and Upchurch, R. P. (1965). The influence of temperature and time on the adsorption of paraquat, diquat, 2,4-D and prometone by clays, charcoal and an anion-exchange resin. *Soil Sci. Soc. Am. Proc.*, **29**, 678–688.

Weber, J. B. and Weed, S. B. (1968). Adsorption and desorption of diquat, paraquat and prometone by montmorillonite and kaolinite clay minerals. *Soil Sci. Soc. Am. Proc.*, **32**, 485–487.

Yamane, V. K. and Green, R. E. (1972). Adsorption of ametryn and atrazine on an oxisol, montmorillonite and charcoal in relation to pH and solubility effects. *Soil Sci. Soc. Am. Proc.*, **36**, 58–64.

# 2 Transport in Solution

*M. Leistra*
Institute for Pesticide Research,
Marijkeweg 22,
6709 PG Wageningen,
The Netherlands



## I. Introduction

Knowledge of the movement of herbicides in soils helps us to understand the performance of soil-applied herbicides against weeds and the safety of the crops. It is also necessary in order to evaluate the risk of leaching through the soil to ground water and surface water and of the exposure of non-target organisms. In the last decade, authorities in various countries have drawn up guidelines for research on pesticides and these pay considerable attention to mobility in soil (Anon., 1973; Anon., 1975, 1978).

Usually, soil-acting herbicides must penetrate at least a few centimetres into the soil to achieve good weed control. If there is little rainfall and the top layer dry, the herbicide will remain very close to the soil surface and its uptake by weeds may be limited. Strong adsorption reduces the availability of the herbicide and also restricts its penetration into the soil, even with sufficient rainfall. Thus in order to know under which soil and climatic conditions herbicide penetration into the soil will be inadequate, data on herbicide adsorption and mobility are needed.

Many soil-acting herbicides rely on depth protection by the soil to avoid damage to the crop, so they must not penetrate in significant concentrations into soil zones that are critical to the crop. The flow of water in the soil and the adsorption of herbicide on the solid phase are the main controlling factors. This mechanism has been described in some detail and has been designated variously as "depth protection" (Gerber and Guth, 1973), "positional selectivity" (Dunham and Crawford, 1973) or "spatial selectivity" (Riley and Morrod, 1976).

In some countries there is concern over the risk that pesticides will leach into ground water and water courses. Therefore it is vital to know how herbicides move with the liquid phase in field soils, when retarded by adsorption. Other important processes such as chemical and biochemical conversion and uptake by plants must also be considered. In the Federal Republic of Germany and in the Netherlands, pesticides are systematically evaluated for the risk of residues leaching from the soil and the use of substances that may lead to contamination of ground water is prohibited in large protection areas around pumping stations for drinking water.

To evaluate the chance that a herbicide will affect non-target organisms, both within and outside the soil system, the level and the duration of exposure to the compound must be known. One of the processes determining the exposure pattern is the movement of herbicides with the liquid phase in soil.

This chapter begins by reviewing experimental information on the movement of herbicides with the liquid phase in soil. This includes soil column experiments, usually with constant water fluxes at fairly high rates, and the more well-defined experiments under field conditions. Complications during movement mainly result from the lack of local equilibrium between concentrations in different soil phases and therefore some conceptual models used for describing non-equilibrium are discussed. Various ways in which herbicide mobility can be characterized are outlined and their mutual connection is stressed. Finally, a standard computation model for herbicide behaviour in field soils is described and the results of a few computer runs are given to demonstrate the possible uses of such models.

## II. Movement at Steady-state Water Flow

### A. Movement Measured in Soil Columns

The rate of water flow through the soil is heterogeneous on the microscopic scale. It is controlled by the size of the pores and it may be almost zero in pore spaces outside the continuous pathways of flow. Within pores, flow rate increases concomitantly with distance from the wall (laminar flow). Differences in flow rate result in the gradual spread of a herbicide band as it moves through the soil. This process is called "convective dispersion". A rapid interchange of substance between various pore sections, which occurs by flow or by transverse diffusion, reduces the extent of dispersion.

When the water fluxes are comparatively low, so that ample time is available for the establishment of adsorption–desorption equilibrium, the spreading in solute distribution can usually be described by normal values of the convective dispersion coefficient (Leistra, 1973). In a range of moderate liquid fluxes, with adsorption–desorption close to equilibrium, the spreading is often enhanced. The peak of the distribution is then found near the expected position or appears in the effluent from a column near the expected time.

When a herbicide solution is fed to soil columns at high liquid fluxes (for example above 100 mm/day), the front part of the distribution is often comparatively steep. The peak of the distribution moves faster than would be expected from equilibrium adsorption; this has been measured for prometryne and fluometuron (Abernathy and Davidson, 1971; Davidson and McDougal, 1973). Similarly, the peak is displaced over a greater distance in the soil than at low liquid fluxes using the same volume of liquid (Davidson and Chang, 1972). The rear of the distribution then shows considerable tailing. The concentrations in the liquid phase move ahead of the corresponding equilibrium contents in the adsorbing phase (Hornsby and Davidson, 1973). In soil column studies using napropamide with water percolating at a high rate under saturated condition (Chu-Huang Wu *et al.*, 1975) a small fraction of the dosage moved very fast. The main part of the distribution was skewed with the front comparatively steep and extensive tailing.

The transport of solutes through highly-aggregated soils was studied by Green *et al.* (1972). At high liquid fluxes, part of the solutes appeared rapidly in the effluent and there was extensive tailing in the effluent curves. Presumably, liquid flow was fast in the larger voids between the aggregates while equilibration with much of the liquid and adsorbing phases by diffusion through the micropores in the aggregates was comparatively slow. The solute that had penetrated into the aggregates after some time was only slowly released, which resulted in considerable tailing. Similarly Davidson and Chang (1972) applied herbicide in solution to soil columns consisting of either small or large aggregates and found at high liquid fluxes the non-equilibrium effects were more marked with the larger than with the smaller aggregates. This was explained by the longer time needed for the solute in the flowing liquid to equilibrate by diffusion with the interior of the large aggregates.

If a considerable time elapses between application of herbicide to the column and the start of leaching, time is available for penetration into small pores and for adsorption onto the solid surfaces. When water is applied at high rates in such situations, the substance will only slowly be released into the flow paths so the main part of it will then move substantially slower than it would under equilibrium conditions.

## B. Modelling Concepts for Non-equilibrium Transport

### 1. *First-order rate equations*

The rate of herbicide adsorption on soil may be assumed to be proportional to the concentration in solution. Similarly, the rate of desorption may be taken to be proportional to the amount on the adsorbing phase. The combined first-order rate equation for a volume element of the soil system reads:

$$\partial(\rho_b S)/\partial t = k_a \varepsilon_l c_l - k_d \rho_b S \qquad (1)$$

where $\rho_b$ is the bulk density of the soil ($kg/m^3$), $\varepsilon_l$ is the volume fraction of liquid, $S$ is the herbicide content of the adsorbing phase (mmol/kg), $t$ is the time (days), $k_a$ and $k_d$ (1/day) are the first-order rate coefficients for adsorption and desorption, respectively. In this way the rate of adsorption–desorption equilibration is assumed to be proportional to the difference between equilibrium and actual adsorption.

The concept of the first-order rate equation for adsorption–desorption was used by Lindstrom *et al.* (1971). It was included in the differential equation for solute transport under steady-state water flow conditions and the resulting equation was solved by a finite-difference procedure. The substance was assumed to be initially distributed in a top layer of the soil column with equilibrium between the phases. At comparatively fast adsorption and desorption during leaching, the computed distributions showed normal spreading, and the peaks of the distributions in the liquid and adsorbing phases were at almost the same position (Fig. 1). When the rate coefficients of adsorption–desorption were of moderate values, the spreading in the distributions was greater. At low equilibration rates the distribution of concentration in the liquid phase moved markedly ahead of that in the adsorbing phase, with the latter distribution showing extensive tailing.

### 2. *Simultaneous fast and slow equilibration*

The concept may be extended to consider separate equilibration with two fractions of the adsorbing surface. One could think of two dominant adsorption mechanisms with markedly different equilibration rates or it is possible that on the adsorbing surface some positions are easily accessible whereas others are difficult to reach. The concept of simultaneous fast and

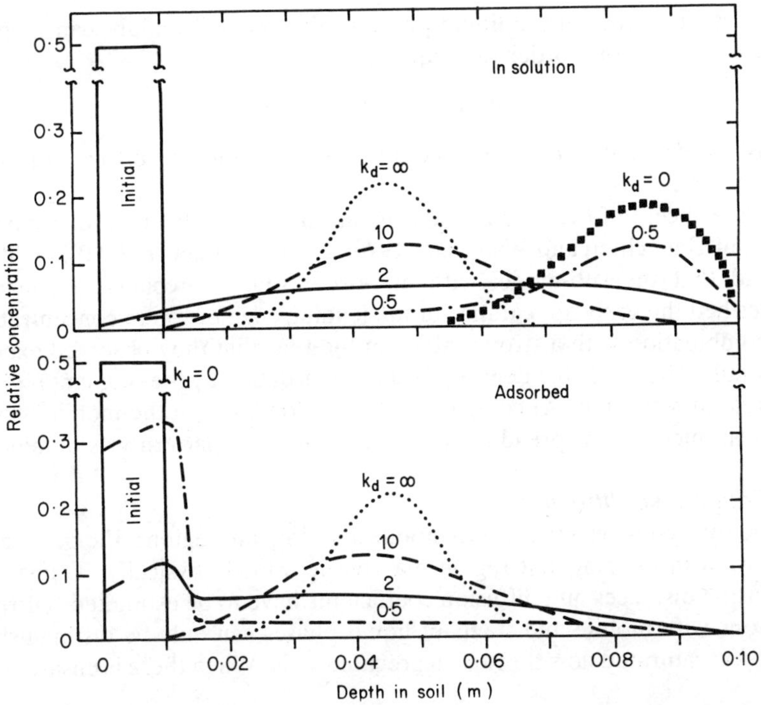


Fig. 1. Computed concentrations of a substance in solution and in the adsorbing phase at various adsorption–desorption rates (after Lindstrom *et al.*, 1971).

slow equilibration processes was used in computations by Leistra and Dekkers (1977). When the substance was introduced in solution to the soil, the calculated movement was comparatively rapid, because the slow-adsorbing phase contributed little to the retardation. If adsorption equilibrium in the top compartment initially was assumed, strong retention near the soil surface was simulated. Rao *et al.* (1979) used a similar submodel with instantaneous equilibration of herbicide in a fraction of the soil system, whereas first-order kinetics were applied to adsorption equilibration with the remaining fraction of the soil.

### 3. *Mobile and stagnant liquid phases*

The liquid phase in soil may be divided into a mobile phase (subscript $m$) and a stagnant phase (subscript $s$), and fractions of the adsorbing phase may be assigned to each of these phases. The interchange by diffusion of a substance between the liquid phase in the two soil regions can then be described by a first-order rate equation:

$$\partial c_{1,s}/\partial t = \alpha (c_{1,m} - c_{1,s}) \tag{2}$$

$c_1$ is the concentration in the liquid phase (mmol/m$^3$) and $\alpha$ is the rate coefficient.

The concept of the mobile and stagnant phases has been used in models by Van Genuchten and Wierenga (1976) and by Gaudet *et al.* (1977). They found that the sub-model with first-order rate interchange adequately described the early appearance of the leading edge of solute concentration in combination with a strong tailing in the rear that they observed experimentally. Evidently the degree of solute retardation by the stagnant regions increased with time. At comparatively low flow rates of the mobile phase, only an increase in spreading in the concentration pattern was obtained.

### 4. *Transverse diffusion*

If the soil system is divided into mobile and stagnant regions, the transverse diffusion in the stagnant region may be described physically. Transverse diffusion distances and diffusion coefficients have to be estimated. Distinct regions with mobile and stagnant liquid phases can only be distinguished easily in saturated flow through aggregated soils. When there is unsaturated

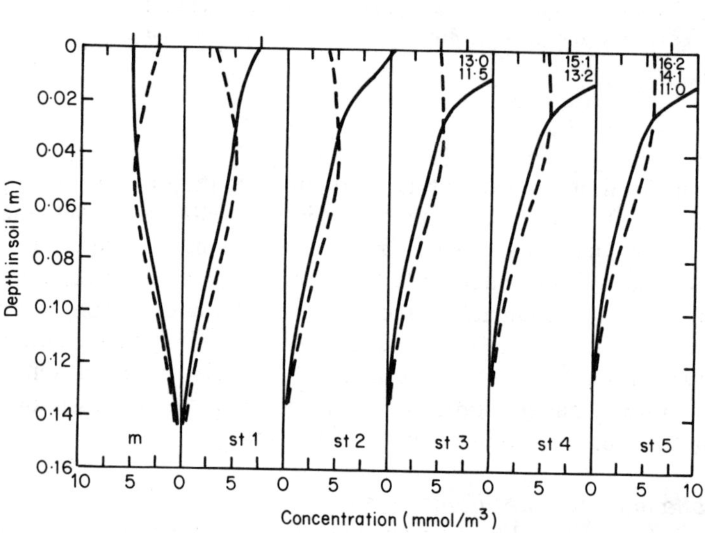


Fig. 2. Concentrations of a pesticide computed for the mobile (m) and for the slices of the stagnant (st) regions in a simplified soil system. Initial conditions in the top compartment: - - - = in solution; ——— = adsorption equilibrium (after Leistra, 1977).

water flow, it is difficult to make a clear-cut division of the soil into mobile and stagnant regions.

The idea of limiting convective flow to the larger sections of the liquid phase and transverse diffusion in a stagnant liquid phase was used by Skopp and Warrick (1974). Poor equilibration between liquid and adsorbing phases at the leading edge can be described well with this concept. The slow release of herbicide by diffusion from stagnant regions explains the considerable tailing measured at high water fluxes.

Leistra (1977) introduced various rainfall patterns and transverse diffusion distances into a computation model with diffusion as the rate-limiting process for pesticide adsorption–desorption. The soil system was divided into vertical slabs with alternately mobile and stagnant liquid phases. When a substance was in solution at the beginning and flow rates were rather high, the stagnant regions only gradually participated in adsorption (Fig. 2). With adsorption equilibrium in a top layer at the beginning, the pesticide was computed to be released only slowly from the stagnant region near the soil surface.

#### 5. *Non-single-valued adsorption–desorption*

In slurry-type experiments it is possible to measure the adsorption on soil of a substance applied in aqueous solution. In desorption experiments, a fraction of the equilibrium solution is replaced by water and the system shaken again. By repeating this procedure several times, a "desorption curve" is obtained that may give higher values for the soil/water partition of the substance than is given by the adsorption curve. This led to the concept of non-single-valued adsorption–desorption proposed by Davidson, Mansell and Baker (1973) and Van Genuchten *et al.* (1974). Unfortunately, these researchers did not measure the final amount of substance in the system, so there was no check on the effects of cumulative analytical errors and of decomposition. In addition, repeated shaking followed by withdrawal of equilibrium solution may induce changes in the composition and surface characteristics of soil organic matter, which could affect adsorption.

Hence the experimental basis for this idea is not yet properly established. In addition it is not compatible with the concept that sorption processes involve the intermittent random capture and release of solutes by the adsorbing surface.

### C. General Remarks

The list of conceptual sub-models for adsorption–desorption dynamics during the movement of substances in soils could be extended. Some already have a fairly long history in different scientific disciplines (Coats and Smith, 1964). Since numerical solution techniques were introduced in studies of the movement of substances in soil, the imagination of research workers has been less restricted by mathematics than before. As the capacity and speed of computer facilities increase, new and more complex conceptual models can be developed.

Ideally sub-models for the adsorption–desorption dynamics of herbicides should be based on sound physico-chemical principles and use parameters whose values can be estimated independently for a range of soil and climatic conditions.

Each sub-model for adsorption–desorption dynamics in soil raises the problem of estimating suitable values for the parameters. In soil column studies the values are usually adjusted until the observed results, or part of them, are fairly well described because it is often not possible to measure or estimate the parameter values independently. When using curve-fitting procedures, a particular set of results can often be described with different sub-models. Rao *et al.* (1979) derived parameter values from a set of results for leaching experiments with 2,4-D and atrazine, but they could not apply these values to explain results for different conditions. When complex submodels are designed, the number of unknown parameters usually increases and the approximation of their values by curve-fitting becomes less satisfactory.

Many simplifications of the geometry of the liquid phase have been used so that the transport mechanism can be exactly analysed mathematically. For example, the pore system has been represented as a bundle of capillary tubes of different size, which are not interconnected. A more realistic idea involves regularly-spaced interconnections between the tubes. A survey of such models of the liquid-filled pore system, used in various disciplines, has been given by Rao (1974). However, even quite complex geometric models are no more than greatly simplified representations of the intricate geometry of the liquid phase in soils.

## III. Movement in Soil Under Field Conditions

### A. Measured Movement in the Soil

In temperate regions the movement of herbicides in soil after application in spring may be expected to be rather limited. Much of the rainfall evaporates from the soil surface or is taken up from the upper layers of the soil by plant roots. The movement of most herbicides is significantly retarded by adsorption. In winter, the rainfall surplus is larger so residues of herbicides may move somewhat more.

Smith and Meggitt (1970) studied the movement and persistence of chloridazon in soils under field conditions after application in spring. Under rainfall conditions which averaged 61 mm/month, the highest concentrations remained in the top 0·05 m layer, while the concentrations decreased to low values at a depth of about 0·15 m. There was little movement in soils with comparatively high organic matter contents (above 6%).

The residues of simazine at various depths in soil after annual applications to an uncropped soil were measured by Clay and Stott (1973). At normal dosages, measurable concentrations were only found in the top few decimetres of the soil. At abnormally high dosages, measurable concentrations penetrated to depths of about 0·6 m. Baumeister and Hurle (1975) traced the distribution of atrazine and simazine in various soils after application in spring to fields cropped with maize or grass. In the autumn the highest concentrations remained in the top 0·05 m; only very low concentrations penetrated to depths of about 0·15 or 0·20 m.

Other researchers have traced the movement of terbutryne and fluometuron in soil under field conditions and correlated it with rainfall totals measured for the sampling intervals (Gerber and Guth, 1973; Burkhard, Eberle and Guth, 1975). Terbutryne showed very little movement during the six months after application in spring. Fluometuron showed some movement following the high amounts of rainfall (772 mm in 182 days). Movement under field conditions was much less than when comparable amounts of water were applied in soil column leaching studies in the laboratory. Evaporation of water from the soil surface reduced downward movement in the field and non-equilibrium adsorption phenomena in laboratory and field may have contributed to the differences.

Uracil herbicides like bromacil and terbacil are comparatively slowly degraded in soil and they are quite mobile at high water fluxes in laboratory studies. Leistra, *et al.* (1975) applied bromacil annually in spring to two orchard soils over a period of at least six years. Concentrations measured just before the next spraying were highest in or around the 0·1 to 0·2 m layer, with a gradual decrease to below 0·1 mg/kg at depths below 0·5 m. Around 1 m depth, the concentrations dropped below 0·01 mg/kg, except for the higher dosage (2.4 kg a.i./ha) in one of the soils (0·02 to 0·04 mg/kg). Marriage *et al.* (1977) measured the distribution of terbacil residues in soil after a series of annual applications (4·5 kg/ha) in an orchard. Two years after the last of the applications, low concentrations of the herbicide were measured down to about 0·6 m depth.

In a similar study, Khan *et al.* (1976) found that after seven annual applications of 4·5 kg diuron/ha to an orchard soil, most of the herbicide stayed in the upper 0·15 m, while low concentrations penetrated to depths of about 0·6 m. The relative distribution of the conversion product 3,4-dichloroaniline in soil was almost the same, although the concentration levels were much lower.

## B. Complications in Herbicide Movement

Transport behaviour in soil sometimes deviates from the expected pattern. One of the possible complications is rapid movement of very small amounts of solute to the subsoil. Under some conditions, part of the water that infiltrates into the soil seems to flow very quickly to the groundwater or to the tile drains. A small fraction of the material in the top layer may then be carried along.

Schiavon and Jacquin (1973a) applied atrazine in spring at a dosage of 1·25 kg a.i./ha to a field with a clay soil and to a field with a clay loam soil. In these soils, which showed considerable swelling and shrinkage, low amounts of atrazine were transported fairly rapidly to depths of a few decimetres. The highest concentrations remained in the top 0·06 m of the soil. They also (Schiavon and Jacquin, 1973b) measured concentrations of atrazine over a nine month period in water from tile drain systems placed 0·9 m deep in the soils. After rainy periods, average concentrations ranging from 6 to 29 $\mu g/dm^3$ were found in the drain water. The total amounts that leached via the drains were estimated to be about 1 and 2% of the dosage for the two fields, respectively. Complications in water and solute movement were promoted by the deep shrinkage of these clayey soils in dry periods.

The leaching of some triazine herbicides and their conversion products via tile drains at a depth of 1·2 to 1·6 m in loamy soils was reported by Muir and Baker (1976, 1978). Following six annual applications of atrazine (2·8 kg/ha) in spring, the concentrations measured in the drainwater taken after rainstorms in summer ranged from 0·06 to 10·8 $\mu g/dm^3$. In the first season after application of cyanazine (3·36 kg/ha) and metribuzin (0·56 kg/ha), these compounds appeared in concentrations ranging from 0·00 to 1·06 and from 0·00 to 1·65 $\mu g/dm^3$, respectively. The first residues seemed to appear after only six days during a period with heavy rainfall. A few metabolites, such as de-ethylated atrazine and cyanazine amide, were found at the same concentration levels as their respective parent compounds. The total discharge of the compounds in the period April to December was estimated to be about 0·22% of the annual dosage or less (Muir and Baker, 1978). Because the slopes in the area were not given, it is difficult to judge whether surface run-off contributed to the contamination of the water in the sampling devices.

Von Stryk and Bolton (1977) monitored atrazine in tile-drain water after applications in ten consecutive years to a clay soil on which maize was grown. Over a four-year period, water was collected from tile drains at a depth of 0·7 m in the soil. The concentrations of atrazine in the drainage water from continuous maize plots ranged from 0·6 to 3·5 $\mu$g/dm$^3$. The total amount of atrazine leached per year was estimated to be about 0·8% or less of the annual dosage.

The rapid movement of a small fraction of the herbicide dosage through the soil is also illustrated by the study on bentazone by Stoller *et al.* (1975). In June, they applied bentazone, which is very weakly adsorbed, to four soils in lysimeters on which soya beans were being grown. During the first two months, very small amounts of bentazone (less than 2% of the dosage) appeared in the leachate from the lysimeters (1 m deep) that were regularly irrigated with small volumes of water. Under natural rainfall conditions, bentazone, (less than 1% of the dosage) only appeared in the leachate from one of the four soils. After an initial period of two months no more bentazone was found in the percolation water.

## C. Initial and Boundary Conditions

Most soil-acting herbicides are applied in a spray liquid containing a wettable powder or an emulsifiable concentrate. The rate at which a herbicide passes to the soil phases around the site of deposition needs to be estimated. Calvet *et al.* (1975) measured the rates of dissolution of three chloro-triazine herbicides from their pure solid state into moving water, in a laboratory system. Within a day substantial amounts of the herbicides had dissolved, the largest amounts for the triazine with the highest water solubility. However, at 10°C it took a few days before the concentrations reached half of the water solubility. Hance and Embling (1979) applied spray solutions of formulated simazine, metribuzin and linuron to layers of moist and air-dry soil aggregates in the laboratory. At intervals, soil solution was collected from moist soil with a pressure membrane apparatus. The solutions collected from moist soil during the first day already contained concentrations that approached the values predicted from the adsorption isotherms. In most instances the concentrations obtained after delayed wetting (24 h) of dry soil were comparatively low during the first few days. It was suggested that after spraying on a soil surface that remains dry for at least a few days, the rate of herbicide release to the flow pathways may be restricted, resulting in less movement upon rainfall than predicted. Dunham and Crawford (1973) suggested the use of scanning electron microscopy to study the state of a herbicide at the soil surface.

## D. Application of Computation Models

Leistra *et al.* (1974) simulated the movement of propyzamide in simplified soil systems under field conditions. The water fluxes (rainfall minus evaporation) were averaged per period and soil water content was kept constant. Measured adsorption coefficients were introduced into the computations. A fair degree of correspondence was found between computed and measured concentration patterns in different soils (Fig. 3). Leistra and Dekkers (1976) have applied a computation model to make preliminary estimates of the risk that pesticides will leach into the sub-soil. The movement and conversion in soil of compounds applied to a field in spring were simulated; uptake by plant roots was included. This allowed them to tabulate computed fractions of the dosage leached from the top metre of the soil at various combinations of adsorption coefficients and conversion rates.

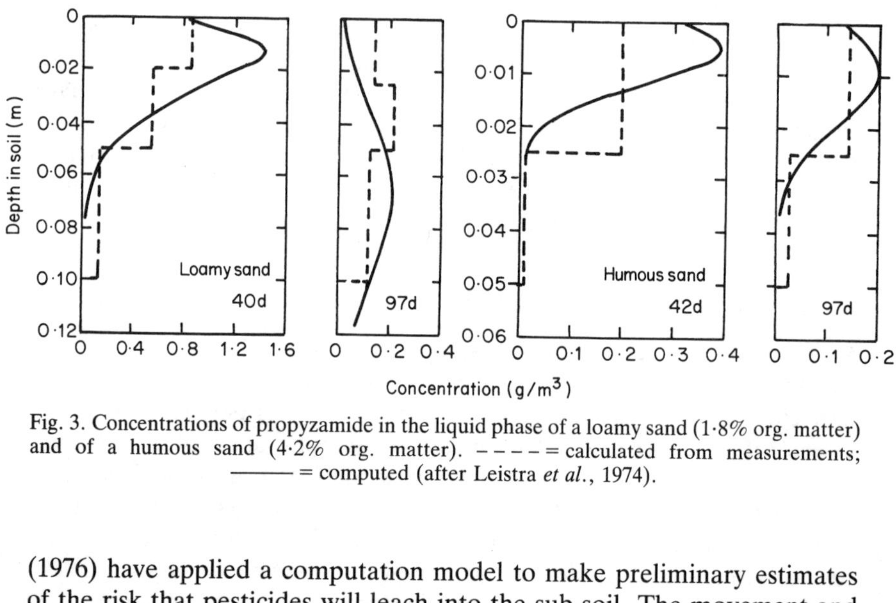


Fig. 3. Concentrations of propyzamide in the liquid phase of a loamy sand (1·8% org. matter) and of a humous sand (4·2% org. matter). – – – – = calculated from measurements; ——— = computed (after Leistra *et al.*, 1974).

Wierenga (1977) has described computer simulation of the transport of substances at transient water flow in soil. He simulated the transport of chloride and nitrate ions at intermittent flood irrigation of soil and obtained fairly good correspondence with measured results. The description of the leaching of solute from soil columns (1.5 m long) by a rapid computation model based on steady-state water flow was surprisingly good.

The transport of a solute with the liquid phase in structured field soils was modelled by Addiscott (1977). He assumed that the liquid phase was divided into a mobile and a stagnant fraction. While the water was being supplied to the soil surface, the solute was carried along with the mobile phase without interchange with the stagnant phase. After the liquid flow stopped, the substance of the mobile and stagnant phases of a computation compartment was instantaneously equilibrated. This model described a very broad distribution of substance in the soil profile and an early appearance of a small fraction of the solute in percolation water, especially under intensive rainfall conditions.

Leistra and Dekkers (1977) introduced patterns of intermittent rainfall into computations including rate equations for adsorption–desorption; these allowed ample time for equilibration in between the showers. The deviation from equilibrium movement in these field simulations was much less than the deviation at the continuous high water fluxes used in many soil column studies. A model with transverse diffusion as the rate-limiting process also showed greatly reduced non-equilibrium effects with an intermittent rainfall pattern (Leistra, 1977). In the field there is often almost adsorption equilibrium for a herbicide in the top layer when rainfall starts. Consequently, it should be expected that at high rainfall intensities by far the greatest part of a herbicide moves more slowly than would follow from equilibrium adsorption (Leistra, 1977; Leistra and Dekkers, 1977).

## E. General Remarks

In many instances, the movement of herbicides in soil under field conditions follows the anticipated general pattern. After application in spring, the substance remains in the upper decimetres of the soil for a considerable time, so that complete or almost complete conversion may occur. More attention should be paid to the possibility that herbicides and conversion products that are still present in the soil system in winter will leach through the soil.

Complications in the transport behaviour of herbicides can be expected under some field conditions. Soils containing much clay may show large cracks upon drying. When intense rain falls on such a soil, part of the water may percolate rapidly via the cracks before the top layer is completely moistened. This water may carry a small amount of the substance along to deeper layers. The main fraction of the substance, which is inside the structure elements, is then moved less than would be expected.

The general picture of herbicide movement in soil under field conditions can be described by computation models and for some purposes the details are not very important. Further research is needed for more accurate simulation. In particular comparisons between computed and measured distributions in soil will show which submodels and basic data need to be improved and when it is necessary to apply non-equilibrium models.

## IV. Basic Experimental Data on Herbicide Mobility

The various procedures used to characterize the mobility of herbicides in soils are only briefly discussed here. More details are given by Gerber and Guth (1973), Guth *et al.* (1976), Anon. (1975, 1978).

### A. Adsorption Isotherms

The adsorption of substances by soil can be measured quickly, for example with the slurry technique. These measurements result in adsorption isotherms and coefficients that can be introduced directly into calculations or computations on movement in soil. The extent of adsorption measured is somewhat dependent on details of the procedure followed. Detailed discussions of adsorption are presented in Chapter I.

### B. Leaching Studies With Soil Columns

The mobility of herbicides can be measured in leaching experiments using soil columns. Various details in the procedure determine the success or failure of the experiment. During coring or filling the column and any subsequent handling, great care should be taken to prevent the collapse of soil structure. When water flow rates through the columns are very high, serious complications in the transport behaviour may occur (Section II) and the use of formulated products (especially granules) may affect the release of the compound. The percolating liquid should be evenly distributed over the soil surface and applied at a low rate (preferably below 50 mm/day). Rainfall simulation is best and total amounts between 200 and 500 mm water could be given. The water applied to the columns should contain a low salt concentration (for example 0·01 M $CaCl_2$ or $CaSO_4$) to limit swelling of loamy and clayey soils. A sand filter at the bottom end promotes the flow of liquid from the column. Detailed descriptions of suitable soil column installations have been given by Poelstra *et al.* (1973) and by Bannink *et al.* (1977) one of which is illustrated in Fig. 4. As well as the concentrations in the effluent fractions, the concentrations in different soil layers at the end of the experiment should be measured.

A comparatively new application is the leaching of soil columns treated with aged residues (Anon., 1975; Guth *et al.*, 1976). Radio-labelled herbicide is first incubated in soil, for example for a few months. Then the soil is placed on top of the soil column, which is irrigated at a low rate. Besides measuring the total radioactivity in fractions of the effluent and in column sections, attempts can be made to quantify the most important conversion products.

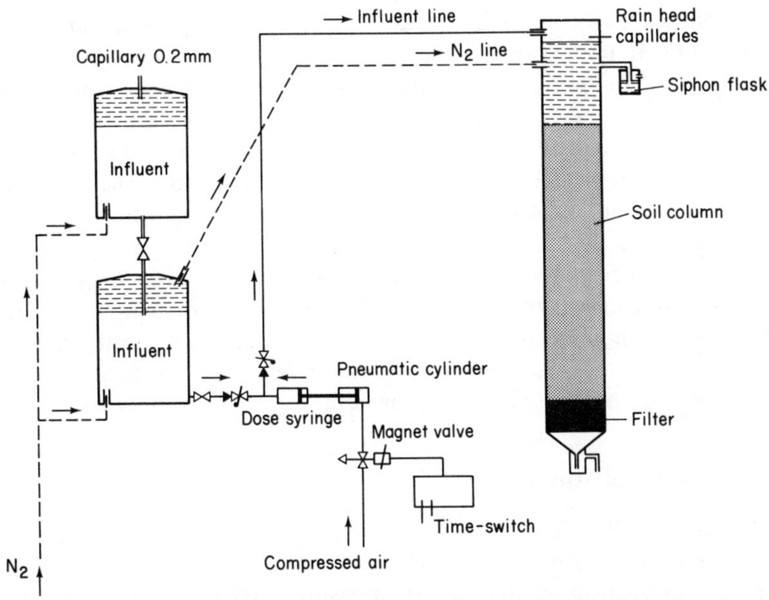


Fig. 4. Soil column installation for studying the mobility of herbicides. $N_2$ to be replaced by air for studies under aerobic conditions (after Bannink *et al.*, 1977).

## C. Soil-layer Chromatography

The relative mobility of a herbicide can be estimated by soil-layer chromatography. The thickness of the layer may vary from about 1 mm (thin layer) to about 5 mm (thick layer). At the very high water flow rates involved, there is usually insufficient time for adsorption equilibration to occur. The rate of release from the application site (dissolution, desorption) may be limited. The results are only qualitative and need to be compared with those obtained for one or more reference compounds.

## D. Field Studies

Studies of the movement of herbicides in soil under field conditions may give useful direct information. For this the soil must be sampled in layers and the concentration patterns must be completely measured in a sufficient number of replicates which means that many samples are needed. Rainfall, evaporation and temperature have to be measured or estimated at frequent intervals both for the interpretation of the experimental observations and to allow the results to be applied to other climatic conditions.

Concentrations of herbicides and their conversion products in sub-soil and groundwater are usually very low (μg/kg or less) which makes high demands on the experimental techniques to be used. Special soil sampling techniques are needed to prevent the intermixing of soil from different layers. Extreme care should be taken to prevent contamination of groundwater samples, especially with topsoil material. Any space between sampling tube and surrounding soil allows water and top soil to move downwards rapidly, especially under high rainfall conditions. When collecting and sampling the effluent from tile-drain systems (usually at low points in the field), there may be contamination from above, for example by surface run-off. The collection of soil solution via suction probes in the profile is often beset with technical problems. Only in fairly wet conditions can part of the soil solution (possibly not representative) be withdrawn.

## E. Lysimeter Studies

Field experiments with soil lysimeters allow the amounts of substance that might leach to the sub-soil to be assessed, because the volume of percolation water and the concentrations can be measured. By growing a crop on the lysimeters, field conditions can be closely simulated.

A procedure for placing large little-disturbed soil monoliths in lysimeters has been described by Brown *et al.* (1974) but the need to use such monoliths can be questioned. Taking large monoliths usually requires heavy equipment and involves extra labour and costs. However, root systems can develop more naturally in such monoliths, as disturbance of the soil often promotes rooting. The condition of the soil structure is especially important at very high liquid fluxes through soil near water-saturation. However, in a lysimeter some disturbance of the soil, especially near the walls, seems unavoidable. Under temperate rainfall conditions, filling lysimeters with sieved soil seems to be acceptable, especially when the sequence of the soil layers is maintained.

An interesting development is the use of radio-labelled substances in lysimeter studies under outdoor conditions, as has been described by Führ *et al.* (1976). Components of the material balance, such as uptake by a crop and leaching from a rooting zone, can be estimated, both for a parent compound and its conversion products.

## F. General Remarks

Basic experimental data on the mobility of herbicides and their conversion products in soils can be obtained from standard trials in laboratory or greenhouse. It is essential that all the relevant details from these trials are recorded. Only then is it possible to apply such results to the varying field conditions in which herbicides are used. Computation models could be developed into useful tools for transferring information from laboratory to field conditions. They need to be checked in a limited number of experiments and subsequently improved to increase the accuracy with which predictions for field conditions can be made. The results from laboratory and greenhouse studies and from computer simulations could be used to establish which compounds require more intensive study under field conditions (field plots, lysimeters, columns).

The procedures for recording experimental details from field experiments need to be greatly extended and improved. The following checklist of details essential for studies on herbicide movement in soils could be helpful in devising well-defined field experiments.

The items of the water balance: rainfall, irrigation, evaporation, transpiration and percolation.
The composition of the soils: organic matter content (organic C), clay and silt (soil textural class) and $CaCO_3$.
Physical properties of the soil: bulk density, solid phase density, moisture contents at the sampling times and a description of soil structure.
Chemical characteristics of the soil: pH; sometimes cation exchange capacity and cationic composition.
Soil classification, for example according to Soil Survey Staff (1975).
Soil cultivations, soil surface conditions, slope of the area. Plant growth, including root development.

Often the characteristics of each soil layer need to be measured individually. Part of the climatic data could be collected from nearby meteorological stations, but a rain gauge should be installed on the site.

In field studies, the conversion rate of the herbicide is usually measured at the same time as movement. The checklist may be extended to include details that should be recorded in combined studies of this type.

Soil temperatures.
Aspects of soil fertility: nitrate contents, total nitrogen, phosphate, potassium and magnesium.
Uses and treatments of the soil, especially in the year before the study.

## V. Computer Simulations with a Standard Model

A standard model for the field soil behaviour of pesticides has been developed by Leistra *et al.* (1980). This model has been adapted and updated for this chapter. It shows the type of basic data needed for a quantitative description and illustrates possible applications of computation models for questions related to herbicide movement in soil, particularly with respect to depth protection of the crop. The computer simulation approach used in this section is just one of the possible procedures for modelling herbicide behaviour in soil. Comparative discussions on mathematical procedures lie beyond the scope of this chapter and the reader is referred to the more specialized literature in soil physics and chemistry.

## A. Equations and Basic Data

### 1. *Water flow*

The liquid flux $J_1[\mathrm{m}^3/(\mathrm{m}^2\ \mathrm{day})]$ in soil along the space co-ordinate $x$ can be described by Darcy's equation

$$J_1 = -K \partial p_h / \partial x . \tag{3}$$

The hydraulic conductivity $K[\mathrm{m}^3/(\mathrm{m\ Pa\ day})]$ is a function of the volume fraction of liquid, $\varepsilon_1$, in the soil. $p_h$ (Pa) is the hydraulic pressure, which equals the sum of tensiometer pressure $p_t$ and gravity pressure $p_g$. The relationship between tensiometer pressure $p_t$ and $\varepsilon_1$ is known as the soil moisture retention curve. Typical hydraulic characteristics for different soil textural classes, derived from a compilation of literature data, have been given by Stroosnijder (1976). The conservation equation for the liquid phase in soil reads

$$\partial \varepsilon_1 / \partial t = -\partial J_1 / \partial x . \tag{4}$$

This equation states for a volume element of the soil system that the rate of change in volume of liquid equals the difference between the rates of inflow and outflow.

### 2. *Transport of solutes*

The convective transport of herbicides in soil can often be described by the following equation for the flux $J_c[\mathrm{mmol}/(\mathrm{m}^2\ \mathrm{day})]$

$$J_c = J_1 c_1 - D_{cd} \partial c_1 / \partial x . \tag{5}$$

The product of liquid flux, $J_1$, and concentration in the liquid phase, $c_1$, represents the "average" convective transport (piston flow). The spreading due to convective dispersion is described by analogy to diffusion, with $D_{cd}[\mathrm{m}^3/(\mathrm{m\ day})]$ being the coefficient of convective dispersion. At low to moderate liquid fluxes, this coefficient is approximately proportional to the liquid flux:
$D_{cd} = L_d |J_1|$; the factor $L_d$ (m) being the dispersion length parameter (Bolt, 1979).

The diffusion of a substance through the liquid phase in soil can be described by Fick's law

$$J_{d1} = -D_{1p} \partial c_1 / \partial x . \tag{6}$$

The coefficient $D_{1p}$[m$^3$/(m day)] for diffusion through the liquid phase is calculated by $D_{1p} = f_1 \varepsilon_1 D_1$, that is the coefficient of diffusion in bulk liquid $D_1$(m$^2$/day) multiplied by the volume fraction of liquid $\varepsilon_1$ and by the tortuosity factor $f_1$(m$^2$/m$^2$). The overall flux equation for the substance then reads

$$J_s = J_1 c_1 - D_{cd}\,\partial c_1/\partial x - D_{1p}\,\partial c_1/\partial x\,, \tag{7}$$

$J_s$ is the total substance flux in mmol/(m$^2$ day).

The partition at equilibrium of a substance (ignoring volatility) between the phases in soil can be described by

$$c_t = \varepsilon_1 c_1 + \rho_b K_{s/1} c_1 \tag{8}$$

The amount of substance per unit volume total soil $c_t$ (mmol/m$^3$) is composed of the amount in the liquid phase $\varepsilon_1 c_1$ and the amount adsorbed $\rho_b K_{s/1} c_1$ . $K_{s/1}$ (m$^3$/kg) is the adsorption coefficient. The rate of change in amount of substance per volume element of the soil system is determined by transport and a sink term

$$\partial c_t/\partial t = -\partial J_s/\partial x - R_c\,. \tag{9}$$

The transport contribution consists of the difference between the rates of transport into and out of the element. A sink term is commonly the rate of conversion of the chemical $R_c$[mmol/(m$^3$ day)].

## B. Lay-out of the Computations

Herbicide movement is described in a one-dimensional, vertical soil system, that is assumed to be homogeneous. The system is divided into 18 computation compartments; the top five compartments are 0·01 m thick, while below 0·05 m compartment size increases each time by a factor of 1·25. The tensiometer pressure at 1·0 m depth is assumed to be in equilibrium with a ground water table at 1·5 m depth. Water flow in the soil system is described using the hydraulic characteristics given by Stroosnijder (1976, Fig. 26 i) for a sandy loam soil. At the start of the computations, the soil moisture profile is in equilibrium with the ground water table.

The weather data are taken from reports of daily observations in the months April and May at De Bilt weather station (KNMI, 1973; 1978). The spring of 1973 was wetter than the average, the rainfall in April and May being 1·5 times and 2·1 times average respectively. The corresponding period in 1978 was drier with 0·8 times and 0·5 times the average rainfall in April and May, respectively. The daily values of rainfall introduced into the computations are given in Fig. 5, together with the cumulative amounts of rainfall for the periods. The evaporation of water from the soil surface

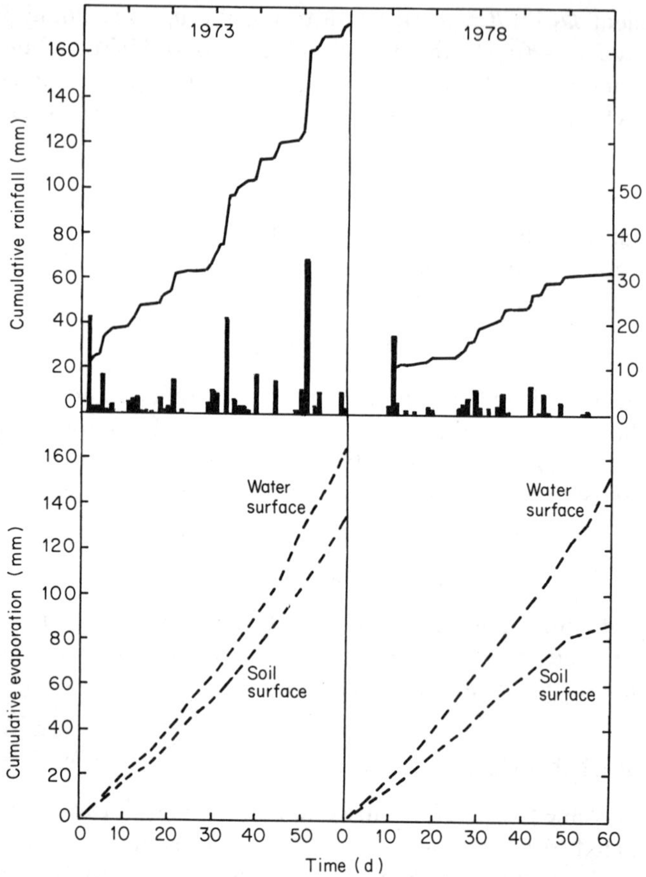


Fig. 5. Rainfall data from the De Bilt weather station, evaporation from a water surface calculated with Penman's method, and evaporation from the soil surface estimated in computations. Day 0 = 1 April in 1973 and 1978, respectively.

was estimated in separate computations by a procedure described by Leistra *et al.* (1980) and the cumulative evaporation patterns are given in Fig. 5. Daily values of evaporation rate are introduced as forcing functions into the main computations.

A simplified pattern of soil temperatures as a function of time is included, based on records for several years for 10 cm depth in soil. Starting from 7°C on April 1, there was a linear increase in temperature to 11°C around May 1 and a further increase to 15°C at the end of May.

Herbicides are assumed to be applied on 1 April at a dosage of 0·91 mmol/m$^2$, which corresponds to 2 kg a.i./ha (molecular mass 220). At the start of the computations, this dosage is assumed to be present in a top layer 0·01 m thick. In separate computer runs, three values of the adsorption coefficient $K_{s/1}$ were introduced: 0·25 $10^{-3}$, 1·0 $10^{-3}$ and 4.0 $10^{-3}$ (mmol/kg)/(mmol/m$^3$) respectively. These values represent herbicide–soil combinations with weak, moderate and rather strong adsorption, respectively. The conversion rate of the herbicides is introduced as a function of soil temperature and of mass fraction of liquid in the soil. Assuming first-order kinetics of decomposition, the rate coefficients $k_r$ under moist conditions range from 0·0087 d$^{-1}$ (half-life 80 days) at 7°C to 0·0173 d$^{-1}$ (half-life 40 days) at 15°C. At mass fractions of liquid below 0·20 kg/kg, these rate coefficients are multiplied by a factor read from a curve through the points: factor 0·5 at 0·05 kg/kg and factor 0·8 at 0·10 kg/kg. The bulk density of the soil is set at $\rho_b = 1200$ kg/m$^3$. A value of 0·02 m is taken for the dispersion length parameter $L_d$, which is in the range of values found for substances in soils.

The equations, parameters and basic relationships are programmed in the computer language CSMP III (IBM, 1975). The concentrations in the liquid phase in adjacent compartments are averaged in the computation of the convective flux, to suppress numerical dispersion. The various rate equations are integrated with Euler's method using a time step $\Delta t$ of 0·0025 day and the program was run on a DEC system-10 computer.

## C. Results and Discussion

The concentration patterns in soil computed for less than average amounts of rainfall (data for April and May in 1978) are shown in Fig. 6. The differences in the concentration levels for the liquid phase caused by the differences in adsorption coefficient are striking and clearly show how the availability of a herbicide decreases as the adsorption coefficient becomes higher.

During the first 20 days, substantial penetration occurs, mainly because of the rainfall (17 mm) on day 11. The results in Fig. 6 clearly show how penetration in the soil is reduced at the comparatively high values of the adsorption coefficient. The cumulative amount of evaporation from soil in the 60-day period exceeded the cumulative amount of rainfall (Fig. 5). Consequently, the concentrations near the soil surface remain highest throughout the period. For the situation with a mobile herbicide, the concentrations near the soil surface even increase in the period from 20 to 60 days.

The concentration patterns computed for the fairly high amounts of

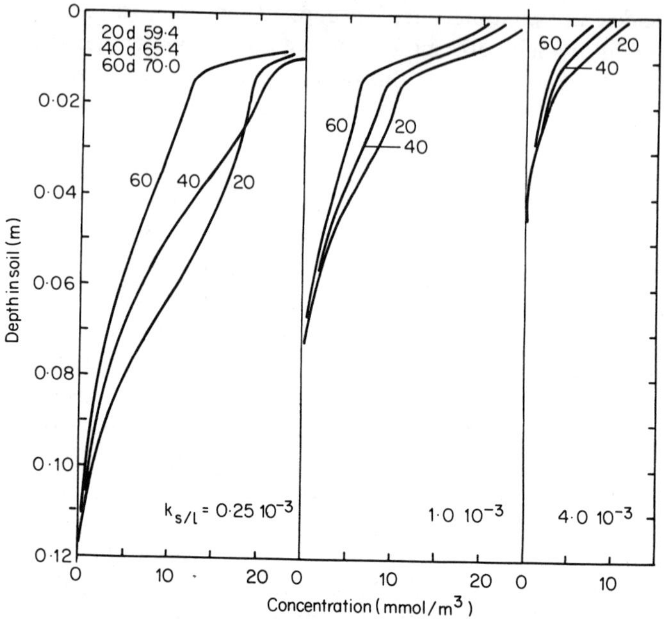


Fig. 6. Computed concentrations of herbicides in the liquid phase at 20, 40 and 60 days after application. Weather conditions for 1978. Adsorption coefficients 0·25 $10^{-3}$, 1·0 $10^{-3}$ and 4·0 $10^{-3}$ $m^3/kg$, respectively.

rainfall (data for 1973) are represented in Fig. 7. The rainfall surplus (see Fig. 5) results in fairly flat distributions with comparatively deep penetration into the soil. At the weakest adsorption, the concentrations below 0·06 m depth are distinctly higher than with low rainfall. The same holds for depths below 0·04 m, when the adsorption coefficient amounts to 1·0 $10^{-3}$ (mmol/kg)/(mmol/$m^3$). These computed results illustrate quantitatively how the depth protection of the crop will diminish as the amounts of rainfall increase.

The influence of a crop on the movement of a substance in soil under spring conditions has been estimated to be small during the first month, increasing gradually during the second month (Leistra, 1978). Consequently, it seems possible to draw some general conclusions from the computed results. Herbicide–soil combinations showing weak adsorption will nearly always present high risk of damage to the crop, even at amounts of rainfall around the average. The other extreme is that when herbicides are quite strongly adsorbed, depth protection will nearly always be adequate. Thus the mechanism of depth protection of moderately adsorbed

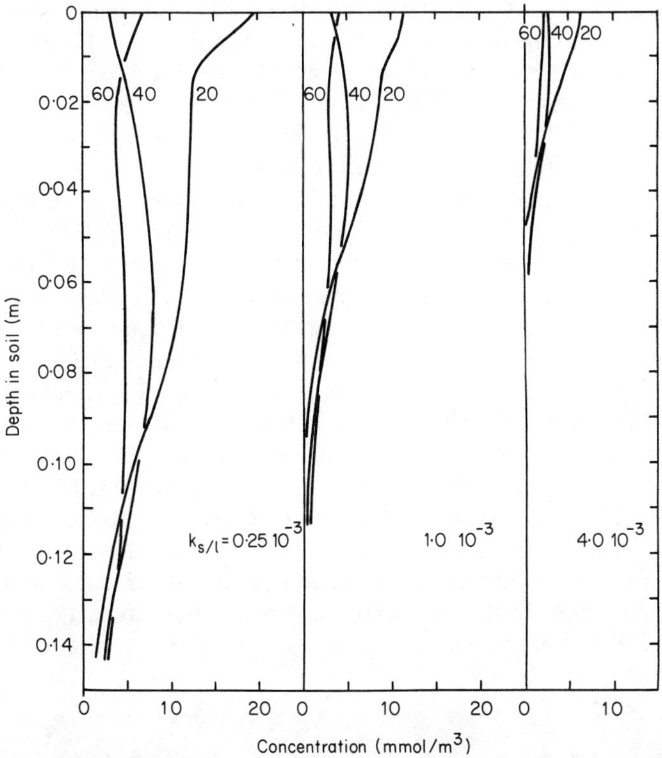


Fig. 7. Computed concentrations of herbicides in the liquid phase at 20, 40 and 60 days after application. Weather conditions for 1973. Adsorption coefficients 0·25 $10^{-3}$, 1·0 $10^{-3}$ and 4·0 $10^{-3}$ $m^3/kg$, respectively.

herbicides needs to be carefully studied. The results of the computations indicate that some penetration into the rooting zone of the crop is always likely, so the crop should also have a certain degree of physiological selectivity.

## D. General Remarks

Only part of the total mechanism of herbicide selectivity to crops is discussed in this section. For a comprehensive description, the complete soil-plant system must be modelled. The elements needed for such comprehensive models are discussed in other chapters in this book. For example, for soil-acting photosynthesis inhibitors the following processes would be included: the development of the root system of the crop; the uptake activity of the roots; the translocation of the herbicide to the chloroplasts, and its conversion during translocation. A start has been made in the development of more comprehensive models by including root development and uptake activity in computations on the movement of pesticides in soils (Leistra and Dekkers, 1976; Leistra, 1978) but much more work is needed before complete sets of selectivity mechanisms can be described quantitatively.

After they have been adequately tested and refined, simulation models could be used to answer various questions. Adsorption coefficients and climatic data could be introduced into models to estimate the dosage that would be needed for safe weed control. Similarly, on the basis of meteorological records, estimates could be made of the frequency with which a dangerous penetration of herbicide in the rooting zone of a crop will occur or in arid regions the optimum amount and timing of sprinkler irrigation in the initial period after herbicide application could be selected. Injury to the crop, such as reported by Von Amsberg and Burtch (1970) after application of chloridazon in sugar beet growing on soils low in organic matter, could thus be prevented. The management of herbicide penetration into the soil by sprinkler irrigation has been reviewed by Horowitz (1972) who points out the important consequences for effectiveness and for the safety to the crop. In the future, the number of field trials and the period of testing of soil-acting herbicides may be significantly reduced by using computation models.

## VI. Conclusions and Recommendations

Conclusions on specific topics are given at the end of the relevant sections in this chapter; some more general conclusions are given below, especially with respect to the evaluation of field behaviour of herbicides in soil.

Various methods of characterizing the mobility of herbicides in soil are available. These methods range from the straightforward adsorption measurements in the laboratory to the time-consuming measurement of movement in field soil. For most of the compounds, adsorption measurement or soil column experiments in the laboratory will be sufficient. The lay-out of experiments could be greatly improved and more attention should be paid to recording all the essential experimental details.

The results of laboratory experiments can be introduced into computer simulations of the field behaviour of herbicides and soil and climatic conditions relevant to the application areas in various parts of the world can be included. First estimates can be made on aspects such as the risk of damage to the crop or the risk of leaching to groundwater. Further development and testing of the models is needed to improve the accuracy of the predictions.

The accumulation of results from numerous poorly-defined field experiments seems a waste of time and skill. A limited number of high-quality field experiments would be a better use of resources. Their main function should be to check whether or not the expectations based on laboratory data, greenhouse data and computer simulations are correct. More extensive field experimentation would then be needed only in instances in which there seems to be risk of crop damage or a risk of leaching from the soil.

Interest in the movement of conversion products of herbicides in soil is increasing, especially in view of possible environmental consequences. Computer simulation models for herbicides and pesticides can be fairly easily extended to include the behaviour of their conversion products (Bromilow and Leistra, 1980). Again, conversion rates and adsorption coefficients can be measured in the laboratory.

Computation models constitute a framework for quantitative thinking and in this way are helpful in the interchange of information and ideas between research workers. They enable the results from related scientific disciplines to be utilized for herbicide work. Further research can be better planned after computed and measured concentration patterns have been compared: the nature and extent of any discrepancies indicate which aspects deserve further attention.

## References

Abernathy, J. R. and Davidson, J. M. (1971). Effect of calcium chloride on prometryne and fluometuron adsorption in soil. *Weed Sci.*, **19**, 517–521.

Addiscott, T. M. (1977). A simple computer model for leaching in structured soils. *J. Soil Sci.*, **28**, 554–563.

Anon. (1973). *Prüfung des Versickerungsverhaltens von Pflanzenschutzmitteln.* Merkblatt Nr. 37, 1 Auflage. Biologische Bundesanstalt für Land- und Forstwirtschaft, Braunschweig, Deutsche Bundesrepublik.

Anon. (1975). Environmental Protection Agency. Guidelines for registering pesticides in the United States. *Federal Register*, **40**, No. 123, 26801–26928.

Anon. (1978). Environmental Protection Agency. Proposed guidelines for registering pesticides in the United States. *Federal Register*, **43**, No. 132, 29696–29741.

Bannink, D. W., Poelstra, P. and Frissel, M. J. (1977). An installation for transport and accumulation studies on undisturbed soil columns under controlled conditions. *Neth. J. agric. Sci.*, **25**, 126–133.

Baumeister, P. and Hurle, K. (1975). Abbau von Atrazin und Simazin unter Freilandbedingungen. *Z. PflKrankh. PflSchutz, Sonderheft*, **7**, 154–167.

Bolt, G. H. (1979). Movement of solutes in soil: Principles of adsorption/exchange chromatography. *In Soil Chemistry B. Physico-chemical Models.* (G. H. Bolt, Ed.) Chapter 9. Elsevier Sci. Publ. Cy. Amsterdam.

Bromilow, R. H. and Leistra, M. (1980). Measured and simulated behaviour of aldicarb and its oxidation products in fallow soils. *Pestic. Sci.*, **11**, (in press).

Brown, K. W., Gerard, C. J., Hipp, B. W. and Ritchie, J. T. (1974). A procedure for placing large undisturbed monoliths in lysimeters. *Soil Sci. Soc. Am. Proc.*, **38**, 981–983.

Burkhard, N., Eberle, D. O. and Guth, J. A. (1975). Model systems for studying the environmental behaviour of pesticides. *Envir. Qual. Saf., Suppl.* **3**, *Pesticides*, 203–213.

Calvet, R., Terce, M. and Le Renard, J. (1975). Cinétique de dissolution dans l'eau de l'atrazine, de la propazine et de la simazine. *Weed Res.*, **15**, 387–392.

Chu-Huang Wu, Buehring, H., Davidson, J. M. and Santelmann, P. W. (1975). Napropamide adsorption, desorption, and movement in soils. *Weed Sci.*, **23**, 454–457.

Clay, D. V. and Stott, K. G. (1973). The persistence and penetration of large doses of simazine in uncropped soil. *Weed Res.*, **13**, 42–50.

Coats, K. R. and Smith, B. D. (1964). Dead-end pore volume and dispersion in porous media. *Soc. Petrol. Engrs. J.*, **4**, 73–84.

Davidson, J. M. and Chang, R. K. (1972). Transport of picloram in relation to soil physical conditions and pore-water velocity. *Soil Sci. Soc. Am. Proc.*, **36**, 257–261.

Davidson, J. M. and McDougal, J. R. (1973). Experimental and predicted movement of three herbicides in a water-saturated soil. *J. envir. Qual.*, **2**, 428–433.

Davidson, J. M., Mansell, R. S. and Baker, D. R. (1973). Herbicide distributions within a soil profile and their dependence upon adsorption–desorption. *Proc. Soil Crop Sci. Soc. Fla.*, **32**, 36–41.

Dunham, R. J. and Crawford, D. V. (1973). Seedbed moisture and the activity of triazines. *Proc. Eur. Weed Res. Coun. Symp. Herbicides-Soil*, 251–261.

Führ, F., Cheng, H. H. and Mittelstaedt, W. (1976). Pesticide balance and metabolism studies with standardized lysimeters. *Landw. Forsch., Sonderheft*, **32**, 272–278.

Gaudet, J. P., Jégat, J., Vachaud, G. and Wierenga, P. J. (1977). Solute transfer, with exchange between mobile and stagnant water, through unsaturated sand. *Soil Sci. Soc. Am. J.*, **41**, 665–670.

Gerber, H. R. and Guth, J. A. (1973). Short theory, techniques and practical importance of leaching and adsorption studies. *Proc. Eur. Weed Res. Coun. Symp. Herbicides-Soil*, 51–69.

Green, R. E., Rao, P. S. C. and Corey, J. C. (1972). Solute transport in aggregated soils: Tracer zone shape in relation to pore-velocity distribution and adsorption. *Proc. 2nd Symp. Fundamentals of transport processes in porous media* (IAHR-ISSS), **2**, 732–752. Univ. of Guelph, Canada.

Guth, J. A., Burkhard, N. and Eberle, D. O (1976). Experimental models for studying the persistence of pesticides in soils. Br. Crop Prot. Coun. Monograph 17, *Persistence of insecticides and herbicides* (K. I. Beynon, Ed.) 137–157.

Hance, R. J. and Embling, S. J. (1979). Effect of soil water content at the time of application on herbicide content in soil solution extracted in a pressure membrane apparatus. *Weed Res.*, **19**, 201–205.

Hornsby, A. G. and Davidson, J. M. (1973). Solution and adsorbed fluometuron concentration distribution in a water-saturated soil: Experimental and predicted evaluation. *Soil Sci. Soc. Am. Proc.*, **37**, 823–828.

Horowitz, M. (1972). Ecology of herbicides under irrigated subtropical conditions. *Proc. 2nd int. Congr. Pestic. Chem.*, (A. S. Tahori, Ed.) **6**, 101–115. Gordon and Breach, New York.

IBM (1975). Continuous system modelling program III (CSMP III). Program Reference Manual, Program No. 5734-XS9. Data Processing Division, White Plains, New York.

Jürgens-Gschwind, S. and Jung, J. (1979). Results of lysimeter trials at the Limburgerhof facility, 1927–1977: The most important findings from 50 years of experiments. *Soil Sci.*, **127**, 146–160.

Khan, S. U., Marriage, P. B. and Saidak, W. J. (1976). Persistence and movement of diuron and 3,4-dichloroaniline in orchard soil. *Weed Sci.*, **24**, 583–586.

KNMI (1973, 1978). *Maandelijks overzicht der weersgesteldheid.* Royal Netherlands Meteorological Institute, De Bilt, The Netherlands.

Leistra, M. (1973). The use of computation models in studies on herbicide behaviour in soil. *Proc. Eur. Weed Res. Coun. Symp. Herbicides-Soil*, 216–226.

Leistra, M. (1977). A model for the transport of pesticides in soil with diffusion-controlled rates of adsorption and desorption. *Agric. Envir.*, **3**, 325–335.

Leistra, M. (1978). Computed redistribution of pesticides in the root zone of an arable crop. *Pl. Soil*, **49**, 569–580.

Leistra, M. and Dekkers, W. A. (1976). Computed leaching of pesticides from soil under field conditions. *Wat. Air Soil Pollution*, **5**, 491–500.

Leistra, M. and Dekkers, W. A. (1977). Computed effects of adsorption kinetics on pesticide movement in soils. *J. Soil Sci.*, **28**, 340–350.

Leistra, M., Bromilow, R. H. and Boesten, J. J. T. I. (1980). Measured and simulated behaviour of oxamyl in fallow soils. *Pestic. Sci.*, **11**, (in press).

Leistra, M., Smelt, J. H., Verlaat, J. G. and Zandvoort, R. (1974). Measured and computed concentration patterns of propyzamide in field soils. *Weed Res.*, **14**, 87–95.

Leistra, M., Smelt, J. H. and Zandvoort, R. (1975). Persistence and mobility of bromacil in orchard soils. *Weed Res.*, **15**, 243–247.

Lindstrom, F. T., Boersma, L. and Stockard, D. (1971). A theory on the mass transport of previously distributed chemicals in a water saturated sorbing porous medium: Isothermal cases. *Soil Sci.*, **112**, 291–300.

Marriage, P. B., Khan, S. U. and Saidak, W. J. (1977). Persistence and movement of terbacil in peach orchard soil after repeated annual applications. *Weed Res.*, **17**, 219–225.

Muir, D. C. G. and Baker, B. E. (1976). Detection of triazine herbicides and their degradation products in tile-drain water from fields under intensive corn (maize) production. *J. agric. Fd. Chem.*, **24**, 122–125.

Muir, D. C. G and Baker, B. E. (1978). The disappearance and movement of three triazine herbicides and several of their degradation products in soil under field conditions. *Weed Res.*, **18**, 111–120.

Poelstra, P., Bannink, D. W. and De Swart, J. G. (1973). An experimental approach to study the migration of pollutant in undisturbed soil columns. *Agricultura* (Leuven), **21**, 147–157.

Rao, P. S. C. (1974). *Pore-geometry effects on solute dispersion in aggregated soils and evaluation of a predictive model.* PhD Thesis, Univ. of Hawaii.

Rao, P. S. C., Davidson, J. M., Jessup, R. E. and Selim, H. M. (1979). Evaluation of conceptual models for describing nonequilibrium adsorption-desorption of pesticides during steady-flow in soils. *Soil Sci. Soc. Am. J.*, **43**, 22–28.

Riley, D. and Morrod, R. S. (1976). Relative importance of factors influencing the activity of herbicides in soil. *Proc. 13th Br. Weed Control Conf.*, 971–980.
Schiavon, M. and Jacquin, F. (1973a). Contribution à l'étude de la migration de deux triazines sous l'infuence des précipitations. *Proc. Eur. Weed Res. Coun. Symp. Herbicides-Soil*, 80–90.
Schiavon, M. and Jacquin, F. (1973b). Etude de la présence d'atrazine dans les eaux de drainage. *C.r. Journées d'Etudes Herbic., Columa*, 35–43.
Skopp, J. and Warrick, A. W. (1974). A two-phase model for the miscible displacement of reactive solutes in soil. *Soil Sci. Soc. Am. Proc.*, **38**, 545–550.
Smith, D. T. and Meggitt, W. F. (1970). Movement and distribution of pyrazon in soil. *Weed Sci.*, **18**, 255–259.
Soil Survey Staff (1975). *Soil taxonomy*. Soil Conservation Service, U.S. Dept. of Agriculture. Handbook No. 436. U.S. Government Printing Office, Washington D.C.
Stoller, E. W., Wax, L. M., Haderlie, L. C. and Slife, F. W. (1975). Bentazon leaching in four Illinois soils. *J. agric. Fd Chem.*, **23**, 682–684.
Stroosnijder, L. (1976). Infiltration and redistribution of water in soils. *Agric. Res. Rep.*, **847**. Pudoc, Wageningen, The Netherlands.
Van Genuchten, M. T. and Wierenga, P. J. (1976). Mass transfer studies in sorbing porous media. I. Analytical solutions. *Soil Sci. Soc. Am. J.*, **40**, 473–480.
Van Genuchten, M. T., Davidson, J. M. and Wierenga, P. J. (1974). An evaluation of kinetic and equilibrium equations for the prediction of pesticide movement through porous media. *Soil Sci. Soc. Am. Proc.*, **38**, 29–35.
Von Amsberg, H. and Burtch, L. (1970). Some factors influencing sugar beet herbicide performance under sprinkler irrigation in the far Western United States.
*Second Int. Meeting on Selective Weed Control in Beet Crops*, Rotterdam, 283–290.
Von Stryk, F. G. and Bolton, E. F. (1977). Atrazine residues in tile-drain water from corn plots as affected by cropping practices and fertility levels. *Can. J. Soil Sci.*, **57**, 249–253.
Wierenga, P. J. (1977). Solute distribution profiles computed with steady-state and transient water movement models. *Soil Sci. Soc. Am. J.*, **41**, 1050–1055.

# 3 Transport in the Vapour Phase

*R. J. Hance*
Agricultural Research Council
Weed Research Organization
Begbroke Hill,
Yarnton, Oxford OX5 1PF U.K.



## I. Introduction

This subject has been reviewed frequently in recent years (Hartley, 1967; Hamaker, 1972; Spencer *et al.*, 1973; Spencer and Cliath, 1975; Plimmer, 1976; Wheatley, 1976). Therefore wherever possible reference will be made to these articles in order to minimize repetition. Despite the activity that these reviews have summarized, there remain gaps in our knowledge in both experimental observation and theory. It is important that we attempt to fill these gaps. As noted in the EPA Guidelines for Registering Pesticides in the US (Anon., 1978) "Pesticide residue movement may cause contamination of the food web (including non-target crops), loss of usable land and water resources to man and habitat loss to wild life". In the case of herbicides applied to the soil, to which this chapter is confined, movement also affects their performance as weed-killers. Whilst we know that many soil herbicides have substantial potentials for movement in the vapour phase, it is not at the moment always clear how large a role such movement plays in bringing them to the target plant, or conversely, in providing a mechanism for loss from the treated area.

## II. Factors Affecting Transfer in the Vapour State

### A. Vapour Pressure

Although the relative intrinsic tendency of pure materials to evaporate is indicated by their vapour pressures, the indiscriminate use of vapour pressure to characterize compounds can be misleading. Vapour pressure is an equilibrium property but as Que Hee and Sutherland (1974) point out, the atmosphere can be regarded as an infinite sink so that volatilization of pesticides is a dynamic process in which kinetics are important. Hence the evaporation rate of a pure compound applied to an inert surface depends not only on its vapour pressure and the temperature, but also on air flow and the ratio of surface area to applied mass of the deposit.

### B. Air–water Partition

The number of factors involved increases in the soil. After application at least some of the material is in solution so solubility, or rather the distribution ratio between water and air, must be considered. For this reason vapour movement can be relatively important for compounds of very low vapour pressure if they also have low water solubility. DDT is a good example. Its vapour pressure at 25° is only about $2{\cdot}7 \times 10^{-5}$ Pa ($2 \times 10^{-7}$ mm) but since its water solubility is only some 0·0012 μg/ml, the water: air partition ratio is around 320, a figure very similar to that for trifluralin, which has a vapour pressure about 1000 times as large. Since pesticides often reach aquatic systems, this ratio can be a useful guide to likely volatility behaviour with regard to environmental contamination and tables of values for a number of pesticides can be found in Hamaker (1972), Wheatley (1976) and the EPA guidelines (Anon., 1978). It must, however, be used with discretion because it is the vapour pressure that determines the magnitude of the quantities that move as vapour.

### C. Adsorption

The other soil process to consider is adsorption. In moist soil this controls the concentration of herbicide in solution hence its partial pressure. Therefore in order to characterize the evaporative tendencies of a soil applied herbicide we need to know something of its adsorption behaviour as well. As is discussed in Chapter 1, adsorption is a function of soil, as well as herbicide, properties. For this reason the EPA guidelines (Anon., 1978) contain a table of estimated tendencies of chemicals to volatilize from a moist soil. This is based on adsorption by a "standard" soil of 2% organic matter and therefore assumes adsorption is mostly determined by soil organic matter. To obtain a rough estimate of this sort it is perhaps not necessary to consider the soil as such. Briggs (1973) has shown that soil adsorption characteristics of non-ionic substances are reasonably well predicted by octanol : water partition ratios ($P$) using the relationship $\log Q = 0{\cdot}52 \log P + 0{\cdot}62$ where $Q$ is given by $\frac{100K}{\%\ \text{organic matter}}$, $K$ being the soil : water partition ratio. This approach might be taken even further as Chiou and Freed (1977) suggested there is also a relationship between vapour pressure and octanol : water partition, at least within groups of chemicals containing the same hydrophilic groups.

At the moment these relationships are not sufficiently well established to allow us to rely on octanol : water partition ratios to predict vapour pressure and adsorption characteristics with confidence. If enough data can be assembled to show that this could be done, then early screening procedures for pesticides would be greatly simplified. There seems to be some reluctance to publish vapour pressure figures and partition ratios even though they are required for registration purposes in many countries. This remark does not apply to all companies, at least as far as vapour pressures are concerned, but little over half the entries in the tables of herbicide properties in the Weed Control Handbook (Fryer and Makepeace, 1977) include figures for vapour pressures at temperatures of 25°C or below. Whilst caution in disseminating many sorts of information is commercially prudent, in this case it may be counterproductive if it delays the possible streamlining of registration requirements.

The discussion so far relates to wet soil. In dry soil adsorption may be even more important. This is because many soil surfaces adsorb water more readily than non-polar compounds but can adsorb the latter if water is absent. Hence as the soil dries, new adsorption sites become available to the herbicide so vapour pressure falls. The relation between adsorption and soil water content varies with the compound. With weakly adsorbed materials adsorption does not increase markedly till the water content is less than that needed to form a monomolecular layer as has been demonstrated for dieldrin (Spencer *et al.*, 1969), lindane (Spencer and Cliath, 1970) and atraton and monuron (Hance, 1977). However, some compounds compete sufficiently well with water for increases in adsorption to occur at higher water contents. For instance Spencer and Cliath (1974) reported that trifluralin vapour pressure began to decrease (hence adsorption increased) at higher soil water contents than those of lindane and dieldrin and Hollingsworth (1977) found that volatilization of metribuzin was less sensitive to changes in soil water than that of trifluralin.

## D. Temperature

Temperature influences vapour movement because of its effects on vapour pressure. The response usually follows the relationship $\log_{10} p = A - B/T$ where $A$ and $B$ are constants, $T$ is the temperature and $p$ is the vapour pressure. The values of $A$ and $B$ in any particular circumstance depend not only on heat of vaporization but also on the heats of solution and adsorption. Sometimes there is the added complication that increase in temperature causes the soil to dry so that adsorption increases and volatilization of the herbicide actually decreases with increasing temperature, an effect demonstrated by Deming (1963) with allidochlor.

# III. Processes Involved in the Transfer of Vapour

## A. Diffusion

This is the consequence of random molecular motion as a result of which material moves from a higher to a lower concentration, in accordance with the second law of thermodynamics. The rate of transfer of material is proportional to the concentration gradient normal to the direction of movement (Fick's law). The basic mathematics of the process are comprehensively described in Hamaker (1972) and Crank (1975).

Vapour phase diffusion in the soil is, of course, controlled by the same factors that control vapour pressure, that is temperature, adsorption and soil water content. Other factors are also involved, however, including soil porosity, hence bulk density, the tortuosity of soil pores and the number of blocked pores.

For molecules the size of herbicides, diffusion as vapour is $10^4$ or more times faster than diffusion in water so that the effect of soil water content on overall diffusion rate depends very much on the air : water partition ratio of a compound. One illustration of this is given by Graham-Bryce's (1968) work with disulfoton and dimethoate. Dimethoate, which has a water : air partition of about $2{\cdot}8 \times 10^8$, diffuses primarily in the liquid phase so its overall diffusion rate increased sharply with increasing soil water content. Disulfoton diffusion was relatively insensitive to water content because its water : air distribution (about $5 \times 10^3$) is such that its rate of transfer is about the same in both phases. Scott and Phillips (1972) made similar observations with 10 herbicides of which only trifluralin (water : air partition $3{\cdot}2 \times 10^2$) showed an increase in diffusion rate as soil water content decreased. Calculations based on these results by Letey and Farmer

(1974) suggested a possible vapour phase component in the diffusion of chlorpropham, diphenamid, prometon and simazine as well as for trifluralin. Jaques and Harvey (1979) reported that benfluralin and profluralin, as well as trifluralin, diffused more at soil water contents below saturation while with dinitramine and fluchloralin vapour and solution diffusion compensated for each other as moisture levels changed.

Soil porosity, together with soil water content, gives a measure of the air space available for vapour diffusion. Working with trifluralin, Bode *et al.* (1973a) found that at a constant water content, but variable bulk density, vapour diffusion decreased by about 50% for each 10% decrease in air-filled porosity. Shearer *et al.* (1973) noted with lindane that vapour diffusion reached a maximum at about 4% water. At lower levels adsorption increased so diffusion was reduced; from 4–20% water there was a slight decrease and above 20% the rate fell rapidly as pores became filled with water. They also suggested that diffusion could occur at air–water and soil–water interfaces. Similar trends were reported for tritiated water and metribuzin by Scott and Paetzold (1978).

Bode *et al.* (1973b) studied the effects of temperature, soil moisture and bulk density on trifluralin diffusion and developed a multiple regression equation to describe the interactions. This work was done in isothermal conditions but in the field there are temperature gradients. It is known that water moves from high to low temperature regions (see for example Jackson, 1973) and undoubtedly the same process will occur with pesticides, but there appears to be no published information on the subject. Similarly the effects of volume changes brought about by diurnal temperature fluctuations have not received attention.

The quantities of herbicide moving by vapour diffusion could be substantial. Bode *et al.* (1973a) obtained values for the apparent vapour diffusion coefficient of trifluralin in soil which varied between $27{\cdot}7 \times 10^{-8}$ and $2{\cdot}5 \times 10^{-8}$ cm$^2$/s. Graham-Bryce (1968) calculated that disulfoton (overall diffusion rate in soil of about $2{\cdot}6 \times 10^{-8}$ cm$^2$/s) if present in soil at a concentration of 110 $\mu$g/g would diffuse to a root of radius 0·02 cm at a rate of 0·156 $\mu$g/h/m root length. Hence vapour diffusion could well be important in moving some herbicides to the target plant. It could also be important as a route of loss of material from the soil. Mayer *et al.* (1974) developed a series of predictive mathematical models for the diffusion controlled volatilization of soil incorporated pesticides which took account of different assumptions about the behaviour of the air at the soil surface. They fitted experimental results reasonably well. However, while in the laboratory boundary conditions can be controlled, this is not possible in the field so that models based on diffusion alone are incomplete.

## B. Movement in Solution to Evaporating Surfaces

A major factor in the field is the movement of dissolved herbicide when there is mass flow of water to the soil surface (the "wick" effect). This replenishes pesticide lost from the surface by volatilization. Hamaker (1972) suggested that pesticide loss by the diffusion route could be calculated from the ratio of diffusion rates of water and pesticide if water losses by diffusive movement were known so that the total loss could be calculated from

$$Qt = \frac{VP}{VP_{H_2O}} \times \frac{D}{D_{H_2O}} \times (Q_{H_2O})_V + C \times (Q_{H_2O})_L$$

where $Q$ is the quantity of water lost as vapour ($V$) and liquid ($L$), $C$ is pesticide concentration in soil solution, $VP$ and $VP_{H_2O}$ are the vapour pressures of pesticide and water, $D$ and $D_{H_2O}$ are diffusion coefficients and $Qt$ is total loss of pesticide. Although this expression provides a method for estimating the rate of loss from knowledge of total water loss, it ignores many factors. Starting with a uniform system, the initial rate of pesticide evaporation will be determined by diffusion in air and the initial concentration. The surface concentration will then be depleted by diffusive loss and increased by pesticide arriving with the water flow. If the solute is less volatile than water, then its concentration at the surface will increase until a steady state is reached when the sum of diffusive loss to the air and diffusive flux in the "wick" are equal to flow transfer in the "wick". In practice the upward flow of water will not be uniform, evaporation is not necessarily confined to the surface and solute can also move downwards in percolating water. Thus the situation is much more complicated than this equation implies.

Spencer and Cliath (1973) examined the relationship between diffusion and mass flow movement by measuring rates of lindane flux from moist soil into air of different relative humidities. They found that the calculation of volatilization rates from the product of water flux and lindane concentration and adding this value to an assumed diffusion rate gave similar rates of loss to those obtained experimentally. For the less soluble dieldrin there was some accumulation at the surface after exposure to low humidities. This treatment, however, can be criticized on similar grounds to that of Hamaker.

Taylor *et al.* (1976) were able to explain rates of volatilization of dieldrin and heptachlor from a maize field by reference to water movement. There were marked diurnal variations with maxima close to noon. By contrast, Harper *et al.* (1976), working with trifluralin and lindane, also found diurnal variations but with day time minima. This, they suggested, was because the surface soil dried out during the day hence adsorption of pesticide increased and volatilization decreased. At night evaporation of water was lower, so the surface soil became wetter, adsorption decreased and pesticide volatilization increased. They estimated that only when the soil moisture content exceeded that equivalent to three molecular layers did pesticide fluxes respond to temperature and air turbulance.

Although these two opposite results can be explained reasonably, given present knowledge they could not have been predicted accurately. To do this, information is needed of the relative strengths of adsorption of water and pesticides on soil mineral and organic surfaces as well as better quantitative understanding of water movement in soil.

## IV. Volatility Losses in the Field

A number of estimates based on laboratory data have been made (see for example Plimmer, 1976; Spencer *et al.*, 1973) but very little field information has been obtained. Table I, taken from Taylor (1977) summarizes much of it. It shows clearly that such losses can be large and that incorporation of the chemical with the soil substantially reduces them although the possibility that some photochemical degradation occurred cannot be excluded. Many more field observations are needed before theory can be used predictively.

For example, even the consequences of incorporating volatile materials in the soil are not completely certain. Kennedy and Talbert (1977) found that although delays of up to seven days in incorporating a range of dinitroaniline herbicides led to losses, the relative persistence of these compounds could not entirely be explained in terms of volatilization. Similarly the effect of soil water content is variable. Anderson (1975) reported no significant vapour loss of di-allate after application to a dry soil, Ishikawa *et al.* (1977) found greater losses of benthiocarb from flooded than non-flooded soil and these are the expected results. Plimmer (1976) quotes several similar observations. However, Hollingsworth (1977) noted that volatilization of trifluralin, but not that of metribuzin, was affected by soil moisture. This presumably is a consequence of different strengths of adsorption compared with that of water. Such an explanation does not help for the observation of Swann and Behrens (1972b) that losses of trifluralin sprayed on wet soil were 4–10 times higher than those when it was flooded into dry soil with enough water to give the same final soil water content.

This raises the question of reversibility with respect to pesticide adsorption and soil water content. Harper *et al.* (1976), on the basis of trifluralin vapour fluxes, suggested that the increase in adsorption as the soil dried

Table I. Summary of management practices, experimental conditions, location and time in field experiments on pesticide volatilization, with general summary of results. Taken from Taylor (1977).

| Compound | Equilibrium vapour pressure (mm Hg at 25°C) | Experimental surface | Pesticide management | Season | Place | Fraction volatilized in stated period | Reference |
|---|---|---|---|---|---|---|---|
| Heptachlor | $3 \times 10^{-4}$ | Bare soil moist | Surface applied | August | Beltsville Maryland | 90% in 2–3 days | Glotfelty (Taylor 1977) |
| | | Vegetation | Surface application | July | Beltsville | 90% in 30 days | Taylor (1977) |
| | | Soil under maize | Incorporated to 7·5 cm depth | May–October | Coshocton Ohio | 7% in 167 days | Taylor *et al.* (1976) |
| Trifluralin | $2{\cdot}4 \times 10^{-4}$ | Bare soil moist | Surface applied | August | Beltsville Maryland | 90% in 2–3 days | Glotfelty (Taylor 1977) |
| | | Soil under soya beans | 90% incorp. to 2·5 cm | June–October | Athens Georgia | 22% in 120 days | White *et al.* (1977) |
| | | Soil under soya beans | Incorporated evenly to 7·5 cm | May–September | Harford New York | 3·4% in 90 days | Taylor (1977) |

| | | | | | | | |
|---|---|---|---|---|---|---|---|
| Chlorpropham | $10^{-5}$ (est) | Soil under soya beans | Surface application (spray emulsion) | May–July | Frederick Maryland | 49% in 50 days | Turner *et al.* (Taylor, 1977) |
| Dieldrin | $2{\cdot}8 \times 10^{-6}$ | Bare soil | Surface application moist, irrigated soil | September–January | Baton Rouge Lousiana | 20% in 50 days | Willis *et al.* (1971) |
| | | Vegetation | Surface application | August | Beltsville | 88% in 30 days | Taylor *et al.* (1977) |
| | | Soil under maize | Incorporated to 7·5 cm depth | May–October | Coshocton Ohio | 3·6% in 167 days | Taylor *et al.* (1976) |

was reversible but Hance and Embling (1979) found that herbicide concentration in soil solution was much lower following application to air dry soil that was wetted 24 h later than following application to a wet soil. The differences existed for at least 96 h after wetting, the length of the experiment, and seemed likely to last much longer. A final complication comes from the suggestion of Kennedy and Talbert (1977) that the effect of water content on volatility is greater in mineral than in organic soils.

Apart from incorporation as soon after application as possible, the other standard method for reducing vapour loss is to formulate the material as granules (e.g. Parochetti *et al.*, 1971; Hance *et al.*, 1973; Parochetti and Hein, 1973; Spencer-Jones and Wilson, 1976). Collins (1975) suggested that although the rate of vapour release from granular matrices is sensitive to wind and temperature, it should usually follow first order kinetics. The results of Hance *et al.* (1973), that the field persistence of tri-allate in granules was essentially the same whether they contained 2·5, 5 or 10% active ingredient, are consistent with this idea. As expected there is a clear relationship between number of granules applied per unit area and herbicide performance but Maas and Popp (1976) point out that the optimum rate can depend on air conditions. In the greenhouse the compound can be deposited and reabsorbed by the soil whereas under turbulent field conditions it is blown away.

Oil formulations have also received some attention. They do not seem to have any effect in dry soil but may reduce losses from wet soil (Weaver *et al.*, 1971; Smith and Wiese, 1973).

## V. Uptake of Herbicide Vapour by Plants

There is an extensive literature on the phytotoxicity of the vapour of herbicides applied to foliage but this is not considered here. The reader is referred to Plimmer (1976) and Wheatley (1973). Gerber (1977) commented that little had been done to investigate volatility by bioassays and this is broadly still true. There is, nevertheless, good evidence that some plants can take in herbicide vapour through their aerial parts. A good example is that of the parasitic weed dodder (*Cuscuta* spp.) which can be killed by chlorpropham whether or not it is in contact with soil (Slater *et al.*, 1969; Dawson, 1979).

Several authors have used twin pots systems of the sort shown in Figs 1 and 2 to allow roots and shoots to be exposed separately to herbicide vapour (Swann and Behrens, 1972a; Koren, 1976; Jordan *et al.*, 1978). The results show that in general vapour can enter via roots and shoots but there are differences between plant species with the same herbicide and

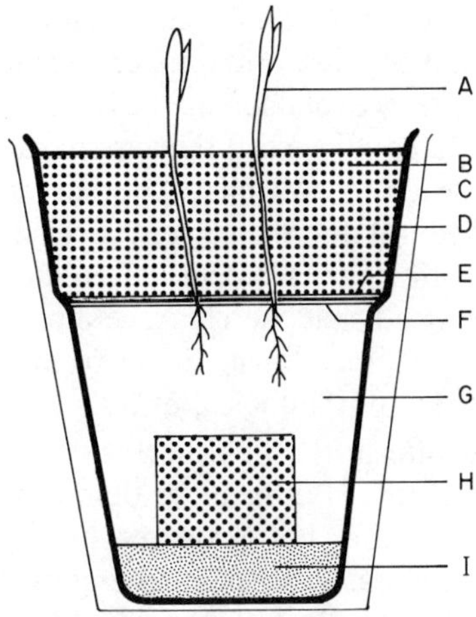


Fig. 1. Schematic diagram of a unit used for assay of trifluralin vapour phytotoxicity to roots. A. grass seedlings, B. untreated soil, C. aluminium-foil covered carton, D. 420-ml dye-pot beaker, E. glass wool, F. galvanized 3-mm screen, G. air space, H. can with treated or untreated soil, and I. water saturated sand. (Swann and Behrens, 1972)

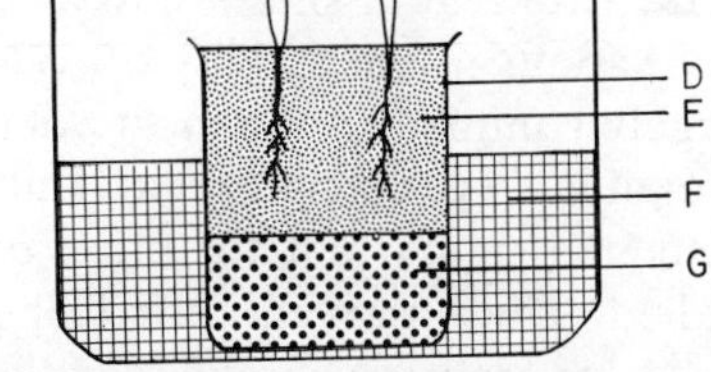


Fig. 2. Schematic diagram of unit used for assay of trifluralin vapour phytotoxicity to shoots. A. aluminium foil, B. 1000-ml beaker, C. grass seedling, D. 100-ml beaker, E. untreated soil, F. treated or untreated soil, G. vermiculite. (Swann and Behrens, 1972)

between herbicides with the same test plant. Yamasue (1975) used closed chambers to demonstrate that vapour of benfluralin and some other dinitroanilines could damage tobacco plants and Ahrens (1978) suggested a routine procedure for evaluating vapour activity by enclosing in plastic bags plants growing in pots above herbicide-treated soils. There are pitfalls in simple experiments such as these, however. Parker (1966) found that root uptake of trifluralin by sorghum (an *S. vulgare* × *S. sudanese* hybrid) seemed to be more effective than shoot uptake. Swann and Behrens (1972a) found the reverse, which they explained on the basis that Parker's apparatus was made of a plastic that could adsorb the vapour, a factor that could also affect the plastic bag system of Ahrens. Another problem was noted by Velmuruga (1974) who found that the solvent vapour used in a dinitramine formulation markedly affected its activity. Length of exposure may be critical. Cox (1974) reported that the chloracetamides had two actions in a wheat shoot assay (i) a "knockdown" effect from which the plant could recover, (ii) an irreversible physiological effect.

It is usually assumed that in the field herbicide vapour enters only the aerial parts of plants but this is difficult to demonstrate. There certainly seems no reason to doubt that herbicide reaching seeds by diffusion as vapour could be taken in as easily as that arriving by diffusion in soil water (Scott and Phillips, 1972). Harvey (1974) argued that some dinitroaniline herbicides must reach the plant as vapour because soil solution concentrations would be too low to account for the observed biological effects. Miller and Nalewaja (1976) showed that though tri-allate vapour was toxic to wild oats (*Avena* spp.), the site of action appeared to be below the soil surface. Ashford (1975), however, had questioned the importance of vapour movement in tri-allate activity as he found that granules placed 10 mm or more from an oat plant had no effect, at least over the experimental period of 10 days. Later work (Ashford and Holroyd, 1977) which compared different granule distributions in field and greenhouse conditions led to the conclusion that, in the greenhouse at least, vapour movement played a part but did not throw light on the problem of site of entry.

Interest in the role of vapour in the activity of soil-applied compounds seems to have been largely confined to the most volatile members of the chloracetamide, dinitroaniline, carbamate and thiocarbamate groups. Whether vapour movement is important for other compounds does not seem to have been studied. The observation of Talbert *et al.* (1971) that cotton could be damaged by prometryne volatilization from soil raises a doubt that this possibility has been underestimated. On the basis of soil–water–air distribution, the quantities of many soil acting herbicides that are present as vapour will be very small. Unfortunately information on the sensitivities of different test species and on the relationships between concentration, time and phytotoxicity is frequently lacking so the significance of herbicide vapour is difficult to assess.

The importance of adsorption by soil is also sometimes overlooked when the more volatile compounds are considered. It is timely that Beestman and Deming (1976) found that tri-allate did not volatilize from soils more rapidly than propachlor and alachlor, compounds whose performance is not normally dependent on incorporation. They suggested that incorporation placed tri-allate in moist soil where it is less adsorbed, hence more available to plants, than in the dry soil at the surface.

Work on vapour phase uptake thus seems to be piecemeal and little information has emerged about sites of uptake under field conditions and the quantities entering the plant by these routes. This presumably reflects the view that in most cases the vapour of soil-acting compounds is of little significance. It would be reassuring if this view could be supported by more extensive evidence than we have at the moment.

## VI. Fate of Herbicide Vapour in the Atmosphere

Once a vapour has left the soil and passed through the crop canopy, it is quickly dispersed in the air as a result of turbulence although its distribution will be restricted if there is a temperature inversion ceiling. Taylor *et al.* (1976) estimated from flux movements that the residence time of a pesticide molecule in a layer of air 1 m thick varied from only 20 s to 5 min, depending on the time of day. Transport processes thereafter, except in a local (< 20 km) scale, are not well understood and the technical difficulties of monitoring the low concentrations are formidable. The principles of the processes are discussed by Wheatley (1973) and Glotfléty and Caro (1975) but there is little experimental evidence with which to evaluate them. The latter authors do, however, point out that it is possible to make rough estimates of transport potential from considerations of water : air partition and the stability of a compound. They quoted air monitoring studies that support this suggestion.

Pesticide vapour in the air may be removed by decomposition, adsorption on aerosols followed by dry deposition, and in precipitation. There is little information available with which to make estimates of the relative importance of these processes. Certainly oxidation, photo-oxidation and photolytic reactions are likely. There is a good deal of laboratory evidence to show that most herbicides are susceptible to such reactions (Crosby, 1976). Hartley (1967) suggested that these reactions increase in importance with height in the atmosphere and play a major part in destroying the waste products of man's activities. Field data are rare but Soderquist *et al.* (1975) describe evidence for the photolysis of trifluralin in the air.

The biological significance of the vapour of soil-applied herbicides and their degradation products to non-target organisms is uncertain. Cox (1974) quoted a Dutch report of apparent propachlor damage to tulips that might have been caused by vapour moving from adjacent fields but instances of vapour activity over distances of several kilometres, as have been reported for some foliar applied compounds, do not seem to have been recorded. There is some justification for the view that, since damage to plants in this way is rare, there is little cause for concern. Similarly, because of the great dilution which occurs, any degradation products formed in the atmosphere would need either to be unusually toxic or else to accumulate in biological tissues if they were to be of importance. Glotfelty and Caro (1975) point out that reactions in aerosols tend to involve oxygenation so products are likely to be polar, water soluble compounds that would not accumulate in tissues or trophic chains. Crews *et al.* (1975) proposed that problems of the atmospheric behaviour of pesticides should be classified in relation to toxicity, persistence, volatility and other relevant physical, chemical and biological properties. Such criteria would put most, if not all, herbicides low in order of priority compared with many other atmospheric pollutants.

## VII. Experimental Techniques

### A. Measurement of Vapour Pressure

The vapour pressures of herbicides range from about 1 Pa to less than $10^{-5}$ Pa at 20°C so that there are considerable difficulties in obtaining reasonable estimates of the vapour pressures of the less volatile materials. The dangers of extrapolation from measurements obtained at high temperatures are generally acknowledged but the suspicion remains that data quoted for the older herbicides, which perhaps are now off-patent, were obtained in this way.

The two most common methods used are the Knudsen effusion method (Hamaker and Kerlinger, 1967) and the gas saturation method (Spencer and Cliath, 1969). In the Knudsen method the loss of vapour through a small hole into a vacuum is measured either by the weight loss of the effusion cell or by measuring the recoil from the stream of effusing gas by means of a torsion balance. Hamaker and Kerlinger (1967) modified the cell so that the vapour was trapped and determined directly. This reduced the sample size needed and, by using a suitably specific analytical method, allowed the errors caused by the evaporation of impurities to be ignored. In the gas saturation method a stream of gas is saturated with vapour which is then removed in a suitable trap so that the amount of material transported by a known volume of gas can be measured. This method is also insensitive to impurities given a suitable analytical procedure. In addition it can be used for measurements in the presence of soil.

These methods should allow absolute determinations of vapour pressure to be made but the EPA (Anon., 1978) recommendation "to make an initial check of the performance of the apparatus by the use of compounds of known vapour pressure" suggests that some workers have found difficulties. If prior calibration of the apparatus of this sort is acceptable, then simpler methods are available. Gückel *et al.* (1973) proposed a method in which pesticide is coated on a roughened glass plate which is suspended from a balance arm in a stream of air of known temperature and flow rate. A standard curve of rate of evaporation against vapour pressure allows the calculation of unknown vapour pressures from evaporation rate measurements. Unfortunately, in addition to the need for a range of reference compounds, volatile impurities cause errors. The method of Grover (1978) does not have the second disadvantage as in essence it is a simple version of that of Spencer and Cliath. Gas flows over a tray of herbicide at a known rate and the vapour is subsequently trapped. Vapour pressures are estimated from the same sort of standard curve as in the method of Gückel *et al.* (1973).

Gas chromatography has been used (Jensen and Schall, 1966) on the assumption that separation is solely a function of vapour pressure. However, for most compounds extrapolation from values obtained at elevated temperature is necessary.

An intriguing possibility is raised by the work of Chmil (1978). He found a linear relationship between the vapour pressure of a homologous series of phenoxyalkanoic acid esters and their *Rf* values in a thin-layer chromatography system.

The basic requirements for information on methods of vapour pressure measurement are easy to define. First, we need to know how reproducible and reliable are the standard effusion and gas saturation methods for making absolute measurements. Secondly, figures are needed with which to assess the possibility of using simpler methods, or even partition coefficients (Chiou and Freed, 1977) to obtain order-of-magnitude estimates of vapour pressure. No doubt there is enough information to answer both these points in the confidential files of the registration authorities of the world. It is a pity it cannot be made available. At the moment we can only agree with Plimmer (1976) that different techniques appear to give quite different results.

## B. Measurement of Vapour Loss from Soil

### 1. *Laboratory measurements*

A variety of methods have been described reflecting the differences in experimental objectives. Perhaps the simplest, if only semi-quantitative, system is that of Cox (1974) in which indicator plants are grown in pots inside chambers containing herbicide-treated soil. Comparison with a standard series using herbicide solutions would allow approximate vapour pressures to be calculated. Other simple systems involve sampling the air over soil held in conical flasks. Weaver *et al.* (1971) used a gas sampling syringe and injected an aliquot directly into a gas chromatograph which is ideal if the compound is sufficiently volatile. Hance *et al.* (1973) obtained an indirect measurement by suspending squares of glass-fibre paper in the flasks which were removed after a suitable time and the absorbed herbicide extracted with a solvent. Both methods are suitable only for equilibrium studies.

For dynamic measurements some sort of gas train system is usual. Parochetti *et al.* (1971), Swann and Behrens (1972b) and Que Hee and Sutherland (1974) used the technique of passing air over a cylinder or petri-dish of treated soil and then through a suitable trap (containing Florisil, xylene or ethylene glycol) to remove the herbicide. The method is quite flexible and allows control of air humidity and temperature so that the effects of soil water content and temperature can be studied. Spencer and Cliath (1974) used a similar method and also one they developed earlier (1972) in which air passes through a column of soil so that measurements of partial pressure can be made. Kearney and Kontson (1976) suggested that the use of polyurethane plugs as vapour traps, first used by Bidleman and Olney (1974) would allow $CO_2$ to be trapped separately so that volatility and metabolism could be studied simultaneously.

Beestman and Deming (1974, 1976) used an indirect method. Soil columns were placed in chambers, air was passed over them, then herbicide remaining in the soil was extracted and analysed. Similar columns were kept sealed so that a correction could be made for losses due to decomposition. The major problem with such a method is that the results contain two sets of analytical error.

Thus although there is no standard method, there is no great difficulty in devising apparatus for the study in the laboratory of factors affecting volatility.

### 2. *Field measurements*

For small plot work it is possible to put some sort of enclosure over the soil and draw air through and into a suitable trap but this mimics a laboratory measurement and excludes many of the normal field variables.

For field scale measurements rates of volatilization must be calculated from estimates of pesticide flux through the air. Parmele *et al.* (1972) showed that vertical pesticide flux rates ($P$) followed reasonably well the relationship

$$P = K_z \frac{(\partial p)}{\partial z}$$

where $K_z$ is the vertical diffusivity coefficient of the pesticide at height $z$. The value of $K_z$ depends on meteorological conditions and is not a function of properties of the vapour, hence can be assumed to be the same as that for water vapour. If lysimeter data and water flux measurements are available, it can be calculated directly, otherwise indirect estimates are possible based on meteorological data. Errors arise because such calculations assume a horizontally uniform surface and their magnitude may depend on the method of calculation used. For instance Harper *et al.* (1976) compared the energy balance and momentum balance methods of estimating $K_z$ and found that the former gave consistently lower results but that the differences were smaller as the soil became drier.

Estimates of flux of pesticide vapour require air sampling at different heights and to be satisfactory any method used should be capable of reliable measurement down to 1 ng/m$^3$. Seiber *et al.* (1978) suggest that the ethylene glycol impinger (Stanley *et al.*, 1971) is a suitable standard method with which to judge others and this sort of device has frequently been used (Caro *et al.*, 1972; Parmele *et al.*, 1972; White *et al.*, 1977; Hollingsworth, 1977). However, any liquid trap has some limit to the volume of air that can be sampled, so solid adsorbents are of interest. Bidleman and Olney (1974) found that plugs of polyurethene foam 4″ diameter and 2″ thick were useful for polychlorinated biphenyls and could easily sample 2000 m$^3$. The later report of Kearney and Kontson (1976) indicates that this material may be a good general adsorbent for pesticides. Johnson *et al.* (1977) have reported that XAD-2 resin is an effective trap for 2,4-D, 5 g of resin being adequate to sample for 24 h at a rate of 10 l/min allowing a sensitivity of 0·01 $\mu$g/m$^3$ of 2,4-D. Grover and Kerr (1978), in a laboratory evaluation, found it was also satisfactory for bromoxynil octanoate, tri-allate and trifluralin.

At the moment a major problem in air sampling is to develop equipment that can distinguish between particulate matter and molecular vapour. The difficulty is to prevent incoming vapour being adsorbed by trapped particles and, conversely, to prevent volatilization from trapped particles.

## VIII. Conclusions

It is often assumed that the fundamental principles of vapour phase movement through and out of the soil are reasonably well understood. This confidence could be misplaced, however, as in some areas it is not supported by convincing evidence. As Hamaker (1972) observed, the mathematics involved can be extremely complicated and even with the availability of powerful computers, the tendency among many workers has been to impose narrow boundaries to their experiments or to make simplifying assumptions of uncertain validity. In addition there is simply a lack of sufficient data to check many of our ideas.

A basic need is for suitable methods for measuring vapour pressures at ambient temperatures that can be used reliably in any laboratory. Uncertainty over such a fundamental property is a clear hindrance to precise work. In addition, the usefulness of simple measurements, such as octanol : water partition, as a rough guide to the likely volatility of pesticides requires evaluation.

There is much to be done if we are to understand fully vapour phase processes in the soil. Perhaps most crucial is to obtain better information on the competitive adsorption between water and pesticides as there seem to be distinct differences between different pesticides and between different soils. The effect of temperature gradients and diurnal fluctuations in the soil on the movement of water and pesticide vapour is a potentially important factor that has received little attention.

The significance of vapour movement in dosage transfer of soil-acting herbicides to the target plant remains uncertain. This is partly because of our imperfect quantitative understanding of the processes occurring in the soil but it is also due to ignorance concerning possible sites of entry into the plant and concentration-time-response relationships.

Although it is known that at least some herbicides may be lost from the soil in large quantities by volatilization, there is a need for many more measurements if the magnitude of such losses is to be reliably predicted from information about soil, crop, climate and pesticide. The need for studies of the fate in the atmosphere of herbicide vapour is perhaps not so acute but it would be comforting to have some data to give reassurance that this attitude is justified.

## References

Ahrens, J. F. (1978). Phytotoxicity of vapours from herbicides on soil in plastic containers. *Abs. Weed Sci. Soc. Amer.*, **34**.

Anderson, J. P. E. (1975). Einfluss von Temperatur und Feuchte auf Verdampfung, Abbau und Festlegung von Diallat im Boden. *Z. PflKrankheit PflSchutz. Sonderheft* **7**, 141–146.

Anon (1978). Environmental Protection Agency: Guidelines for registering pesticides in United States, *Federal Register*, **43**, No 132.

Ashford, R. (1975). Triallate granule spacing effect on oat. *Weed Sci.*, **23**, 470–472.

Ashford, R. and Holroyd, J. (1977). Post-emergence effects of different distribution patterns of tri-allate granules on *Avena fatua* L. and wheat plants. *Weed Res.*, **17**, 415–422.

Beestman, G. B. and Deming, J. M. (1974). Dissipation of acetanilide herbicides from soils. *Agron. J.*, **66**, 308–311.

Beestman, G. B. and Deming, J. M. (1976). Triallate mobility in soils. *Weed Sci.*, **24**, 541–544.

Bidleman, T. F. and Olney, C. F. (1974). High volume collection of atmospheric polychlorinated biphenyls. *Bull. environm. Contamin. Toxicol.*, **11**, 442–450.

Bode, L. E., Day, C. L., Gebhardt, M. R. and Goering, C. E. (1973a). Mechanism of trifluralin diffusion in silt loam soil. *Weed Sci.*, **21**, 480–484.

Bode, L. E., Day, C. L., Gebhardt, M. R. and Goering, C. E. (1973b). Prediction of trifluralin diffusion coefficients. *Weed Sci.*, **21**, 485–489.

Briggs, G. G. (1973). A simple relationship between soil adsorption of organic chemicals and their octanol: water partition coefficients. *Proc. 7th Br. Insecticide and Fungicide Conf.*, pp. 83–86.

Caro, J. H., Taylor, A. W. and Lemon, E. R. (1972). Measurement of pesticide concentrations in the air overlying a treated field. *Proc. Symp. Identification and Measurement of Environmental Pollutants*, Nat. Res. Council, Canada, pp. 72–77.

Chiou, C. T. and Freed, V. H. (1977). Assessments of volatility and partition coefficient from physical chemical properties. In *Terrestrial Microcosms and Environmental Chemistry* (J. M. Witt and J. W. Gillett, Eds) pp. 17–22. National Science Foundation/RA, 79-0026 Washington DC.

Chmil, V. D. (1975). Chromatographic methods for assessing the biological activity of herbicides In *Materialy 10ge Mezhdunarodnogo Simpoziuma Stron-chlenov SEV, Pushchino, SSSR*, Vol. 1, 162–167 (Weed Abs (1978) **27**, 535).

Collins, R. L. (1975). Microgranules and microencapsulation of pesticides. *Proc. 1975 Int. Controlled Release Pesticide Symposium*, Dayton, Ohio 105, 112.

Cox, T. I. (1974). Experiments on vapour activity of alachlor and related herbicides. *Weed Res.*, **14**, 379–383.

Crank, J. (1975). *Mathematics of Diffusion.* 2nd Ed., Clarendon Press, Oxford.

Crews, W. B., Brewer, J. W. and Peterson, T. J. (1975). Modeling of atmospheric behaviour: a submodel of the dynamics of pesticides. In *Environmental Dynamics of Pesticides* (R. S. Haque and V. H. Freed, Eds) pp. 17–43, Plenum.

Crosby, D. G. (1976). Herbicide Photodecomposition. In *Herbicides; Chemistry Degradation and Mode of Action* (P. C. Kearney and D. D. Kaufman, Eds) Vol. 2, pp. 836–890, Dekker.

Dawson, J. H. (1979). Movement of chlorpropham after application to soil for control of dodder (*Cuscuta* spp.) *Weed Science*, **27**, 274–277.

Deming, J. M. (1963). Determination of Volatility Losses of $C^{14}$-CDAA from Soil Surfaces, *Weeds*, **11**, 91–95.

Fryer, J. D. and Makepeace, R. J. (1977). *Weed Control Handbook Vol. 1: Principles, Including Plant Growth Regulators.* Blackwell Scientific Publns, Oxford.

Gerber, H. R. (1977). Biotests for herbicide development. In *Crop Protection Agents—Their Biological Evaluation* (N. R. McFarlane, Ed.) pp. 307–321, Academic Press, London and New York.

Glotfelty, D. E. and Caro, J. H. (1975). Introduction, transport and fate of persistent pesticides in the atmosphere. American Chemical Society Symposium Series 17: *Removal of Trace Contaminants from the Air*, pp. 42–62.

Graham-Bryce, I. J. (1968). Movement of systemic insecticides through soil to plant roots. Soc. Chem. Ind. Monograph 29: *Physicochemical and Biophysical Factors Affecting the Activity of Pesticides*. 251–263.

Grover, R. (1975). A method for determining the volatility of herbicides. *Weed Sci.*, **23**, 529–532.

Grover, R. and Kerr, L. A. (1978). Evaluation of silica gel and XAD-4 as adsorbents for herbicides. *Journal of Environmental Science and Health B*, **13**, 311–322.

Guckel, W., Synnatschke, G. and Rittig, R. (1973). A method for determining the volatility of active ingredients used in plant protection. *Pestic. Sci.*, **4**, 137–148.

Hamaker, J. W. (1972). Diffusion and volatilization. In *Organic Chemicals in the Soil Environment*, Vol. 1 (C. A. I. Goring, and J. W. Hamaker, Eds) pp. 341–397, Dekker.

Hamaker, J. W. and Kerlinger, H. O. (1969). Vapour pressure of pesticides. In *Pesticidal Formulations Research* (R. F. Gould, Ed.) pp. 39–54,American Chemical Society Advances in Chemistry Series, No. 86.

Hance, R. J. (1977). The adsorption of atraton and monuron by soils at different water contents. *Weed Res.*, **17**, 197–202.

Hance, R. J. and Embling, S. J. (1979). Effect of soil water content at the time of application on herbicide concentration in soil solutions extracted in a pressure membrane apparatus. *Weed Res.*, **19**, 201–205.

Hance, R. J., Holroyd, J., McKone, C. E. (1973). Some aspects of tri-allate volatility. *Pestic. Sci.*, **4**, 13–17.

Harper, L. A., White, A. W. Jr., Bruce, R. R., Thomas, A. W. and Leonard, R. A. (1976). Soil and microclimate effects on trifluralin volatilization. *J. Environm. Quality*, **5**, 236–242.

Hartley, G. S. (1969). Evaporation of pesticides. In *Pesticidal Formulations Research* (R. F. Gould, Ed.) American Chemical Society, Advances in Chemistry Series No. 86, 115–134.

Harvey, R. G. (1974). Soil adsorption and volatility of dinitroanaline herbicides. *Weed Sci.*, **22**, 120–124.

Hollingsworth, E. B. (1977). Volatility of trifluralin and metribuzin. *Proc. 30 Mtg Southern Weed Sci. Soc.*, 369.

Ishikawa, K., Nakamura, Y. and Kuwatsuka, S. (1977). Volatilization of benthiocarb herbicide from the aqueous solution and soil. *J. Pestic. Sci.*, **2**, 127–134.

Jackson, R. D. (1973). Diurnal changes in soil water content during drying. In *Field Soil Water Regime* (chairman R. Russell Bruce, Ed.) pp. 37–55, Soil Sci. Soc. Am. Special Publication Series No. 5, Madison.

Jacques, G. L. and Harvey, R. G. (1979). Adsorption and diffusion of dinitroaniline herbicides in soils. *Weed Sci.*, **23**, 450–455.

Jensen, D. J. and Schall, E. D. (1966). Determination of vapour pressures of some phenoxyacetic herbicides by gas-liquid chromatography. *J. agric. Fd. Chem.*, **14**, 123–126.

Johnson, E. R., Yu, T. C. and Montgomery, M. L. (1977). Trapping and analysis of atmospheric residues of 2,4-D. *Bull. environm. Contamination and Toxicology*, **17**, 369–372.

Jordan, T. N., Baker, R. S. and Barrentine, W. L. (1978). Comparative toxicity of several dinitro-aniline herbicides. *Weed Sci.*, **26**, 72–75.

Kearney, P. C. and Kontson, A. (1976). A simple system to simultaneously measure volatilization and metabolism of pesticides from soils. *J. agric. Fd. Chem.*, **24**, 424–426.

Kennedy, J. M. and Talbert, R. E. (1977). Comparative persistence of dinitroanaline type herbicides on the soil surface. *Weed Sci.*, **25**, 373–381.

Koren, E. (1976). A new method for evaluating the herbicidal potential of the gaseous phase of herbicides. *Phytoparasitica*, **4**, 144–145.

Letey, J. and Farmer, W. J. (1974). Movement of pesticides in soil. In *Pesticides in Soil and Water*. pp. 67–97, Soil Sci. Soc. Am., Madison.

Mass, W. and Popp. F. (1976). Granular fomulations of dichlobenil. *British Crop Protection Council Monograph* **18**, *Granular Pesticides* (S. A. Evans, Ed.) 53–58.

Mayer, R., Farmer, W. J. and Letey, J. (1974). Models for predicting volatilization of soil-incorporated pesticides. *Soil Sci. Soc. Am. Proc.*, **38**, 563–568.

Miller, S. D., Nalewaja, J. D. (1976). Phytotoxicity of tri-allate vapours to wild oat. *Weed Sci.*, **24**, 134–136.

Parker, C. (1966). The importance of shoot entry in the action of herbicides applied to the soil. *Weeds.*, **14**, 117–121.

Parmele, L. H., Lemon, E. R. and Taylor, A. W. (1972). Micrometeorological measurement of pesticide vapour flux from bare soil and corn under field conditions. *Water, Air and Soil Pollution*, **1**, 433–451.

Parochetti, J. V., Hein, E. R. and Colby, S. R. (1971). Volatility of dichlobenil. *Weed Sci.*, **19**, 28–31.

Parochetti, J. V. and Hein, E. R. (1973). Volatility and photodecomposition of trifluralin, benefin and nitralin. *Weed Sci.*, **21**, 469–473.

Plimmer, J. R. (1976). Volatility. In *Herbicides: Chemistry Degradation and Mode of Action*. Vol. 2. (P. C. Kearney, and D. D. Kaufman, Eds.) pp. 892–934, Dekker.

Que Hee, G. S. and Sutherland, R. G. (1974). Volatilization of various esters and salts of 2,4-D. *Weed Sci.*, **22**, 313–318.

Scott, H. D. and Paetzold, R. F. (1978). Effects of soil moisture on the diffusion coefficients and activation energies of tritiated water, chloride and metribuzin. *Soil Sci. Soc. Am. J.*, **42**, 23–27.

Scott, H. D. and Phillips, R. E. (1972). Diffusion of selected herbicides in soil. *Soil Sci. Soc. Am. Proc.*, **36**, 714–719.

Seiber, J. H., Woodrow, J. E., Shafik, T. M. and Enos, H. F. (1975). Determination of pesticides and transformation products in air. In *Environmental Dynamics of Pesticides* (R. S. Haque, and V. H. Freed, Eds.) pp. 17–43, Plenum.

Shearer, R. C., Letey, J., Farmer, W. J. and Klute, A. (1973). Lindane diffusion in soil. *Soil. Sci. Soc. Am. Proc.*, **37**, 189–193.

Slater, C. H., Dawson, J. H., Furtick, W. R. and Appleby, A. P. (1969). Effects of chlorpropham vapours on dodder seedlings. *Weed Sci.*, **17**, 238–241.

Smith, D. T., Wiese, A. F. (1973). Delayed incorporation of trifluralin and nitralin. *Weed Sci.*, **21**, 163–165.

Soderquist, C. J., Crosby, D. G., Moilanen, K. W., Seiber, J. N. and Woodrow, J. E. (1975). Occurence of trifluralin and its photoproducts in air. *J. agric. Fd. Chem.*, **23**, 304–309.

Spencer, W. F. and Cliath, M. M. (1969). Vapour density of dieldrin. *Environ. Sci. Tech.*, **3**, 670–674.

Spencer, W. F. and Cliath, M. M. (1970). Desorption of lindane from soil as related to vapor pressure. *Soil Sci. Soc. Am. Proc.*, **34**, 574–578.

Spencer, W. F. and Cliath, M. M. (1972). Volatility of DDT and related compounds. *J. agric. Fd. Chem.*, **20**, 645–649.

Spencer, W. F. and Cliath, M. M. (1973). Pesticide volatilization as related to water loss from soil. *J. environ. Qual.*, **2**, 284–288.

Spencer, W. F. and Cliath, M. M. (1974). Factors affecting vapor loss of trifluralin from soil. *J. agric. Fd. Chem.*, **22**, 987–991.

Spencer, W. F. and Cliath, M. M. (1975). Vaporization of chemicals. In *Environmental Dynamics of Pesticides* (R. S. Haque, and V. H. Freed, Eds.) pp. 61–78, Plenum.

Spencer, W. F. Cliath, M. M. and Farmer, W. J. (1969). Vapor density of soil applied dieldrin as related to soil-water content, temperature and dieldrin concentration. *Soil Sci. Soc. Am. Proc.*, **33**, 509–511.

Spencer, W. F., Farmer, W. J. and Cliath, M. M. (1973). Pesticide volatilization. *Residue Rev.* **49**, 1–47.

Spencer-Jones, D. H. and Wilson, D. (1976). Some Effects of Formulation on the Biological Performance of Dichlobenil. *British Crop Protection Council Monograph 18, Granular Pesticides* (S. A. Evans, Ed.) 59–66.

Stanley, C. W., Barney, J. E., Helton, M. R. and Yobs, A. R. (1971). Measurements of atmospheric levels of pesticides. *Environ. Sci. Technol.*, **5**, 430–435.

Swann, C. W. and Behrens, R. (1972a). Phytotoxicity of trifluralin vapors from soil. *Weed Sci.*, **20**, 143–146.

Swann, C. W. and Behrens, R. (1972b). Trifluralin vapor emission from soil. *Weed Sci.*, **20**, 147–149.

Talbert, R. E., Smith, D. R. and Franz, R. E. (1971). Volatilization, leaching and adsorption of prometryne in relation to selectivity in cotton. *Weeds*, **19**, 6–11.

Taylor, A. W. (1977). Post application volatilization of pesticides under field conditions. "70th Meeting Air Pollution Control Association Toronto, June '77," 45·3.

Taylor, A. W., Glotfelty, D. E. Glass, B. C., Freeman, M. P. and Edwards, W. M. (1976). Volatilization of dieldrin and heptachlor from a maize field. *J. agric. Fd. Chem.*, **24**, 625–631.

Taylor, A. W., Glotfelty, D. E., Turner, B. C., Silver, R. E., Freeman, H. P. and Weiss, A. (1977). Volatilization of dieldrin and heptachlor residues from field vegetation. *J. agric. Fd. Chem.*, **25**, 542–548.

Velmurugu, V. (1974). Studies on mode of action and selectivity of dinitramine. MSc Thesis UC N Wales, Bangor.

Weaver, D. N. (1971). The effect of carriers and carrier additives on the short term persistence of trifluralin. *Diss. Abs. Int. B.*, **32**, (1) 39.

Weaver, D. N., Meyer, R. E. and Merkle, M. G. (1971). Paraffin oil and granules as carriers of trifluralin. *Agron. J.*, **63**, 705–708.

Wheatley, G. A. (1973). Pesticides in the atmosphere. In *Environmental Pollution by Pesticides* (C. A. Edwards, Ed.) pp. 366–408, Plenum, New York.

Wheatley, G. A. (1976). Physical loss and redistribution of pesticides in the vapour phase. *British Crop Protection Council Monograph* 17, *Persistence of Insecticides and Herbicides*, (K. I. Beynon, Ed.) 117–125.

White, A. W., Jr, Harper, L. A., Leonard, R. A. and Turnbull, J. W. (1977). Trifluralin volatilization losses from a soybean field. *J. environ. Qual.*, **6**, 105–110.

Willis, G. H., Parr, J. F., Smith, S. and Carroll, B. R. (1972). Volatilization of dieldrin from fallow soil as affected by different soil water regimes. *J. environ. Qual.*, **1**, 193–196.

Yamasue, Y. (1975). Foliar effects on tobacco (*Nicotiana tabacum* L.) of benefin (*N*-butyl-*N*-ethyl-*ααα*trifluoro-2,6,dinitro-*p*-toluidine) vaporizing from soils. *Diss. Abs. Int. B* **35** (10) 4734–5.

# 4 Persistence and its Prediction

*Karl Hurle*
Institut für Phytomedizin,
Universität Hohenheim,
7000 Stuttgart 70,
Bundesrepublik Deutschland

*Allan Walker*
National Vegetable Research Station,
Wellesbourne,
Warwick CV35 9EF, England

---



## I. Introduction

The ideal soil-applied herbicide would persist long enough to give an acceptable period of weed control but not so long that soil residues after crop harvest limit the nature of subsequent crops which can be grown. Unfortunately persistence is not a fixed property of the chemical but is influenced by factors such as soil type and the weather conditions after application. Since these factors vary from site to site and from year to year, the results from any field study of persistence are specific to one particular location and season. In order to estimate or predict persistence of a chemical in different situations, we must know how its rate of degradation is affected by soil and environmental factors. Although there is an extensive literature on herbicide persistence in soil in the field, much of it is of limited value for general predictive purposes. Rarely have measurements been made at sufficient sites to allow rates of loss or residues to be related to soil or climatic variables, and in general experiments have not been designed with the intention of kinetic analysis of the data.

## II. Kinetics of Degradation

Degradation of some compounds is characterized by an initial lag-period during which little or no change in concentration occurs. This is followed by rapid degradation which often appears linear with time, but detailed analysis may show that it follows first-order kinetics (Hurle, 1973). Herbicides showing this degradation pattern include 2,4-D, MCPA, dichlorprop, DNOC, dalapon, TCA, chloridazon, propham, chlorpropham and endothal, all compounds of short persistence in soil. Because of the lag-phase, presentation of half-lives for such compounds is not very meaningful.

With other compounds, there is no lag-period, and the rate of degradation is more or less proportional to concentration so that the results can often be interpreted using first-order kinetics. The equation for a first-order reaction is:

$$C = C_0 e^{-kt} \tag{1}$$

where $C$ is the concentration after time $t$, $C_0$ is the initial concentration and $k$ the rate constant. Thus a plot of the logarithm of concentration against time gives a straight line with slope proportional to the rate constant. Writing $t_{\frac{1}{2}}$ as the time taken for 50% degradation or "half-life", equation (1) becomes:

$$t_{\frac{1}{2}} = 0{\cdot}6932/k \tag{2}$$

The half-life in a first-order reaction is independent of initial concentration but this is not so for other orders of reaction. Use of this specific term should therefore be reserved for instances where first-order kinetics apply.

The half-life concept is a valuable tool in comparing rates of herbicide degradation and because of the simplicity of first-order reaction kinetics, this rate law is widely used. In many instances, however, its use may be an oversimplification. Burschel and Freed (1959) and Zimdahl *et al.* (1970) commented that since the quantity of herbicide in the soil is very small in relation to the other components, we might expect herbicide concentration to be rate limiting so that first-order reaction kinetics should apply. However, soil is a complex biological and chemical medium. The kinetics of adsorption and desorption might affect rates of loss by controlling availability for degradation, the activities of soil microorganisms may vary with time depending on the availability of nutrients and other energy sources, and there may be competing reaction sequences within the degradation process. It is therefore not surprising that deviations from simple first-order reaction kinetics are observed.

Rate laws other than first-order have been considered. Hance and McKone (1971) showed that neither zero-order, half-order, first-order nor Michaelis–Menten kinetics described precisely the breakdown of atrazine, linuron or picloram in the laboratory. Empirical curve fitting has also been suggested and Hamaker (1972) proposed a power–rate equation:

$$C = (C_0^{(1-n)} + (n-1)kt)^{1/(1-n)} \tag{3}$$

where $C$ is the concentration after time $t$, $C_0$ is the initial concentration, $n$ the apparent order of reaction and $k$ the rate constant. This equation was used by Kempson-Jones and Hance (1979) in studies of the kinetics of linuron and metribuzin degradation. In only 10 of 40 separate experiments did the value 1 fall within the 95% confidence limits of the determined value of $n$. In the remaining experiments the apparent reaction order was greater than 1, and in eight of these, it was greater than 4. Hamaker (1972) reported similar deviations from first-order reaction kinetics when he used equation (3) to calculate apparent orders of reaction from previously published degradation data. Other empirical rate equations have been proposed by Edwards (1972) and Hamaker (1972).

When degradation is followed for extended periods, many compounds show rates of degradation that are disproportionately slow at lower residual concentrations. Hamaker and Goring (1976) suggested a "two compartment" model to explain this. The pesticide is considered to be divided between available and unavailable fractions, with only that which is available subject to degradation. Appropriate rate constants control movement into and out of the unavailable pool. Freshly added chemical is mainly in the available state and the initial rate of degradation is rapid. However, the rate falls off as the pesticide is transferred to the unavailable state and eventually, rate of release from the unavailable pool controls the rate of degradation. Hamaker (1976) and Hamaker and Goring (1976) demonstrated that this model could describe the kinetics of degradation of a number of soil-applied pesticides.

There is little doubt that rate equations are valuable for describing pesticide degradation, but it seems unlikely that a universal equation will be found which applies to all compounds and soils. When rate equations are used, their approximate nature must always be kept in mind.

## III. Factors Influencing Rates of Loss

### A. Concentration and Rate of Application

As we have suggested, the first-order rate law often is not strictly followed. With first-order kinetics, rate constants should be independent of initial concentration but with atrazine (Armstrong *et al.*, 1967), atrazine, linuron, picloram (Hance and McKone, 1971), simazine and prometryne (Walker, 1976a), rates of loss decreased as initial concentration increased. With some compounds, a lag-phase was demonstrated at increased initial concentrations where normally none was apparent, e.g. picloram (Grover, 1967; Hamaker *et al.*, 1968), aminotriazole (Riepma, 1962) and tri-allate (Banting, 1967).

With compounds which normally show a lag-phase, initial concentration may also have an effect, although this varies with the compound. The duration of the lag-phase for TCA was unaffected by increasing concentration (Torstensson, 1976); with 2,4-D it was independent of concentration when this was low but extended at higher doses; for DNOC, the duration of the lag-period was proportional to initial concentration (Hurle, 1973; 1981).

Hance and McKone (1971) suggested that reduced degradation rates at higher initial concentrations might result from a limitation in the number of reaction sites in the soil, and Hurle (1981) suggested that toxic effects on microorganisms or enzyme inhibition might be involved.

If these effects occurred in the field, unusually high rates of application of some compounds (e.g. from accidental overdosing or spillage) might persist for disproportionately long periods. There are few reports in the literature concerning persistence in the field of different initial doses and most of these involved a relatively narrow range. Holly and Roberts (1963) reported no differences in rates of loss between applications of simazine at 0·6 and 2·4 kg/ha, and Schweizer (1976) showed that the rate of loss of ethofumesate was independent of initial dose within the range 2·2 to 9·0 kg/ha. Results of Keys and Friesen (1968) suggested that rates of picloram degradation were essentially the same at all doses from 0·03 to 3·36 kg/ha. However, Roadhouse and Birk (1961) found that simazine residues expressed as a percentage of the amounts applied increased with increasing dose in the range 0·6 to 22·4 kg/ha. Clay and Stott (1973) found no difference in rates of loss of simazine between initial rates of 2·8 and 5·6 kg/ha, but their results suggested a considerably reduced rate of loss at 22·4 kg/ha. Hance *et al.* (1976) measured the disappearance from soil of 112 kg/ha simazine and linuron over a period of five years. They compared their results with those of Fryer and Kirkland (1970) for the same herbicides on an adjacent site, but at much lower doses. At the high dose, the calculated half-life for simazine was about 78 weeks and that of linuron about 30 weeks, whereas at the lower doses half-lives were 12 to 40 weeks for simazine and 8 to 25 weeks for linuron. The authors considered that because of the large variation involved in both sets of data, it was not possible to ascribe any significance to the differences observed but they concluded that use of very high doses did not prevent dissipation processes.

Despite these conflicting results it seems that only at unusually high rates of application is persistence in the field increased disproportionately. Edwards (1972) reviewed the literature on insecticides and reached a similar conclusion.

## B. Soil Type

The influence of soil type on herbicide persistence is not well understood. Because soil microorganisms are usually involved in degradation, soil organic matter might be expected to have some effect since microbial activity is often higher in more organic soils. However, adsorption of most herbicides also increases with an increase in soil organic matter and since adsorption reduces the amount of herbicide available in the soil solution, it might provide protection from degradation. For these reasons, Hamaker (1972) suggested that an increase in organic matter might increase rates of degradation in mineral soils up to a limiting value, above which the rate of loss would be retarded. Reports in the literature have given conflicting results. The persistence of 2,4-D, TCA, chlorpropham and monuron decreased progressively from a light sandy soil through a silt loam to an organic soil (Ogle and Warren, 1954), and simazine (Zurawski and Piss, 1971), dicamba (Donaldson and Foy, 1965) and linuron (Walker and Thompson, 1977) degraded more rapidly in soils with higher organic contents. On the other hand, Harris *et al.* (1969) reported a positive correlation between fenac residues and soil organic matter content as did Smith and Meggitt (1970) with chloridazon and Briska *et al.* (1973) with simazine. In a study of simazine persistence at 26 sites throughout the United Kingdom, Davison and Clay (1972) reported no significant relationship between persistence and any soil factor, although there was a trend towards lower residues with lower soil pH. The rates of loss of picloram from soil have also been assessed in different soils in both laboratory tests (Meikle *et al.*, 1973) and field trials (Hamaker *et al.*, 1967) and in neither case was any soil property related with the rate of loss. In laboratory experiments, only 27% of the variations in the rate of loss could be accounted for by a multiple linear regression with soil properties.

## C. Herbicide Adsorption

As mentioned above, adsorption of herbicides may provide protection from degradation by reducing the concentration in the soil solution. However, since the density of microorganisms near colloidal surfaces is greater than in the soil solution (Burns, 1978) and since adsorption sites may catalyse non-biological reactions, adsorbed herbicide may be degraded more rapidly. There have been some studies of the effects of adsorbents added to the soil and of the natural adsorptive properties of the soil on degradation rates.

Microbial degradation of DNOC was prevented by addition of 1·25% activated charcoal to soil, whereas degradation of endothal was unaffected by 6·2% (Helweg, 1969). Decomposition of maleic hydrazide was progressively reduced by increasing concentrations (0·02 to 0·5%) of activated carbon (Helweg, 1975). Süss *et al.* (1972) could detect no microbial breakdown of monolinuron in soil containing 6% of filter ash and Hurle (1978) reported increased persistence of atrazine and DNOC in soil when straw ash was added. Further studies showed a linear relationship between Freundlich *k*-values and atrazine half-lives in straw-ash amended soils (Hurle, 1981). In unamended soils, increasing half-lives with increased adsorption have been reported for simazine, atrazine, propazine and terbuthylazine (Burkhard and Guth, 1980). On the other hand, for linuron, first-order rate constants for degradation were positively correlated with adsorption constants in 18 different soils (Walker and Thompson, 1977).

Moyer *et al.* (1972) found lower degradation rates for atrazine and chlorthiamid but not linuron in soil amended with activated charcoal although increased adsorption of all three compounds was observed. With montmorillonite, persistence of these compounds increased, but increased adsorption was only observed for atrazine and chlorthiamid. In nutrient solutions, diquat was adsorbed by kaolinite, but adsorption had no effect on its degradation (Weber and Coble, 1968). In the presence of montmorillonite, however, the rate of degradation was proportional to the amount in solution. It is well documented that paraquat is inactivated in soils by adsorption, and is degraded only slowly once adsorbed (Fryer *et al.*, 1975; Calderbank and Slade, 1976).

Clay minerals used as carriers may catalyse degradation of pesticides in dust formulations (Polon, 1973) and chloro-*s*-triazine hydrolysis when adsorbed on clay minerals has been observed (Russell *et al.*, 1968). Armstrong *et al.* (1967) reported faster degradation of atrazine at pH 3·9 in sterile soil than in solution and again suggested adsorption-catalysed hydrolysis. However, later studies (Armstrong and Chesters, 1968) suggested that adsorption does not necessarily lead to catalysed hydrolysis but that specific adsorption bonds are required for this to occur.

The effects of soil organic matter on the rates of herbicide loss were reviewed above. In general, organic matter is the most important adsorbent for herbicides in soil and some studies have been made of the effects of its removal on rates of degradation. Hance (1974), for example, reported that degradation rates of atrazine and linuron in soil samples from which fractions of the organic matter had been removed were not related to adsorption, and slowest degradation was observed in samples with least adsorption containing little organic matter. It is difficult to interpret such results however, since pre-treatment of soils to extract organic matter will affect the surface properties of the organic matter which remains and will have some effect on soil microorganisms. There are similar difficulties in interpreting results of Süss and Wagner (1970) and Süss *et al.* (1972). They treated a clay mineral (not defined) and a peat with monolinuron, chloridazon and simazine, and after 72 h equilibration, extracted part of the herbicide with organic solvents. The adsorbents plus "strongly adsorbed" herbicide were then mixed with soil and rates of herbicide degradation compared with those of herbicides added directly to the soil. The "strongly adsorbed" herbicide degraded more rapidly in all treatments.

The results are therefore conflicting and it is clear that adsorption does not always protect a chemical from degradation, nor does it always lead to increased rates of loss.

## D. Soil pH

Soil pH may affect degradation directly if the stability of the chemical is pH dependent, and indirectly via its effect on adsorption or on the composition of the soil microflora.

With simazine, Nearpass (1965) found faster degradation in two soils at pH 5·4 and 3·9 than at 6·8 and 7·0. Atrazine degradation was more rapid at pH 5·5 than in the same soil adjusted to pH 7·5 (Best and Weber, 1974), and this is consistent with results of Hiltbold and Buchanan (1977) working with a soil adjusted to pH levels of 5·6 and 7. Walker and Thompson (1977) found a significant negative correlation between rates of simazine degradation and pH in 18 unamended soils. In all of these examples, the authors suggested increased rates of hydrolysis at lower soil pH. However, Hance (1979) working with two soils adjusted to four different pH levels in the range 5 to 8 reported only slightly increased decomposition rates for atrazine with decreasing pH in one of the soils, whereas in the other, rates of loss decreased. Best and Weber (1974) found slower rates of degradation for prometryne at lower pH, and Ladlie *et al.* (1976a; 1976b) observed less degradation of metribuzin with decreasing pH. These results were explained by increased adsorption at low pH and hence reduced availability for degradation.

Corbin and Upchurch (1967) investigated the rates of degradation of several herbicides in two high organic matter soils adjusted to four pH levels in the range 4·3 to 7·5. They reported maximum degradation of 2,4-D and dicamba at pH 5·3, of dalapon and aminotriazole at pH 6·5, of vernolate at pH 7·5, but no effect of pH on rates of loss of diuron or chloramben. They suggested that maximum rates of loss would occur at pH levels favourable for growth of the specific microorganisms involved in degradation.

Working with natural soils of pH 3·8, 4·2, 4·8 and 6·3, Torstensson (1975) found most rapid degradation of 2,4-D and MCPA in soil with highest pH, and fermenter studies showed most rapid degradation of these compounds between pH 5·5 and 8·5. Hance (1979) also found increased degradation rates with increasing pH for dichlorprop, linuron and propyzamide. For linuron, these results are consistent with those of Walker and Thompson (1977), but with propyzamide, the latter found no correlation between pH and rates of loss in 18 natural soils. Savage (1973) measured trifluralin and nitralin residues in soil from 250 sites and observed that higher residues were associated with low pH. In laboratory experiments, nitralin was more persistent at pH 4·8 than at 6·8, but degradation of trifluralin was not affected.

These results show that soil pH can influence degradation rates, but we have little understanding of the principles involved, and more information is required before general conclusions can be made. More knowledge of the pH dependence of herbicide hydrolysis in aqueous systems would be helpful. Such information for some compounds is probably available in the files of Industry but unfortunately, not published. The experimental approach used by most workers of adjusting the pH of soil by additions of acid or alkali has been criticized (Hamaker, 1972) since such treatments cause changes in soil characteristics other than pH. However, when appropriate equilibration periods are used, the method may suffice. Use of more natural model systems would be preferred, for example by changing pH in field plots through liming for several years and using these soils in both laboratory and field studies.

## E. Soil Amendments

It is generally assumed that the rate of degradation of most herbicides is influenced by soil microbial activity. Since addition of easily degradable organic substrates and mineral nutrients results in a spontaneous increase in microbiological activity, one would expect enhanced herbicide degradation by such treatments. However, this does not always occur.

Microbial nutrient broths added to soil accelerated degradation of diphenamid, monuron, atrazine, dicamba, chlorpropham and chloramben (McClure, 1970). Glucose enhanced degradation of atrazine (McCormick and Hiltbold, 1966) and metribuzin (Savage, 1977); degradation of dalapon, however, was slower (Burge, 1969). Ground vetch tops increased atrazine and diuron decomposition (McCormick and Hiltbold, 1966) and maize and bean straw enhanced degradation of bromacil and terbacil (Wolf and Martin, 1974). Additions of sewage sludge inhibited degradation of atrazine, diuron and linuron but dairy manure increased rates of loss even though both amendments stimulated microbial activity (Doyle *et al.*, 1978). In a study with two soils, Hance (1973) reported that farmyard manure or straw accelerated atrazine degradation in one soil but not the other, whereas mineral fertilizers (N, P and K) or fertilizers plus straw accelerated degradation in both soils. With linuron, none of these treatments had a significant effect in either soil. In similar experiments, Hurle (1981) also observed that straw enhanced degradation of atrazine, but nitrogen inhibited rates of loss. The effect of straw plus nitrogen was dependent on the straw/nitrogen ratio. DNOC degradation was not affected by any of the treatments.

Once again, no general conclusions can be made. In order to understand more fully the results obtained, we require information on the specific involvement of microorganisms in degradation of particular compounds. Aspects of this are discussed in Chapter 6 of this book.

## F. Temperature and Moisture

Since the rates of non-biological reactions and biological processes are increased by increasing temperature, rates of herbicide degradation should also be increased. Hamaker (1972) reviewed the literature for herbicides and other soil-applied pesticides and found that this is generally the case.

The dependence of the rate constant, $k$, of a chemical reaction on temperature can be expressed by the Arrhenius equation:

$$k = A_0 \, e^{-(E_a/RT)} \tag{4}$$

where $A_0$ is a constant, $R$ the gas constant, $T$ the absolute temperature and $E_a$ the activation energy. With first-order reactions, the dependence of half-lives on temperature can be expressed by:

$$\log H_1 - \log H_2 = \frac{E_a}{4{\cdot}575}\left(\frac{1}{T_1} - \frac{1}{T_2}\right) \tag{5}$$

where $H_1$ and $H_2$ are the half-lives at absolute temperatures $T_1$ and $T_2$. Equation (5) has been satisfactorily applied to degradation data for a number of soil-applied herbicides (Zimdahl *et al.*, 1970; Usoroh and Hance, 1974; Hyzack and Zimdahl, 1974; Gingerich and Zimdahl, 1976; Walker, 1978; Kibler, 1979) and values for the activation energy derived. It is however difficult to ascribe any real significance to values of $E_a$ since first-order kinetics are generally only an approximation of the true kinetics of degradation.

Many degradation studies reported in the literature include only one temperature and it would be more useful if such studies were carried out at several temperatures to cover the range likely to occur in the field environment. This would permit assessment of the general validity of the Arrhenius equation to describe temperature effects on degradation and allow the relative temperature dependence of degradation of different herbicides to be determined.

Adequate water as well as temperature is essential for microbiological activity, but in addition, water acts as a solvent and transport agent, a reaction medium for both biological and non-biological processes and is a reagent in hydrolytic reactions. There is evidence that herbicide degradation rates are increased under moist soil conditions which, in most instances, probably reflects increased biological activity. Results from studies with herbicides and other pesticides were again reviewed by Hamaker (1972) who commented on the general lack of information available on this aspect. In recent years, reports of studies with a wide range of compounds have been published (Meikle *et al.*, 1973; Usoroh and Hance, 1974; Walker, 1976a; 1976b; 1976c; Zimdahl and Gwynn, 1977; Walker, 1978; Kibler, 1979), and these have shown the expected effect of increased degradation rates with increasing soil moisture up to field capacity. Since moisture conditions in the surface soil in the field are often limiting during the summer months, further investigations of the moisture dependence of herbicide degradation would be justified.

Increased degradation rates at increased temperature and soil moisture content have been confirmed in numerous field studies. Holly and Roberts (1963) measured persistence of a number of triazine herbicides at three locations in the United Kingdom and concluded that within this restricted range of soils, variations in persistence between seasons at any one site were greater than those between sites. In similar experiments at various field locations in Nebraska, Burnside *et al.* (1963) and Burnside *et al.* (1969) showed that simazine, atrazine, propazine, linuron and monuron were less persistent at sites in the wetter eastern parts of the state compared with the relatively dry western areas. Harris *et al.* (1969) measured persistence of atrazine and chlorfenac at 12 locations in the USA and demonstrated that both compounds were more persistent in the cooler northern and the dryer western states than in the semi-tropical southern states. Similar results were obtained for picloram by Hamaker *et al.* (1967). Its persistence in 18 states and provinces in the USA and Canada was expressed in terms of half-order kinetics. In general, the lowest rate constants were calculated for cool, dry states and the highest in hot, wet states and they could be significantly correlated with rainfall and temperature.

Variations in persistence as a result of different weather conditions have also been demonstrated following herbicide application to the same site at different times of the year. Walker (1970b) measured soil residues of propyzamide following spring, early summer or autumn application.

When applied at the beginning of April, there was little loss of herbicide during the first 4 to 8 weeks followed by a period of relatively rapid decline. When applied in late spring or early summer, rapid loss commenced soon after application. Low temperatures in April and early May presumably prevented degradation, a conclusion supported by the pattern of loss recorded after autumn application. When applied in September, 15 to 20% of the amount applied was lost during the subsequent 4 to 8 weeks but there was little further change in soil residues throughout the winter months until soil temperatures rose again the following spring. Similar variations in summer and winter losses of other soil-applied herbicides have been reported (Burnside *et al.*, 1961; Talbert and Fletchall, 1964; Morris and Penney, 1971; Smith and Emmond, 1975; Schweizer, 1976).

## G. Distribution in Soil

Some relatively volatile herbicides such as trifluralin and EPTC must be incorporated into soil to avoid rapid loss. Persistence of trifluralin is greatly increased if it is incorporated and persistence increases with depth of incorporation (Savage and Barrantine, 1969; Menges and Tamez, 1974). Shallow incorporation or subsurface placement of some herbicides which are normally left on the soil surface can also increase their persistence. Schweizer and Holstun (1966) reported that residues of prometryne were more phytotoxic to a subsequent oat crop when the herbicide had been applied sub-surface than when surface-applied and Walker (1976a) demonstrated that prometryne persistence also increased following its incorporation into the surface 2 to 3 cm. Diuron, fluometuron, propham (Horowitz and Herzlinger, 1974), propachlor (Menges and Tamez, 1973) and chlorpropham (Parochetti and Warren, 1968) behave in a similar way.

In these examples, it was suggested that incorporation increased persistence by reducing losses due to volatility, photochemical degradation or both, and with compounds which are not prone to losses by either mechanism, incorporation should have little effect. Walker (1976a) showed that this was so for simazine and Regan and Krempler (1974) observed no effect from depth of placement within the top 10 cm of soil on persistence of simazine or atrazine.

Since soil organic matter, temperature and aeration are more favourable for microbial activity in topsoil than in subsoil, degradation rates might decrease if a herbicide should penetrate beyond the cultivated layer into the subsoil. In laboratory experiments, reduced rates of loss in subsoils compared with soil from the plough layer have been reported for a number of herbicides including atrazine (Roeth *et al.*, 1969), dicamba (Hahn *et al.*, 1969) and asulam (Babiker and Duncan, 1977).

In field tests, Harris *et al.* (1969) measured the persistence of atrazine and chlorfenac at depths of 7–8, 22–23 and 37–38 cm at 12 different sites. Although the data showed considerable variation and recoveries from the different depths at some locations were the same, on average 60% more atrazine and 40% more chlorfenac remained at 37 to 38 cm than at 7 to 8 cm. Lavy *et al.* (1973) confirmed reduced rates of atrazine degradation in the field in subsoils compared with topsoils. When buried at a depth of 15 cm, no atrazine could be detected after 5 months; at 40 cm, none could be detected after 15 months but at 90 cm measurable quantities were still present after 47 months.

## H. Repeat Treatment

Most experiments on herbicide persistence in field soils have examined behaviour of a single application at a rate similar to that used for selective weed control. In some agricultural and horticultural situations the same herbicide may be applied to the same soil each year for a number of years so the possibility of accumulation of residues arises.

Parka and Tepe (1969) sampled soils from 107 sites in the USA which had received trifluralin applications for the preceding 1, 2, 3 or 4 years. They found no evidence of differences in soil residue levels which could be related to the number of times the herbicide had been applied. Similar results were reported for trifluralin and nitralin residues in soils from 250 sites in Mississippi (Savage, 1973).

Fryer and Kirkland (1970) applied MCPA, simazine, linuron and triallate at their normal use rates each year to plots on which an appropriate crop was grown, and at twice the recommended rate, two times each year to bare plots. The experiments continued from 1963 to 1968 and in no case was there any evidence of significant build-up of residues. For example, when simazine was applied at 3.36 kg/ha in spring and in autumn each year, the maximum residue found 6 months after autumn application was 1·90 kg/ha and after spring application was 0·56 kg/ha. Similar results for simazine were reported by Laar (1964) who determined only 0·92 kg/ha remaining in the soil 6 months after the ninth application of 5 kg/ha in 5 years.

Marriage *et al.* (1975) determined soil residues of simazine, atrazine, linuron and diuron following their application at 4·5 kg/ha in each of nine consecutive years in a peach orchard. The maximum residues measured in the autumn following spring application varied considerably between the herbicides and ranged from 7·3 kg/ha for diuron to 0·4 kg/ha for atrazine. Maximum residues for linuron and simazine were 3·8 and 1·6 kg/ha, respectively. After the final herbicide application, soil residues dissipated during the subsequent 1 to 3 years depending upon the chemical. Horowitz *et al.* (1974) also found no significant build-up following application of 10 different compounds to the same plots for four successive years.

Thus, when used at the recommended rates there is no evidence that repeat treatment with herbicides over several years leads to an accumulation of residues in the soil. This suggests that use of herbicides does not adversely affect degradation rates of subsequent applications, which was confirmed for three herbicides by Hance *et al.* (1978). They compared persistence of linuron, simazine and tri-allate in field plots that had received the relevant herbicide for the preceeding 14 years or had received no herbicide during this time. In no case was persistence affected. Similar results for simazine and monolinuron were reported by Paeschke *et al.* (1978).

With compounds which are used as microbial energy sources, and to which organisms become adapted, repeat treatments may degrade more rapidly than the first few applications if the interval to retreatment is not too long. A shortening of the lag-period and hence reduced persistence has been demonstrated in laboratory experiments for 2,4-D (Audus, 1949; 1951; 1960; 1964; Newman and Thomas, 1949), MCPA (Audus, 1949; 1960; 1964), dichlorprop (Brownbridge, 1956), DNOC (Jensen, 1966; Hurle and Pfefferkorn, 1972), chloridazon (Drescher and Otto, 1969; Engvild and Jensen, 1969), dalapon (Thiegs, 1955; Leasure, 1964; Kearney *et al.*, 1965), endothal (Jensen, 1965; Horowitz, 1966), propham (Kaufman and Blake, 1973) and chlorpropham (Kaufman and Kearney, 1976). Cross adaptation has been demonstrated with 2,4-D and MCPA (Tortensson, 1975). MCPB was shown to degrade faster in soil previously treated with MCPA (Kirkland and Fryer, 1972) as did dicamba in soil previously treated with 2,4-D (Kaufman, 1975).

Similar increased rates of loss of compounds such as TCA, MCPA and 2,4-D following repeated use have also been reported in field experiments (Kirkland, 1967; Fryer and Kirkland, 1970; Hurle and Rademacher, 1970; Torstensson, 1977).

## I. Cropping

The effects of a crop on herbicide persistence are difficult to assess because the presence of a crop will have indirect effects. For example, soil temperature and moisture levels will be different in cropped compared with fallow soils and there is little doubt that the presence of roots will affect soil microbial activity.

Although a crop might absorb herbicide from the soil it seems unlikely that the amounts removed in this way will be of practical importance. Sikka and Davis (1966) showed that 20 to 25% of the atrazine uniformly incorporated into soil in pots in the glasshouse could be absorbed by corn, sorghum and johnsongrass. In the field, where a herbicide is sprayed on the soil surface, much lower amounts are likely to be taken into the plant since the bulk of the plant root system will be in untreated soil. Nevertheless, in field trials, atrazine residues declined more rapidly in plots cropped with corn than in fallow plots (Sikka and Davis, 1966) and simazine residues were also lower in cropped than fallow plots (Zurawski and Ploszynski, 1968; Manorik and Malichenko, 1970). Birk and Roadhouse (1964) on the other hand found soil residues of atrazine much greater in cropped than fallow plots. The corn plots were much drier than the fallow plots for most of the growing season and the authors suggested that this resulted in reduced microbial activity in the cropped plots. Burnside *et al.* (1971) could find no difference between simazine and atrazine residues in fallow compared with cropped plots when irrigation was regularly applied, and cropping had no effect on the persistence of tri-allate, simazine or linuron in field trials in the United Kingdom (Hance *et al.*, 1978).

Apparently the effects from a crop on the rate of herbicide loss will depend on indirect soil and environmental factors which are likely to vary between soils, sites and seasons.

## J. Formulation

The formulation of a pesticide may have a marked effect on its persistence. Edwards (1972) reviewed the literature for insecticides and concluded that the relative order of persistence was granules > emulsions > miscible liquids > wettable powders. In general, granular herbicides are also more persistent than other formulations and this is particularly so with volatile compounds such as tri-allate (Hance *et al.*, 1973), propham (Siegman *et al.*, 1975), dichlobenil (Spencer-Jones and Wilson, 1976) and EPTC (Schreiber, 1976). Most of these herbicides are normally incorporated into soil to avoid rapid loss but formulation as granules permits application to the soil surface. The effects obtained can depend on weather conditions after application. Siegman *et al.* (1975) showed that granular formulations of propham gave advantages only when weather conditions were favourable for excessive volatilization from wettable powder formulations.

Relatively non-volatile compounds can persist longer when formulated as granules although again the effects are variable suggesting an interaction with weather conditions. For example, in some experiments granular atrazine was shown to persist longer than a wettable powder formulation (Buchholz, 1965; Frank, 1966), but in others, formulation had no effect (Meggitt, 1962; Regan and Krempler, 1974). However, because of the increased possibility of residue carry-over, granular atrazine is no longer marketed for selective weed control (LeBaron, 1970).

Controlled release granular formulations of herbicides can be particularly useful when relatively short persistence compounds are used in situations where long-term weed control is required but where repeat treatment is not economic (Collins *et al.,* 1973). The factors influencing the rate of release of herbicide from such formulations have been discussed by Furmidge (1972) and Osgerby (1972), and some of the practical uses of these materials were reviewed in detail by Wilkins (1976).

## K. Pesticide Combinations

The question whether the persistence of a pesticide in soil is affected when other pesticides are present is of considerable importance since there is increasing use of tank-mixes and complex spray programmes. If pesticides have an impact on soil microorganisms, degradation may be affected, and physicochemical effects such as competition for degradation sites could occur. There is evidence that persistence of a number of herbicides may be changed when used in combination with other pesticides although most of the experiments have been conducted under laboratory conditions; there have been few studies to determine the significance of these changes in the field.

In laboratory tests, persistence was prolonged for chlorpropham in the presence of quintozene (Walker, 1970a) and carbaryl (Kaufman *et al.*, 1970); proximpham with diuron, fenuron, propham, chlorpropham and lenacil (Hamroll and Jumar, 1973); linuron and monolinuron with methiocarb, aldicarb, carbaryl, propoxur, parathion, monuron, TMTD and diallate (Poschenrieder *et al.*, 1975; Wallnöfer *et al.*, 1977); isoproturon with PCMC (*p*-chlorophenyl-*N*-methylcarbamate), benomyl and maneb (Repiquet and Fournier, 1977); prometryne with thiuram (Wagenbreth and Kluge, 1971); propanil with PCMC (Kaufman *et al.*, 1971); dalapon with aminotriazole (Kaufmann, 1966); 2,4-D with parathion, carbaryl, dinoseb

(Repiquet and Fournier, 1977) and DNOC (Hurle, 1981); chlortoluron with demeton-*S*-methyl-sulphoxide (Kibler, 1979).

In all instances where different concentrations have been used, persistence of the compound under examination increased with increasing concentration of the second pesticide. When degradation curves have been presented, it is noticeable that the kinetics of degradation were not changed. For example, persistence of simazine was extended in the presence of DNOC, dinoseb-acetate and captan, but degradation still approximated to first-order kinetics, and with 2,4-D, the lag-period was extended in the presence of DNOC, but the degradation phase was not affected (Hurle, 1981).

No change in rate of degradation was found for aminotriazole in the presence of dalapon (Kaufman, 1966); di-allate with chlorpyrifos, azinphos-methyl, chloridazon, dimethoate or thiophanate (Domsch and Anderson, 1979); tri-allate with dinoseb, bentazone, dichlorprop, 2,4,5-T, 2,4-D, chlormequat, cholinechloride, tridemorph or thiophanate (Domsch and Anderson, 1979); 2,4-D with dicamba, dichlorprop, difenzoquat, TCA, 2,4,5-T (Smith, 1979) or benomyl (Repiquet and Fournier, 1977); 2,4,5-T with 2,4-D (Smith, 1979); simazine with parathion, sulphur or glyphosate (Kibler, 1979); chlortoluron with mecoprop or carbendazim (Kibler, 1979).

Increased degradation has been reported for simazine with three and ten times the normal field rate of paraquat, and for chlortoluron with a commercial fungicide formulation containing thiophanate-methyl, maneb and captafol (Kibler, 1979).

Where degradation was enhanced or inhibited, the investigators suggested that an impact of the second pesticide on microbiological processes was responsible. The mode of action, however, has generally not been investigated, with the exception of some insecticidal methylcarbamates which increase persistence of carbamate herbicides by competitive enzyme inhibition (Kaufman *et al.*, 1970; 1971; Blake and Kaufman, 1975; Engelhardt and Wallnöfer, 1975).

Our knowledge of degradation mechanisms and possible side effects of pesticides in soil is still limited so that, at present, we are unable to predict whether a given compound will affect persistence of another. There is a need for more information on the interactions between pesticides, and in particular, on the mechanisms by which certain compounds modify degradation rates of others. Because of the large number of possible combinations, studies should concentrate on those most frequently used. Any major effects found under laboratory conditions should be investigated further in the field to determine whether they are of real significance in the practical situation.

It is possible that combinations with desirable effects on herbicide degradation rates could be used deliberately to prolong persistence of very short-lived compounds as has been demonstrated for chlorpropham with the microbial inhibitor PCMC (Dawson, 1969).

## IV. Prediction of Persistence

A number of attempts have been made to quantify persistence in such a way that more generalized predictions of losses can be made. In spite of the complex of factors which can influence rates of degradation, the few attempts which have been made have given surprisingly good results (Goring *et al.*, 1975). Since few studies have involved soil-applied herbicides, studies with other pesticides will also be reviewed. However, before discussing these examples, it is worth examining the variability which can occur in measurements of pesticide residues in samples of field soils since it is within the context of this variability that the accuracy required from predictive models must be assessed.

### A. Variability in Measurements of Herbicide Residues

In laboratory experiments, a well sieved and therefore relatively uniform soil sample can be used; the herbicide can be distributed uniformly through the soil; incubation conditions can be controlled. In the field none of these factors can be controlled with precision. Much of the increased variation in field compared with laboratory experiments will result from inequalities in the initial distribution of herbicide on the soil surface. Some examples of the size of these errors can be taken from the data of Fryer and Kirkland (1970). Over six years of repeated treatment of plots with four different herbicides, initial recoveries of the nominally applied rate varied from 42 to 100% with MCPA, 37 to 90% with tri-allate, 60 to 104% with simazine and 32 to 144% with linuron. Similar variation was observed for atrazine, chlorfenac (Harris *et al.*, 1969), simazine, prometryne and propyzamide (Walker, 1976a; 1976c). Hence to assess quantitatively the kinetics of degradation under field conditions, detailed measurements of the amounts present initially are essential.

Fryer and Kirkland (1970) also measured the point to point variation in deposition of the spray application. In experiments with linuron, the mean deposition on several 12·9-cm$^2$ filter papers placed at random on the soil surface was 94, 65, 73, 81 and 94% of the theoretical quantity in different tests and the coefficient of variation was 16, 60, 23, 17 and 11% respectively. Mechanical incorporation of herbicides applied to the soil surface can increase this variation (Walker *et al.*, 1976; Wauchope *et al.*, 1977).

Variations in residual concentrations from one position to another within a treated area tend to increase with time, presumably because of variations in soil properties which affect degradation rates (Hance *et al.*, 1976) and differences in redistribution of herbicide both vertically and laterally within the treated area (Fryer and Kirkland, 1970; Clay and Stott, 1973). Fryer and Kirkland showed that four weeks after application of simazine, residues measured in 20 individual soil cores gave a coefficient of variation about the mean of 50%. Greater variability was reported by Taylor *et al.* (1971). The dieldrin content of 108 individual cores taken two years after application to small field plots showed a 50-fold range. They calculated that the coefficient of variation between bulk samples made by combining different numbers of cores could not be reduced practically below 20% which would be achieved by bulking 12 cores per plot.

Few other workers have assessed levels of variability but coefficients of variation in residues measured in samples from replicate field plots from 5 to 80% (Dawson *et al.*, 1968), 5 to 78% (Hance *et al.*, 1976), 3 to 66% (Smith, 1975), and 4 to 40% (Walker, 1978) have been reported.

## B. Theoretical Equations

Hill *et al.* (1955) observing that degradation of a number of substituted urea herbicides followed first-order kinetics, derived the following expression to show how residues would build up in soil following repeated annual use:

$$R = \frac{AX(1 - X^n)}{(1 - X)} \tag{6}$$

where $R$ is the residue at the end of the $n$th year, $X$ is the proportion of the applied dose remaining after 1 year and $A$ is the amount applied each year. They calculated the theoretical build-up curves following repeat application of 1 kg/ha of pesticides with different rates of loss (Fig. 1) and concluded that only repeated use of highly stable compounds would lead to residue accumulation. Even if only 50% were lost each year, the maximum residue one year after the $n$th application would equal the annual application rate.

A similar equation was used by Hamaker (1966) to predict disappearance data for aldrin reported by Lichtenstein and Schultz (1965). Lichtenstein and Schultz studied persistence of aldrin following a single application of 28 kg/ha and five annual applications of 5·6 kg/ha. Results from the single application suggested that 24% of the aldrin was converted to dieldrin and that the dieldrin subsequently degraded according to first-order kinetics

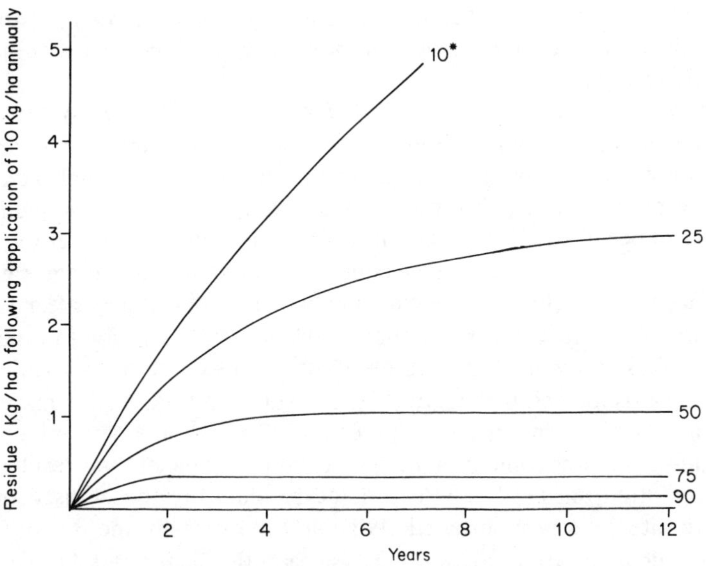


Fig. 1. Pesticide residues in soil following repeated annual application of compounds with different rates of loss. ⋆, percentage lost in one year. (From Hill *et al.*, 1955).

with a half-life of 4·2 years. With this situation equation (6) modifies to:

$$R = \frac{EAX(1 - X^n)}{(1 - X)} \tag{7}$$

where $A = 3{\cdot}2\ \mu g/g$ (5·6 kg/ha corrected for soil sampling depth and bulk density), $E = 0{\cdot}24$ (the efficiency of conversion of aldrin to dieldrin) and $X = 0{\cdot}8479$ (fraction of the amount applied remaining after 1 year corresponding to a half-life of 4·2 years). The results calculated by Hamaker (1966) using this expression are shown in Table I.

Table I. Accumulated residues of dieldrin following 5 annual applications of aldrin at 5·6 kg/ha (from Hamaker, 1966).

| | μg/g in soil after time (years) | | | | |
|---|---|---|---|---|---|
| | 1 | 2 | 3 | 4 | 5 |
| Calculated | 0·66 | 1·22 | 1·67 | 2·10 | 2·45 |
| Observed | 0·76 | 1·2 | 1·4 | 1·8 | 2·0 |

From *Organic Pesticides in the Environment*, Advances in Chemistry Series 60, American Chemical Society

Calculated and observed concentrations were in good agreement and the calculations showed that residues from repeat annual application followed a predictable pattern.

Similar results were reported by Dekker *et al.* (1965). They measured the rate of loss of aldrin from soil at one site in Illinois following its application at rates from 1·12 to 3·36 kg/ha. Residues of aldrin plus dieldrin declined rapidly while aldrin was the main component but subsequent loss of dieldrin occurred more slowly. A series of straight lines was empirically fitted to the data to represent maximum, minimum and average rates of loss. The average situation, for example, was a residue of 20% after 1 year, 15% after 2 years and a subsequent half-life of 3 years for the dieldrin residue. Dekker *et al.* then measured aldrin plus dieldrin residues in soil from 35 fields located throughout Illinois which had recieved from 3 to 13 applications of aldrin during the preceding 13 years, and determined how well their equations could predict these residues. Some of their results have been replotted in Fig. 2. The calculated residues are those based on the average rates of loss mentioned above and the straight line is the line of best fit calculated by regression analysis ($r = 0{\cdot}903$, $p < 0{\cdot}001$). The calculated intercept is 0·030 and the slope 0·960 and these are not significantly different from 0 or 1·0, respectively. The degree of correspondence between observed and predicted residues is very impressive when it is considered that no allowance was made for variations in soil type or weather conditions.

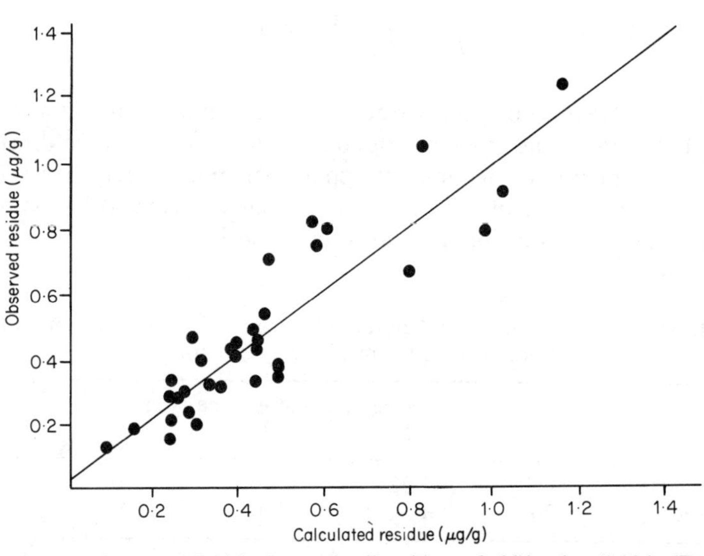


Fig. 2. Calculated compared with observed soil residues of aldrin plus dieldrin (Replotted from Dekker *et al.*, 1965).

These examples suggest that further use of relationships similar to equations (6) and (7) should be applied to data from long-term experiments involving herbicide persistence. Such attempts would provide further information on the possibilities that repeated herbicide use either restricts or enhances the rate of degradation of subsequent applications.

## C. Empirical Equations

Briggs (1976) proposed a method of predicting the relative rates of degradation of pesticides in soils based on their physical and chemical properties. The underlying concept of this method was that rates of pesticide loss are a function of the availability of the chemical for degradation (A) and its inherent degradability (D). Availability and degradability were assigned numerical values and the product A × D was then used to assess the rate of degradation. Briggs proposed 14 availability classes, determined by the organic matter/water partition coefficients ($Q$) of the pesticides and further showed that the value of $Q$ could be estimated from the octanol/water partition coefficient of the compound. A compound with low $Q$-value was assigned to availability class 1 and a compound with a very high value was assigned to class 14.

Degradability was based on the stability of the functional groups in the pesticide molecule so that a readily degradable group was assigned a low number on the scale 1 to 10 whereas a stable group was assigned a high number.

Although it is unlikely that an empirical model of this type could be used to predict persistence in a specific situation, it shows considerable promise as a technique for the early general classification of the likely persistence of newer compounds. It could be useful in deciding which groups of compounds should be further developed or rejected as potential pesticides.

## D. Regression Equations

Where herbicide residues in soil have been measured at a number of different sites, the data lend themselves to regression analysis. There have been some attempts to fit regression equations to soil residue data, not all of which have been successful.

Davison and Clay (1972) studied simazine persistence at 26 sites in the United Kingdom and attempted to correlate residues with soil and climatic factors. They could not identify any single soil property with which residues could be correlated and attempts to fit multiple regression equations were also unsuccessful. On the other hand, Smith and Meggitt (1970) showed that the rate of loss of chloridazon during the first two months after application was positively correlated with rainfall and negatively correlated with soil organic matter content. The relationship which best fits the data is

$$\mu\text{g/g lost per month} = 0{\cdot}51 - 0{\cdot}07\ (\%\text{OM}) + 0{\cdot}43\ (\text{cm rain/month}) \qquad (8)$$

This equation is probably of limited predictive ability since its use would be specific to a particular initial rate of application (5·3 kg/ha) and also to a two-month time period. Recalculation of the residues in terms of rate constants before regression analysis could possibly have produced an equation of more general applicability.

The most successful attempt to fit regression equations to soil persistence data was reported by Hamaker *et al.* (1967) for picloram. They assessed picloram persistence at 207 sites in North America and the data were transformed to equivalent half-order rate constants. An average rate constant was calculated for each state or province and these average values were related to various soil and climatic variables. Inclusion of terms for rainfall, temperature, rate of application, soil pH, organic matter and clay content in a multiple linear regression gave a correlation coefficient of 0·70 ($P < 0{\cdot}01$). An equation based on the two climatic variables alone gave a correlation coefficient only marginally poorer and this equation had the form

$$k_{\frac{1}{2}} = 0{\cdot}0119\,X_1 + 0{\cdot}0056\,X_2 - 0{\cdot}049 \qquad (9)$$

in which $X_1$ = annual rainfall (inches) and $X_2$ = days with temperature over 90°F (32·2°C). The usefulness of this equation was subsequently tested by Goring and Hamaker (1971) against other published persistence data for picloram and the rate constants predicted by equation (9) were all within the 95% confidence limits of those observed.

## E. Simulation Models

Walker (1973, 1974) developed a mathematical model for simulation of herbicide persistence in soil under field conditions which defined more precisely the variable field environment. The model combined laboratory measurements of the effects of soil temperature and soil moisture content on the rate of herbicide degradation with the fluctuations in surface soil moisture and temperature in the field. The field environment was predicted empirically from measurements of rainfall, evaporation and 10-cm soil temperature. Results from experiments to test the model with a number of herbicides have been reported (Walker, 1976a; 1976b; 1976c; 1976d; 1978; Walker and Bond, 1978; Walker and Smith, 1979; Kibler, 1979), and although the model showed limitations with some herbicides, its general predictive ability was good.

An example of the type of result obtained is shown for simazine in Fig. 3. Walker (1976b) measured the rate of loss of simazine at different controlled temperatures and moisture contents in untreated soil from a site adjacent to the long-term experiments of Fryer and Kirkland (1970).

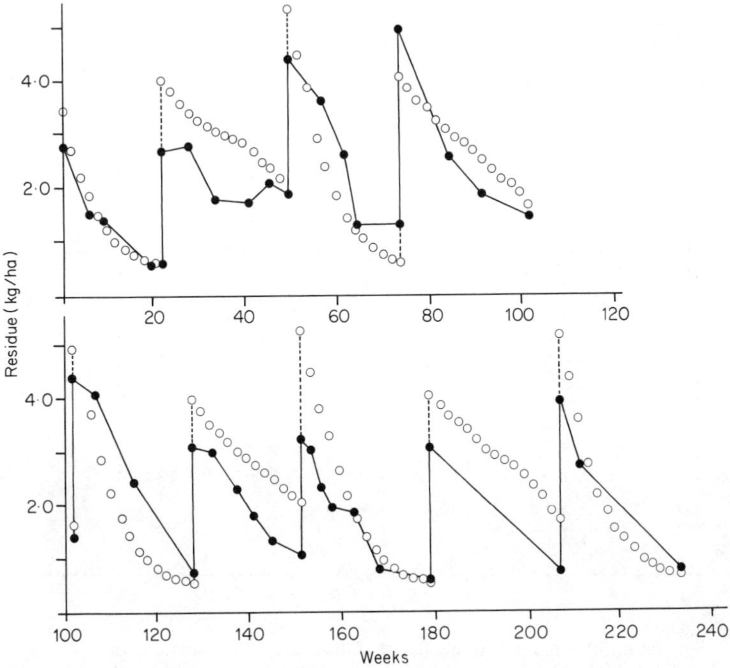


Fig. 3. Simulation of simazine persistence in long-term experiments. ●, Observed; ○, Simulated; time 0 = May 27, 1964. (From Walker, 1976b).

Relationships derived from these data were used in the computer program with the relevant weather records for the years 1964 to 1968 to simulate the persistence data published by Fryer and Kirkland. The simulated disappearance curves (Fig. 3) were of similar shape to those observed and the predicted residue at the end of the fifth year and after nine applications of simazine was 0·58 kg/ha—very close to the determined residue of 0·63 kg/ha.

In similar experiments with asulam in Canada (Smith and Walker, 1977), the laboratory data suggested an optimum temperature of 25°C and no degradation at soil moisture contents below 25% of field capacity. These features were included in the computer program and the results (Fig. 4) suggested that their inclusion to a large extent was successful; the model satisfactorily predicted asulam persistence in the field in five out of six experiments.

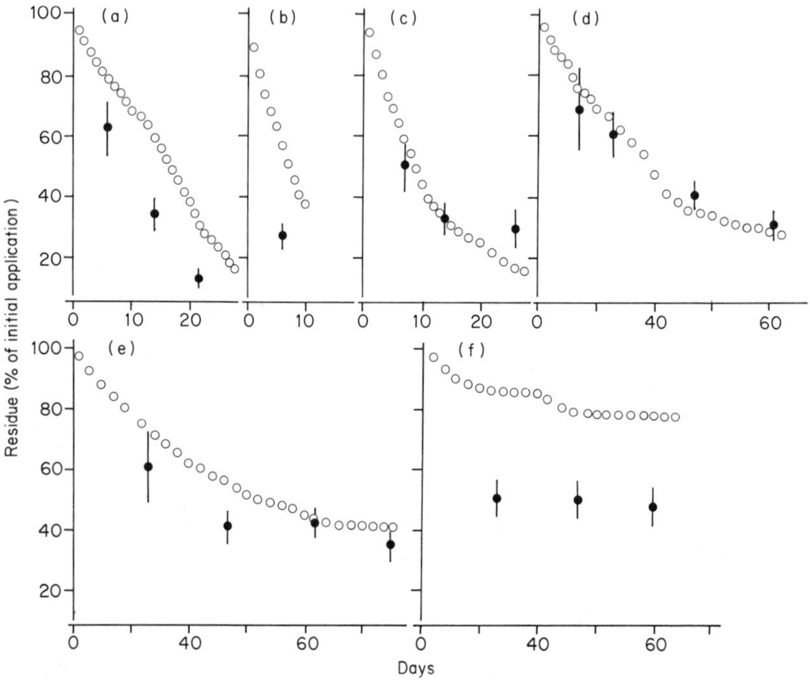


Fig. 4. Simulation of asulam persistence. ●, Observed; ○, Simulated. (a), 14 May; (b), 4 June; (c), 2 July; (d), 30 July; (e), 3 September; (f), 8 October. (From Smith and Walker, 1977).

These two examples demonstrate the usefulness of the computer modeling approach to prediction of persistence but there are a number of limitations in the model described by Walker (1974). The model was based on first-order kinetics and in many instances this is only an approximation to the true kinetics; the model did not allow for losses through volatility or photochemical degradation; the model was highly specific to one soil. In addition, most of the experiments to test the model were made in one soil type and under the temperate weather conditions experienced in the United Kingdom where extremes of temperature and soil moisture are rarely encountered. However, it has been shown to predict adequately the persistence of a number of herbicides in different years, and so more work in this area is justified. In particular, studies in different soils and climates are essential to test the true validity of this technique. Given a valid model, expected behaviour under average and extreme conditions could be predicted from a limited amount of experimental data; persistence of a compound when applied at different times of the year, in different years or in different geographical areas could be assessed; in conjunction with data on crop susceptibilities it could be used to indicate safe cropping sequences for particular herbicide use patterns; in cases of crop damage it could also be used retrospectively to assess whether the residues present early in the life of the crop were likely to have been the cause.

## V. Significance of Residues to Succeeding Crops

There are many reports in the literature of herbicide carry-over problems, some of which were reviewed by Sheets and Harris (1965). In most of these reports, the responses of crops sown at intervals following herbicide application were assessed but measurements of herbicide residues in the soil at the time of sowing or when phytotoxicity symptoms developed were not made (e.g. Lallukka *et al.*, 1974; Burnside and Schultz, 1978). Results from such studies are useful in identifying potential residue problems, but they do not allow quantitative interpretation in terms of residue levels causing crop damage and are therefore of limited value for advisory purposes.

There are a number of problems in obtaining more quantitative information since the absolute values for residual concentrations in the soil will be determined by sampling depth and phytotoxicities of residues will depend on soil type, their distribution in the soil (Eagle, 1978), and climatic factors. However, in recent years, some attempts have been made to measure crop responses to defined soil concentrations of herbicides with the hope that such measurements will allow better interpretation of residue analyses. For example, Caverly (1978) incorporated lenacil, linuron and trifluralin into soil at various rates at three field locations and sowed 16 crops in the treated plots. When symptoms of phytotoxicity developed, the soil was sampled (0–15 cm) and residual concentrations were measured. Some of the results, averaged across the three sites, are shown in Table II. Other results for more limited herbicide–crop combinations have been reported (Miller *et al.*, 1975; Schweizer, 1975; Walker and Bond, 1978; Eagle, 1978). More information of this type is required for other soil-applied herbicides in a range of soil types to allow minimum phytotoxic concentration ranges to be defined.

Table II. Minimum residues causing visible symptoms of phytotoxicity to various crops. (Mean of three sites, recalculated from Caverley, 1978).

| Crop | Herbicide residue (mg/kg in 0 to 15 cm) | | |
|---|---|---|---|
| | Lenacil | Linuron | Trifluralin |
| Turnip | 0·04 | 0·25 | 0·43 |
| Lettuce | 0·07[b] | 0·28 | 1·06[b] |
| Wheat | 0·09[b] | 0·24[b] | 0·19 |
| Barley | 0·10[a] | 0·39[b] | 0·14 |
| Oat | 0·10[a] | 0·33[b] | 0·11[b] |
| Rape | 0·11 | 0·24 | 0·95[b] |
| Pea | 0·13[b] | 0·91 | 0·82 |
| Ryegrass | 0·18[b] | 0·49 | 0·10 |
| Onion | 0·18[b] | 0·77[b] | 0·66 |
| Maize | 0·18[a] | 0·41[a] | 0·12 |
| Carrot | 0·18 | 0·91 | 2·45 |
| Parsnip | 0·18 | 0·91 | 2·45 |
| Potato | 0·21 | 0·67 | 0·82 |
| Dwarf bean | 0·23 | 0·38 | 0·82 |
| Red beet | 0·83 | 0·49 | 0·10 |
| Linseed | 0·94 | 1·00 | 1·72 |

[a] Data from one site only
[b] Data from two sites

## VI. Alleviation of Residue Problems

Although all herbicides degrade in soil, because of variations in the rates of loss between compounds, between soils and between different climatic conditions, occasional residue problems will inevitably occur. Various techniques which can help to minimize them have been discussed by Burnside (1974).

The obvious way to minimize phytotoxicity from residues is to avoid sowing highly sensitive crops, but to do this, more data are needed on relative crop susceptibilities such as those in Table II. Current label recommendations for a number of soil-applied herbicides (e.g. simazine, lenacil, propyzamide) suggest that the soil should be ploughed before any succeeding crop is sown. Burnside (1972) showed that soil residues of nitralin and trifluralin were non-phytotoxic to a subsequent oat crop if the soil was ploughed and Stoinev (1967) showed that deep cultivation reduced phytotoxicity of simazine and atrazine residues to wheat. Eagle (1978) applied 11 herbicides at 1·6 kg/ha at different field locations. The soil was subsequently ploughed and cultivated or just cultivated prior to sowing wheat or barley. The amount of damage caused by every herbicide was reduced by ploughing. The use of activated charcoal to enhance herbicide selectivity is discussed in Chapter 8 and this material has also been used to alleviate soil residue problems. The amounts of carbon required are generally relatively high. For example, Zemanek (1964) showed that 500 kg/ha was required to inactivate atrazine residues completely, although Ostrowski (1968) suggested that for both simazine and diuron, a ratio of 50:1 activated charcoal:herbicide residue would give complete inactivation. Andersen (1968) reported that between 200 and 400 units of charcoal were required to deactivate a unit residue of simazine whereas with linuron the ratio was only 100:1. Other charcoal:residue ratios have been suggested by Ahrens (1964), Smith, D. W. (1968) and Bovey and Miller (1969). Variations in these ratios are likely because all charcoals are not equally adsorptive (Jordan and Smith, 1971) and there are marked variations in the extent to which any one charcoal adsorbs different herbicides (Coffey and Warren, 1969). Andersen (1968) also reported that the amount of charcoal required would depend on its efficiency of incorporation into soil and on soil organic matter content. One problem in the use of charcoal which seems to have been overlooked is its effect on the activity of subsequent herbicide treatments.

Other soil additives such as farmyard manure (Joshi and Datta, 1975), green manures (Beckmann and Pestemer, 1975) and microbial nutrient broths (McClure, 1970) have also been suggested as possible means of increasing herbicide degradation rates, and effects from such amendments were discussed earlier. Insufficient work appears to have been done to assess their true practical value and it is probable that their effects will be small compared with the natural variations in degradation rates which result from fluctuations in temperature and soil moisture.

## VII. Registration Requirements

Information on persistence under field conditions is an essential requirement for registration of pesticides and registration authorities in various countries have guidelines outlining the types of data required. In some instances, these guidelines are general and flexible whereas in others they are more specific. Examples of both will be discussed.

The United Kingdom Pesticides Safety Precautions Scheme (Anon., 1971) does not detail specific requirements for soil residue data, but in a section dealing with residues in crops it states that "at the same time, the possible persistence of a pesticide in the soil should not be overlooked". The implication is that soil residue data similar to those required for crops may be necessary. The guidelines request data from a number of sites, covering a range of rates and methods of application. They request that residue measurements be made immediately after application and on at least four occasions up to and including crop harvest. Particular emphasis is placed on the need for sufficient replication to give a true estimate of variability and the guidelines request full weather records with specific mention of any abnormal weather patterns. If repeat applications are likely, data on persistence of these are also required.

Almost identical guidelines have been adopted by the Council of Europe (Anon., 1973) with the modification that "trials should be designed to cover a wide range of representative growing areas", although the number of sites is not specified. Although of a fairly general nature these guidelines appear realistic and take account of variations in soil type, weather conditions, methods and rates of application and different use patterns. An important feature of the United Kingdom guidelines is that they suggest informal consultation with the registration authorities at an early stage to ensure that the data submitted are "as complete as can reasonably be expected", therefore suggesting a flexibility in data requirements depending on the chemical and its proposed use.

The Environmental Protection Agency (EPA) in the United States (Anon., 1978) has proposed guidelines which are more specific and are designed to cover almost all variables. In contrast with recommendations in other areas of pesticide–soil interactions, the proposed guidelines do not suggest a preferred method but they do specify sampling times, the need for sufficient replication to allow statistical evaluation of the data and the need for complete weather records for the period of the tests. Data are required from a specified number of field locations, which varies according to the proposed use of the chemical (Table III) and the tests must continue

Table III. Number of sites and time periods for which field dissipation data are required by EPA (Anon., 1978).

| Use area | Number of sites | Maximum duration (months) |
|---|---|---|
| Field and vegetable crops | 4 | 18 |
| Tree fruit and nut crops | 2 | 12 |
| Pasture land | 3 | 12 |
| Domestic outdoor, parks, ornamental and turf | 2 | 4 |
| Rights of way, shelterbelts | 2 | 2 |

until 90% of the pesticide has been degraded or up to a maximum time limit which varies with the proposed use (Table III). The reasons for these variations in site number and test duration are not clear and it seems unlikely that a true measure of persistence of a compound can be obtained by measuring residues from two sites for only two months.

A long-term experiment involving persistence of repeat treatments is required if less than 50% loss occurs prior to a recommended subsequent application, and it is suggested that such a study continues for a maximum of three years, or two years beyond the appropriate maximum period specified in Table III. Additional data are also required if the chemical will be used either as a combined formulation or tank mixture with a second compound. A laboratory or field experiment to measure the dissipation rates of each compound alone and in the proposed mixture must be done. This requirement is realistic since a number of interactions between pesticides which affect persistence have been identified (Kaufman *et al.,* 1970; Kaufman, 1977).

In general, registration authorities request data that are both relevant and necessary in order to ensure that pesticide residues in the soil will dissipate within acceptable time limits. It may be argued, however, that since laboratory studies of both rates and pathways of degradation are also required, extensive field evaluation of persistence should not be necessary for compounds which are degraded rapidly. Such information should only be required for compounds which prove to be relatively stable in laboratory tests and therefore to have the potential for extended persistence in the field. A similar suggestion has recently been made by Hemingway (1978).

## VIII. Conclusions and Suggestions for Future Research

A knowledge of the rates of herbicide dissipation from soil is important for agronomic and environmental reasons. With most current herbicides there is little cause for concern. It has been demonstrated with a number of compounds that repeated annual use at normal rates of application does not lead to a build up of residues or cause measurable effects on other soil properties. In the section dealing with registration requirements, it was suggested that in some instances, extended laboratory studies could preclude the need for field persistence data. From a research point of view, however, it is essential that both short and long-term studies of persistence, particularly with new compounds should continue. This is firstly to ensure that residue problems will not occur and secondly to provide a body of valid scientific evidence with which unrealistic objections to the use of herbicides can be counteracted.

There are a number of specific areas where future research should be concentrated:

(i) There is a need for more quantitative interpretation of soil residue data to allow better use of the information generated. At present, the extensive persistence data for many herbicides remain specific to the sites and seasons of collection. A compilation of data from different sites and seasons together with soil characteristics and weather records may allow development of predictive regression equations as described for picloram by Hamaker *et al.* (1967). With this in mind, it would be helpful if all reports of herbicide persistence could include standard information such as soil properties (% sand, silt, clay and organic matter, pH, exchange capacity, available potassium and phosphorus, soil moisture content at field capacity and permanent wilting point) together with records of monthly mean maximum and minimum air temperatures, total rainfall per month plus the number of days with rain and the monthly total potential evaporation.

(ii) Another way to make fuller use of persistence data is the approach suggested by Briggs (1976) of using existing information to develop general relationships between rates of degradation and the physical and chemical properties of different compounds. This approach should be explored and developed further since such relationships would be of particular value during early screening of new compounds.

(iii) The use of computer modelling provides an ideal method of handling the large amounts of data required to characterize fully the variable field environment and attempts to predict persistence in the field from degradation rates measured in the laboratory have given promising results (Walker, 1976a; 1976b; 1976c; 1978; Kibler, 1979). Further development and evaluation of such models in different soils and widely varying weather conditions is required. A valid model could be of great potential benefit and allow estimates to be made of the likely variations in persistence when a herbicide is used at different times of the year, in different years and at different locations.

(iv) The success of predictive models depends to a large extent on a correct description of the kinetics of degradation. Up to now, most models have been based on first-order reaction kinetics which generally only approximate to the real situation. Use of current models is also restricted by our limited understanding of the influence of soil properties on the rates of herbicide degradation. Further studies of these two aspects of herbicide degradation are required. Such studies are best made under laboratory conditions where effects from one or a limited number of variables can be examined in systems where effects from others are minimized. However, close attention must be paid to methodology since any undesired change in the chemical or microbiological properties of soil will undoubtedly affect rates of loss. It has been suggested, for example, that on some occasions, deviations from first-order kinetics may occur as a result of artifacts from the method of incubation (Kempson-Jones and Hance, 1979). Aspects of methodology are discussed in more detail by Guth and Torstensson in other chapters of this book.

(v) There is also a need for more information on the variability in herbicide residue measurements. Does this result primarily from errors in initial application or is inherent variability within soil responsible? It is essential for accurate assessment of persistence under field conditions that care is taken in initial application, that sufficient samples are taken to obtain a representative sample from each plot and that treatments are replicated sufficiently to allow correct statistical analysis of the results. A measure of variability is also essential when considering the accuracy required from predictive models.

(vi) In order to interpret soil residue data in practical terms, more quantitative data on relative crop susceptibilities are required.

Many of these suggestions require data from numerous field sites with variations in soil type and weather patterns. Such studies are probably beyond the scope of individual research laboratories and more collaboration between researchers in different countries would permit generation of the required data from sufficient sites within reasonable periods of time. Presentation of standard soil and weather data as mentioned in (i) above would be a first step in this direction.

## References

Ahrens, J. F. (1964). Antidote for herbicides. *Front. Pl. Sci.*, **17**, 5–7.

Andersen, A. H. (1968). The inactivation of simazine and linuron in soil by charcoal. *Weed Res.*, **8**, 58–60.

Anon. (1971). Ministry of Agriculture Fisheries and Food, Pesticides Safety Precautions Scheme, Revised March 1971.

Anon. (1973). Agricultural Pesticides: Advice and guidance given to manufacturers, national authorities and others concerned with the registration of pesticides. *Council of Europe, Strasbourg.*

Anon. (1978). Environmental Protection Agency, Proposed guidelines for registration of pesticides in the United States. *Federal Register*, Vol. **43**, No. 132.

Armstrong, D. E. and Chesters, G. (1968). Adsorption catalyzed chemical hydrolysis of atrazine. *Environ. Sci. Technol.*, **2**, 683–689.

Armstrong, D. E., Chesters, G. and Harris, R. F. (1967). Atrazine hydrolysis in soil. *Soil Sci. Soc. Amer. Proc.*, **31**, 61–66.

Audus, L. J. (1949). The biological detoxification of 2,4-dichlorophenoxyacetic acid in soil. *Pl. Soil*, **2**, 31–36.

Audus, L. J. (1951). The biological detoxification of hormone herbicides in soil. *Pl. Soil*, **3**, 170–192.

Audus, L. J. (1960). Microbiological breakdown of herbicides in soils. In *Herbicides and the Soil.* (E. K. Woodford and G. R. Sagar, Eds.) pp. 1–19. Blackwell Scientific Publications, Oxford.

Audus, L. J. (1964). Herbicide behaviour in the soil. II. Interactions with soil microorganisms. *In* "The Physiology and Biochemistry of Herbicides." (L. J. Audus, Ed.) pp. 163–206. Academic Press, London and New York.

Babiker, A. J. T. and Duncan, H. J. (1977). Influence of soil depth on asulam adsorption and degradation. *Soil Biol. Biochem.*, **9**, 197–201.

Banting, J. D. (1967). Factors affecting the activity of di-allate and tri-allate. *Weed Res.*, **7**, 302–315.

Beckmann, E. O. and Pestemer, W. (1975). The effect of various organic manures on herbicide decomposition and biological activity in soils. *Landw. Forsch.*, **28**, 24–33. *Weed Abstracts*, No. 2439, 1976).

Best, J. A. and Weber, J. B. (1974). Disappearance of *s*-triazines as affected by soil pH using a balance sheet approach. *Weed Sci.*, **22**, 364–373.

Birk, L. A. and Roadhouse, F. E. B. (1964). Penetration of and persistence in soil of the herbicide atrazine. *Can. J. Pl. Sci.*, **44**, 21–27.

Blake, J. and Kaufman, D. D. (1975). Characterisation of acylanilide-hydrolysing enzymes from *Fusarium oxysporum* Schlecht. *Pestic. Biochem. Physiol.*, **5**, 305–313.

Bovey, R. W. and Miller, F. R. (1969). Effect of activated carbon on the phytotoxicity of herbicides in a tropical soil. *Weed Sci.*, **17**, 189–192.

Briggs, G. G. (1976). Degradation in soils. In *Persistence of Insecticides and Herbicides. British Crop Protection Council. Monograph No. 17*, 41–54.

Briska, A., Cencelj, J., Hocevar, J., Macek, J. and Sisakovic, V. (1973). (Research on soil contamination by herbicides). *Agrohemiya*, **1**, 37–43. (*Weed Abstracts*, No. 1391, 1974).

Brownbridge, N. (1956). Ph.D. Thesis, London University. (From Audus, 1964).

Buchholtz, K. P. (1965). Factors influencing oat injury from triazine residues in soil. *Weeds*, **13**, 362–367.

Burge, W. D. (1969). Populations of dalapon-decomposing bacteria in soil as influenced by additions of dalapon and other carbon sources. *Appl. Microbiol.*, **17**, 545–550.

Burkhard, N. and Guth, J. A. (1980), Chemical hydrolysis of 2-chloro-4,6-bis-(alkylamino)-1,3,5-triazine herbicides and their breakdown in soil under the influence of adsorption. *Pestic. Sci.*, **11**, (In press).

Burns, R. G. (1978). Enzyme activity in soil: some theoretical and practical considerations. In *Soil Enzymes* (R. G. Burns, Ed.) pp. 295–340. Academic Press, London and New York.

Burnside, O. C. (1972). Tolerance of soybean cultivars to weed competition and herbicides. *Weed Sci.*, **20**, 294–297.

Burnside, O. C. (1974). Prevention and detoxification of pesticide residues in soils. In *Pesticides in Soil and Water* (W. D. Guenzi, Ed.) pp. 387–412. Soil Science Society of America, Madison, Wisconsin.

Burnside, O. C. and Schultz, M. E. (1978). Soil persistence of herbicides for corn, sorghum and soybeans during the year of application. *Weed Sci.*, **26**, 108–115.

Burnside, O. C., Schmidt, E. L. and Behrens, R. (1961). Dissipation of simazine from the soil. *Weeds*, **9**, 477–484.

Burnside, O. C., Wicks, G. A. and Fenster, C. R. (1963). Dissipation and leaching of monuron, simazine and atrazine in Nebraska soils. *Weeds*, **11**, 209–213.
Burnside, O. C., Fenster, C. R. and Wicks, G. A. (1971). Soil persistence of repeated annual applications of atrazine. *Weed Sci.*, **19**, 290–293.
Burnside, O. C., Fenster, C. R., Wicks, G. A. and Drew, J. V. (1969). Effect of soil and climate on herbicide dissipation. *Weed Sci.*, **17**, 241–245.
Burschel, P. and Freed, V. H. (1959). The decomposition of herbicides in soils. *Weeds*, **7**, 157–161.
Calderbank, A. and Slade, P. (1976). Diquat and paraquat. In *Herbicides—Chemistry, Degradation and Mode of Action.* (P. C. Kearney and D. D. Kaufman, Eds.) pp. 501–540. Dekker, New York.
Caverley, D. J. (1978). Establishment of crop damage levels in soils for the herbicides lenacil, linuron and trifluralin. *Proc. 1978 Br. Crop Prot. Conf.—Weeds*, 549–555.
Clay, D. V. and Stott, K. G. (1973). The persistence and penetration of large doses of simazine in uncropped soil. *Weed Res.*, **13**, 42–50.
Coffey, D. L. and Warren, G. F. (1969). Inactivation of herbicides by activated charcoal and other adsorbents. *Weed Sci.*, **17**, 16–19.
Collins, R. L., Doglia, S., Mazak, R. A. and Samulski, E. T. (1973). Controlled release of herbicides—theory. *Weed Sci.*, **21**, 1–5.
Corbin, F. T. and Upchurch, R. P. (1967). Influence of pH on detoxification of herbicides in soil. *Weeds*, **15**, 370–377.
Davison, J. G. and Clay, D. V. (1972). The persistence of some horticultural herbicides in soil. *Rep. Weed Res. Org.* 1970–71, 58–65.
Dawson, J. H. (1969). Longevity of dodder control by soil-applied herbicides in the greenhouse. *Weed Sci.*, **17**, 295–298.
Dawson, J. H., Bruns, V. F. and Clore, W. J. (1968). Residual monuron, diuron and simazine in a vineyard soil. *Weed Sci.*, **16**, 63–65.
Dekker, G. C., Bruce, W. N. and Bigger, J. H. (1965). The accumulation and dissipation of residues resulting from use of aldrin in soils. *J. econ. Ent.*, **58**, 266–271.
Domsch, K. H. and Anderson, J. P. E. (1979). Mikrobieller Abbau, Festlegung und Verdampfung der Thiolcarbamate Diallat und Triallat im Boden. *Deutsche Forschungsgemeinschaft—Forschungsbericht Herbizide.* pp. 63–77. Harald Boldt Verlag, Boppard.
Donaldson, T. W. and Foy, C. L. (1965). The phytotoxicity and persistence in soils of benzoic acid herbicides. *Weeds*, **13**, 195–202.
Doyle, R. C., Kaufman, D. D. and Burt, G. W. (1978). Effect of dairy manure and sewage sludge on $^{14}$C-pesticide degradation in soil. *J. agric. Fd Chem.*, **26**, 987–989.
Drescher, N. and Otto, S. (1969). Über den Abbau von 1-Phenyl-4-amino-5-chlor-pyridazon-6 (Pyrazon) im Boden. *Z. PflKrankh. PflPath. PflSchutz*, **76**, 27–33.
Eagle, D. J. (1978). Interpretation of soil analysis for herbicide residues. *Proc. 1978 Br. Crop Prot. Conf.—Weeds*, pp. 535–639.
Edwards, C. A. (1972). Insecticides. In *Organic Chemicals in the Soil Environment.* Vol. 1. (C. A. I. Goring and J. W. Hamaker, Eds.) pp. 513–568. Dekker, New York.
Engelhardt, G. and Wallnöfer, P. R. (1975). Inhibition of phenylamide hydrolysis of *Bacillus sphaericus* with methylcarbamate and organophosphorus insecticides. *Appl. Microbiol.*, **29**, 717–721.

Engvild, K. C. and Jensen, H. L. (1969). Microbiological decomposition of the herbicide pyrazon. *Soil Biol. Biochem.*, **1**, 295–300.

Frank, R. (1966). Atrazine carryover in production of sugar beet in Southwestern Ontario. *Weeds*, **14**, 82–85.

Fryer, J. D. and Kirkland, K. (1970). Field experiments to investigate long-term effects of repeated applications of MCPA, tri-allate, simazine and linuron: report after 6 years. *Weed Res.*, **10**, 133–158.

Fryer, J. D., Hance, R. J. and Ludwig, J. W. (1975). Long-term persistence of paraquat in a sandy loam soil. *Weed Res.*, **15**, 189–194.

Furmidge, C. G. L. (1972). General principles governing the behaviour of granular formulations. *Pestic. Sci.*, **3**, 745–751.

Gingerich, L. L. and Zimdahl, R. L. (1976). Soil persistence of isopropalin and oryzalin. *Weed Sci.*, **24**, 431–434.

Goring, C. A. I. and Hamaker, J. W. (1971). The degradation and movement of picloram in soil and water. *Down to Earth*, **27**, No. 1, 12–15.

Goring, C. A. I., Laskowski, D. A., Hamaker, J. W. and Meikle, R. W. (1975). Principles of pesticide degradation in soils. In *Environmental Dynamics of Pesticides*. (R. Haque and V. H. Freed, Eds.) pp. 135–172. Plenum Press, New York.

Grover, R. (1967). Studies on the degradation of 4-amino-3,5,6-trichloropicolinic acid in soil. *Weed Res.*, **7**, 61–67.

Hahn, R. R., Burnside, O. C. and Lavy, T. L. (1969). Dissipation and phytotoxicity of dicamba. *Weed Sci.*, **17**, 3–8.

Hamaker, J. W. (1966). Mathematical prediction of cumulative levels of pesticides in soil. In *Organic Pesticides in the Environment*. (A. A. Rosen and H. F. Kraybill, Eds.) pp. 122–131. Advances in chemistry series No. 60, American Chemical Society, Washington D.C.

Hamaker, J. W. (1972). Decomposition: quantitative aspects. In *Organic Chemicals in the Soil Environment*. (C. A. I. Goring and J. W. Hamaker, Eds.) pp. 253–340. Dekker, New York.

Hamaker, J. W. (1976). The application of mathematical modelling to the soil persistence and accumulation of pesticides. In *Persistence of Insecticides and Herbicides. British Crop Protection Council. Monograph No. 17*, 181–199.

Hamaker, J. W. and Goring, C. A. I. (1976). Turnover of pesticide residues in soil. In *Bound and Conjugated Pesticide Residues*. (D. D. Kaufman, G. G. Still and G. D. Paulson, Eds.) pp. 219–243. Advances in chemistry series No. 29, American Chemical Society, Washington D.C.

Hamaker, J. W., Youngson, C. R. and Goring, C. A. I. (1967). Prediction of the persistence and activity of TORDON herbicide in soils under field conditions. *Down to Earth*, **23**, No. 2, 30–36.

Hamaker, J. W., Youngson, C. R. and Goring, C. A. I. (1968). Rate of detoxification of 4-amino-3,5,6-trichloropicolinic acid in soil. *Weed Res.*, **8**, 46–57.

Hamroll, B. and Jumar, A. (1973). Modellunterschungen zur Abbaugeschwindigkeit des Herbizids Proximpham im Boden in Gegenwart herbizider Zweitkomponenten. *Chem Tech.*, **25**, 423–424.

Hance, R. J. (1973). The effect of nutrients on the decomposition of the herbicides atrazine and linuron incubated with soil. *Pestic. Sci.*, **4**, 817–822.

Hance, R. J. (1974). Soil organic matter and the adsorption and decomposition of the herbicides atrazine and linuron. *Soil Biol. Biochem.*, **6**, 39–42.

Hance, R. J. (1979). Effect of pH on the degradation of atrazine, dichlorprop, linuron and propyzamide in soil. *Pestic. Sci.*, **10**, 83–86.

Hance, R. J. and McKone, C. E. (1971). Effect of concentration on the decomposition rates in soil of atrazine, linuron and picloram. *Pestic. Sci.*, **3**, 31–34.

Hance, R. J., Holroyd, J. and McKone, C. E. (1973). Aspects of tri-allate volatility. *Pestic Sci.*, **4**, 13–17.

Hance, R. J., Smith, P., Byast, T. H. and Cotterill, E. G. (1976). Variability in the persistence and movement in soils of abnormally high rates of simazine and linuron; some wider implications. *Proc. 1976 Br. Crop Prot. Conf.—Weeds*, pp. 643–648.

Hance, R. J., Smith, P. D., Cotterill, E. G. and Reid, D. C. (1978). Herbicide persistence: effects of plant cover, previous history of the soil and cultivation. *Med. Fac. Landbouww. Rijksuniv. Gent*, **43**, 1127–1134.

Harris, C. I., Woolson, E. A. and Hummer, B. E. (1969). Dissipation of herbicides at three soil depths. *Weed Sci.*, **17**, 27–31.

Helweg, A. (1969). Influence of charcoal-adsorbed herbicides on microorganisms. *Weed Res.*, **9**, 254–257.

Helweg, A. (1975). Degradation of $^{14}$C-maleic hydrazide in soil as influenced by adsorption on activated carbon. *Weed Res.*, **15**, 129–133.

Hemingway, R. J. (1978). Some aspects of the registration of herbicides. *Proc. 1978 Br. Crop Prot. Conf.—Weeds*, 1013–1017.

Hill, G. D., McGahen, J. W., Baker, H. M., Finnerty, D. W. and Bingeman, C. W. (1955). The fate of substituted urea herbicides in agricultural soils. *Agron. J.*, **47**, 93–104.

Hiltbold, A. E. and Buchanan, G. A. (1977). Influence of soil pH on persistence of atrazine in the field. *Weed Sci.*, **25**, 515–520.

Holly, K. and Roberts, H. A. (1963). Persistence of phytotoxic residues of triazine herbicides in soil. *Weed Res.*, **3**, 1–10.

Horowitz, M. (1966). Breakdown of endothal in soil. *Weed Res.*, **6**, 168–171.

Horowitz, M. and Herzlinger, G. (1974). Soil conditions affecting dissipation of diuron, fluometuron and propham from the soil surface. *Weed Res.*, **14**, 257–264.

Horowitz, M., Blumenfeld, T., Herzlinger, G. and Hulin, N. (1974). Effects of repeated applications of ten soil-active herbicides on weed population, residue accumulation and nitrification. *Weed Res.*, **14**, 97–109.

Hurle, K. (1973). Lag-Phase Studien mit DNOC und 2,4-D im Boden. *Proceedings European Weed Research Council Symposium; Herbicides and the Soil.* pp. 151–162.

Hurle, K. (1978). Einfluss des Strohverbrennens auf Aktivität, Sorption und Abbau von Herbiziden im Boden. *Med. Fac. Landbouww. Rijksuniv. Gent*, **43**, 1097–1107.

Hurle, K. (1981). Unterschungen zum Abbau von Herbiziden in Boden. *Acta Phytomedica*, (in press).

Hurle, K. and Pfefferkorn, V. (1972). Experiments on the rate of decomposition of DNOC in soils: the effect of pre-treatments. *Proc. 11th Br. Weed Control Conf.*, 806–810.

Hurle, K. and Rademacher, B. (1970). Unterschungen über den Einfluss langjährig wiederholter Anwendung von DNOC und 2,4-D auf ihren Abbau im Boden. *Weed Res.*, **10**, 159–164.

Hyzack, D. L. and Zimdahl, R. L. (1974). Rate of degradation of metribuzin and two analogues in soil. *Weed Sci.*, **22**, 75–79.

Jensen, H. L. (1965). Bilogisk sønderdeling af ukrudtsmidler i jordbunden. III. Endothal. *Tidsskr. f. Planteavl.*, **68**, 553–571.

Jensen, H. L. (1966). Biologisk sønderdeling af ukrudtsmidler i jordbunden. IV. Dinitro-ortho-cresol. *Tidsskr. f. Planteavl.*, **70**, 149–159.

Jordan, P. D. and Smith, L. W. (1971). Adsorption and deactivation of atrazine and diuron by charcoals. *Weed Sci.*, **19**, 541–544.

Joshi, O. P. and Datta, N. P. (1975). Persistence and movement of simazine in soil. *J. Indian Soc. Soil Sci.*, **23**, 263–265.

Kaufman, D. D. (1966). Microbial degradation of herbicide combinations: amitrol and dalapon. *Weeds*, **14**, 130–133.

Kaufman, D. D. (1975). Persistence and biodegradation of pesticide combinations. *Roczniki glebozn.*, **26**, 55–64.

Kaufman, D. D. (1977). Biodegradation and persistence of several acetamide, acylanilide, azide, carbamate and organo phosphate pesticide combinations. *Soil Boil. Biochem.*, **9**, 49–57.

Kaufman, D. D. and Blake, J. (1973). Microbial degradation of several acetamide, acylanilide, carbamate, toluidine and urea pesticides. *Soil Biol. Biochem.*, **5**, 297–308.

Kaufman, D. D. and Kearney, P. C. (1976). Microbial transformations in the soil. In *Herbicides—Physiology, Biochemistry, Ecology*. (L. J. Audus, Ed.) Vol. 2, pp. 29–64. Academic Press, London and New York.

Kaufman, D. D., Blake, J. and Miller, D. E. (1971). Methyl carbamates affect acylanilide herbicide residues in soil. *J. agric. Fd Chem.*, **19**, 204–206.

Kaufman, D. D., Kearney, P. C., Endt, D. W. Von and Miller, D. E. (1970). Methylcarbamate inhibition of phenylcarbamate metabolism in soil. *J. agric. Fd Chem.*, **18**, 513–519.

Kearney, P. C., Harris, C. I., Kaufman, D. D. and Sheets, T. J. (1965). Behaviour and fate of chlorinated aliphatic acids in soils. *Adv. Pest Contr. Res.*, **6**, 1–30.

Kempson-Jones, G. F. and Hance, R. J. (1979). Kinetics of linuron and metribuzin degradation in soil. *Pestic. Sci.*, **10** (in press).

Keys, C. H. and Friesen, H. A. (1968). Persistence of picloram activity in soil. *Weed Sci.*, **16**, 341–343.

Kibler, E. (1979). Freiland- und Laborversuche zur Abbaukinetik von Chlortoluron und Simazin in Böden in Gegenwart anderer Pflanzenschutzmittel. *Diss. Univ. Hohenheim*, pp. 86.

Kirkland, K. (1967). Inactivation of MCPA in soil. *Weed Res.*, **7**, 364–367.

Kirkland, K. and Fryer, J. D. (1972). Degradation of several herbicides in a soil previously treated with MCPA. *Weed Res.*, **12**, 90–95.

Laar, H. J. Van de (1964). The breakdown of simazine in soil. *Beplantingen en Boomkwekerij*, **20**, 14–17. (*Weed Abstracts* No. 1875, 1964).

Ladlie, J. S., Meggitt, W. F. and Penner, D. (1976a). Effect of soil pH on microbial degradation, adsorption and mobility of metribuzin. *Weed Sci.*, **24**, 477–481.

Ladlie, J. S., Meggitt, W. F. and Penner, D. (1976b). Role of pH on metribuzin dissipation in field soils. *Weed Sci.*, **24**, 508–511.

Lallukka, R., Lode, O., Petersen, E. J. and Svensson, A. (1974). Herbicides in the soil—effects on crop plants. *Reports and Dissertations, Department of Plant Husbandry, Agricultural College of Sweden*, No. 10 pp. 117. (*Weed Abstracts*, No. 501, 1978).

Lavy, T. L., Roeth, F. W. and Fenster, C. R. (1973). Degradation of 2,4-D and atrazine at three soil depths in the field. *J. environ. Qual.* **2**, 132–137.
Leasure, J. K. (1964). The halogenated aliphatic acids. *J. agric. Fd Chem.*, **12**, 40–43.
LeBaron, H. M. (1970). Ways and means to influence the activity and the persistence of triazine herbicides in soils. *Residue Rev.*, **32**, 311–353.
Lichtenstein, E. P. and Schulz, K. R. (1965). Residues of aldrin and heptachlor in soils and their translocation into various crops. *J. agric. Fd. Chem.*, **13**, 57–63.
Manorik, A. V. and Malichenko, S. M. (1970). The role of the soil and plants in the inactivation of simazine and atrazine. *Agrokhimija.*, **7**, 114–119. (*Weed Abstracts*, No. 1872, 1971).
Marriage, P. B., Saidak, W. J. and Stryk, F. G. Von (1975). Residues of atrazine, simazine, linuron and diuron after repeated annual applications in a peach orchard. *Weed Res.*, **15**, 373–379.
McClure, G. W. (1970). Accelerated degradation of herbicides in soil by the application of microbial nutrient broths. *Contr. Boyce Thompson Inst. Pl. Res.*, **24**, 235–240.
McCormick, L. L. and Hiltbold, A. E. (1966). Microbiological decomposition of atrazine and diuron in soil. *Weeds*, **14**, 77–82.
Meggitt, W. F. (1962). Residual studies on triazine herbicides. *Proc. N. cent. Weed Control Conf.*, **19**, 82.
Meikle, R. W., Youngson, C. R., Hedlung, R. T., Goring, C. A. I., Hamaker, J. W. and Addington, W. W. (1973). Measurement and prediction of picloram disappearance rates from soil. *Weed Sci.*, **21**, 549–555.
Menges, R. M. and Tamez, S. (1973). Effect of soil incorporation on selectivity, movement and persistence of herbicides in onion plantings. *J. Amer. Soc. hort. Sci.*, **98**, 390–393.
Menges, R. M. and Tamez, S. (1974). Movement and persistence of bensulide and trifluralin in irrigated soil. *Weed Sci.*, **22**, 67–71.
Miller, J. H., Keeley, P. E., Carter, C. H. and Thullen, R. J. (1975). Soil persistence of trifluralin, benefin and nitralin. *Weed Sci.*, **23**, 211–214.
Morris, R. F. and Penney, B. G. (1971). Persistence of linuron residues in soils at Kelligrews, Newfoundland. *Can. J. Pl. Sci.*, **51**, 242–245.
Moyer, J. R., Hance, R. J. and McKone, C. E. (1972). The effect of adsorbents on the rate of degradation of herbicides incubated with soil. *Soil Biol. Biochem.*, **4**, 307–311.
Nearpass, D. C. (1965). Effects of soil acidity on the adsorption, penetration and persistence of simazine. *Weeds*, **13**, 341–346.
Newman, A. S. and Thomas, J. R. (1949). Decomposition of 2,4-dichlorophenoxyacetic acid in soil and liquid media. *Soil Sci. Soc. Amer. Proc.*, **14**, 160–164.
Ogle, R. E. and Warren, G. F. (1954). Fate and activity of herbicides in soils. *Weeds*, **3**, 257–273.
Osgerby, J. M. (1972). An assessment of the consequences of the controlled release of pesticides from granular formulations. *Pestic. Sci.*, **3**, 753–769.
Ostrowski, J. (1968). The elimination of herbicide carryover by their adsorption on activated carbon. *Roczniki glebozn.*, **19**, 235-241. (*Weed Abstracts*, No. 1962, 1969).
Paeschke, R. R., Ebing, W. and Heitefuss, R. (1978). Abbau von Herbiziden und

deren Einfluss auf die biologische Aktivität im Boden nach erstmaliger oder mehrjähriger Anwendung. *Z. PflKrankh. PflSchutz*, **85**, 280–297.

Parka, S. J. and Tepe, J. B. (1969). The disappearance of trifluralin from field soils. *Weed Sci.*, **17**, 119–122.

Parochetti, J. V. and Warren, G. F. (1968). Biological activity and dissipation of IPC and CIPC. *Weed Sci.*, **16**, 13–15.

Polon, J. A. (1973). Formulation of pesticidal dusts, wettable powders and granules. In *Pesticide Formulations*. (W. Van Valkenburg, Ed.) pp. 143–234. Marcel Dekker, New York.

Poschenrieder, G., Wallnöfer, P. and Beck, Th. (1975). Phenylamid- und Carbamat-Interaktionen in Böden. *Z. PflKrankh. PflSchutz*, **82**, 398–405.

Regan, A. and Krempler, F. (1974). Triazine residues in the soil. *Arh. Hig.*, **25**, 319. (*Weed Abstracts*, No. 1834, 1976).

Repiquet, C. and Fournier, J-C. (1977). Étude en laboratoire de la degradation du 2,4-D et de l'isoproturon en presence d'autres produits de traitement. *Comptes desjournées d'etudes sur les herbicides, Columa*, 801–810.

Riepma, P. (1962). Preliminary observations on the breakdown of 3-amino-1,2,4-triazole in soil. *Weed Res.*, **2**, 41–50.

Roadhouse, F. E. and Birk, L. A. (1961). Penetration of and persistence in soil of the herbicide 2-chloro-4,6-bisethylamino-s-triazine (simazine). *Can. J. Pl. Sci.*, **41**, 252–260.

Roeth, F. W., Lavy, T. L. and Burnside, O. C. (1969). Atrazine degradation in two soil profiles. *Weed Sci.*, **17**, 202–205.

Russell, J. D., Cruz, M., White, J. L., Bailey, G. W., Payne, W. R., Pope, J. D. and Teasley, J. I. (1968). Mode of chemical degradation of triazines by montmorillonite. *Science*, **160**, 1340–1342.

Savage, K. E. (1973). Nitralin and trifluralin persistence in soil. *Weed Sci.*, **21**, 285–288.

Savage, K. E. (1977). Metribuzin persistence in soil. *Weed Sci.*, **25**, 55–59.

Savage, K. E. and Barrentine, W. L. (1969). Trifluralin persistence as affected by depth of soil incorporation. *Weed Sci.*, **17**, 349–352.

Schreiber, M. M. (1976). Efficacy and persistence of starch encapsulated EPTC. *Proc. N. cent. Weed Control Conf.*, 143.

Schweizer, E. E. (1975). Crop response to residues of ethofumesate. *Weed Sci.*, **23**, 409–413.

Schweizer, E. E. (1976). Persistence and movement of ethofumesate in soil. *Weed Res.*, **16**, 37–42.

Schweizer, E. E. and Holstun, J. T. (1966). Persistence of five cotton herbicides in four southern soils. *Weeds*, **14**, 22–26.

Sheets, T. J. and Harris, C. I. (1965). Herbicide residues in soils and their phytotoxicity to crops grown in rotation. *Residue Rev.*, **11**, 119–140.

Siegmann A., Narkis, M. and Tirosh, N. (1975). Slow release herbicides from polymeric granules. *Proceedings, Israel-France Symposium, Behaviour of Pesticides in Soil.* pp. 11–12. Department des Publications Scientifique.

Sikka, H. C. and Davis, D. E. (1966). Dissipation of atrazine from soil by corn, sorghum and johnsongrass. *Weeds*, **14**, 289–293.

Smith, A. E. (1975). Field persistence studies with herbicides in prairie soils. In *Environmental Quality and Safety*. Suppl. Vol. III. (F. Coulston and F. Korte, Eds.) pp. 266–270. George Thieme, Stuttgart.

Smith, A. E. (1979). Soil persistence experiments with $^{14}$C-2,4-D in herbicidal mixtures, and field persistence studies with tri-allate, and trifluralin both singly and combined. *Weed Res.*, **19**, 165–170.

Smith, A. E. and Emmond, G. S. (1975). Persistence of linuron in Saskatchewan soils. *Can. J. Soil Sci.*, **55**, 145–148.

Smith, A. E. and Walker, A. (1977). A quantitative study of asulam persistence in soil. *Pestic. Sci.*, **8**, 449–456.

Smith, D. T. and Meggitt, W. F. (1970). Persistence and degradation of pyrazon in soil. *Weed Sci.*, **18**, 260-264.

Smith, D. W. (1968). Effect of activated carbon on picloram in soil. *Abstr. 1968 Mtg Weed Sci. Soc. Am.*, 74.

Spencer-Jones, D. H. and Wilson, D. (1976). Some effects of formulation on the biological performance of dichlobenil. In *Granular Pesticides. British Crop Protection Council Monograph No. 18*, pp. 59–66.

Stoinev, K. (1967). Effects of tillage on penetration into soil and residual effect of simazine and atrazine. *Rast. Nauki*, **3**, 101–112. (*Weed Abstracts*, No. 1934, 1968).

Süss, A. and Wagner, A. (1970). Die Adsorption von Herbiziden an verschiedene Bodenkomponenten. *Z. PflKrankh. PflSchutz*, Sonderheft VI, 33–41.

Süss, A., Eben, C. and Siegmund, H. (1972). Verhalten von adsorbierten Herbiziden im Boden und deren Verfügbarkeit fur die Pflanze. *Z. PflKrankh. PflSchutz*, Sonderheft VI, 65–74.

Talbert, R. E. and Fletchall, O. H. (1964). Inactivation of simazine and atrazine in the field. *Weeds*, **12**, 33–37.

Taylor, A. W., Freeman, H. P. and Edwards, W. M. (1971). Sample variability and the measurement of dieldrin content of a soil in the field. *J. agric. Fd Chem.*, **19**, 832–836.

Thiegs, B. J. (1955). The stability of dalapon in soils. *Down to Earth*, **11**, 2–4.

Torstensson, L. (1976). Effect of some factors on the persistence of TCA in arable soil. *Proc. 17th Swedish Weed Conf.*, K8–L13.

Torstensson, N. T. L. (1975). Degradation of 2,4-D and MCPA in soils of low pH. In *Environmental Quality and Safety*. Suppl. Vol III. (F. Coulston and F. Korte, Eds.) pp. 262–265. Georg Thieme, Stuttgart.

Torstensson, N. T. L. (1977). Effect of repeated applications on the degradation of some herbicides. *Proc. European Weed Research Society Symposium; Different methods of weed control and their integration*, pp. 113–120.

Usoroh, N. J. and Hance, R. J. (1974). The effect of temperature and water content on the rate of degradation of the herbicide linuron in soil. *Weed Res.*, **14**, 19–21.

Wagenbreth, D. and Kluge, E. (1971) Der Einfluss von Fungiziden (Thiuram, Vapam) auf den Triazinabbau im Boden. *Arch. Pflanzenschutz*, **7**, 451–459.

Walker, A. (1970a). Effects of quintozene on the persistence and phytotoxicity of chlorpropham and sulfallate in soil. *Hort Res.*, **10**, 45–49.

Walker, A. (1970b). Persistence of pronamide in soil. *Pestic. Sci* **1**, 237–239.

Walker, A. (1973). Use of a simulation model to predict herbicide persistence in the field. *Proc. European Weed Research Council Symposium; Herbicides and the Soil*, pp. 240–250.

Walker, A. (1974). A simulation model for prediction of herbicide persistence. *J. environ. Qual.*, **3**, 396–401.

Walker, A. (1976a). Simulation of herbicide persistence in soil. I. Simazine and prometryne. *Pestic. Sci.*, **7**, 41–49.

Walker, A. (1976b). Simulation of herbicide persistence in soil. II. Simazine and linuron in long-term experiments. *Pestic. Sci.*, **7**, 50–58.

Walker, A. (1976c). Simulation of herbicide persistence in soil. III. Propyzamide in different soil types. *Pestic. Sci.*, **7**, 59–64.

Walker, A. (1976d). Effect of varying weather conditions on persistence of three herbicides in soil. *Proc. 1976 Br. Crop Prot. Conf.—Weeds*. 635–642.

Walker, A. (1978). Simulation of the persistence of eight soil-applied herbicides. *Weed Res.*, **18**, 305–313.

Walker, A. and Bond, W. (1978). Simulation of the persistence of metamitron activity in soil. *Proc. 1978 Br. Crop Prot. Conf.—Weeds*. 565–572.

Walker, A. and Smith, A. E. (1979). Persistence of 2,4,5-T in a heavy clay soil. *Pestic. Sci.*, **10**, 151–157.

Walker, A. and Thompson, J. A. (1977). The degradation of simazine, linuron and propyzamide in different soils. *Weeds Res.*, **17**, 399–405.

Walker, A., Farrant, D. M., Bryant, J. H. and Brown, P. A. (1976). The efficiency of herbicide incorporation into soil with different implements. *Weed Res.*, **16**, 391–397.

Wallnöfer, P., Poschenrieder, G. and Engelhardt, G. (1977). Verhalten von Pestizidkombinationen im Boden. *Z. PflKrankh. PflSchutz*, Sonderheft VIII, 199–207.

Wauchope, R. D., Chandler, J. M. and Savage, K. E. (1977). Soil sample variation and herbicide incorporation uniformity. *Weed Sci.*, **25**, 193–196.

Weber, J. B. and Coble, H. D. (1968). Microbial decomposition of diquat adsorbed on montmorillonite and kaolinite clays. *J. agric. Fd Chem.*, **16**, 475–478.

Wilkins, R. M. (1976). Aspects of the chemical and physical control of persistence with special reference to the use of polymers. In *The Persistence of Insecticides and Herbicides. British Crop Protection Council Monograph No. 17*, pp. 201–216.

Wolf, D. C. and Martin, J. P. (1974). Microbial degradation of 2-carbon-14 bromacil and terbacil. *Soil Sci. Soc. Amer. Proc.*, **38**, 921–925.

Zemanek, J. (1964). A study of various means of inactivating atrazine in the soil. *Rostl. Vyroba*, **37**, 959–972. (*Weed Abstracts*, No. 958, 1965).

Zimdahl, R. L. and Gwynn, S. M. (1977). Soil degradation of three dinitroanilines. *Weed Sci.*, **25**, 247–251.

Zimdahl, R. L., Freed, V. H., Montgommery, M. L. and Furtick, W. R. (1970). The degradation of triazine and uracil herbicides in soils. *Weed Res.*, **10**, 18–26.

Zurawski, H. and Piss, J. (1971). Laboratory investigations on the rate of inactivation of simazine. *Roczniki glebozn.*, 127–136. (*Weed Abstracts*, No. 245, 1973).

Zurawski, H. and Ploszynski, M. (1968). Investigations on the disappearance of simazine from light soil. *Proc. 9th Br. Weed Control Conf.*, pp. 115–118.

# 5 The Study of Transformations

*Johann A. Guth*
CIBA-GEIGY Ltd.,
Agricultural Division,
Basle, Switzerland



## I. Introduction

Knowledge of transformation of herbicides in soil is of importance both to agricultural practice and environmental safety. Herbicide transformations in soil can be caused by three main types of process: photochemical, chemical and microbial. Numerous experimental procedures have been described in the literature to study the significance of these processes for the transformation of a particular herbicide in soil. Advantages and disadvantages of the existing laboratory model systems and the available field testing methods are the first main subjects to be discussed in this chapter.

The interpretation of results obtained with this methodology is another aspect which will be considered. Whether the transformations observed in one laboratory soil occur also in other soils and whether laboratory results are representative for field situations are the questions of main interest in this respect.

Finally, present problem areas, such as biological activity of the test soils, soil bound residues, and the influence of other chemicals and of the rhizosphere on herbicide transformation, are recommended for future research activities with special emphasis on registration requirements and scientific interests.

## II. Registration Requirements

Recently the environmental safety aspect has received much attention and this has led to registration requirements. The first written guide available was the PR 70–15 issued by the United States Department of Agriculture in 1970 which was the basis for the environmental chemistry guidelines issued by the US Environmental Protection Agency (EPA) in 1975 (Anon., 1975), later revised in 1978 (Anon., 1978). The data requirements on soil

Table I. Requirements for soil metabolism studies under laboratory and field conditions as given in the EPA guidelines (Anon., 1978).

| | Aerobic soil metabolism | Field dissipation |
|---|---|---|
| 1. Compound | Radiolabelled | Not specified |
| 2. Objective | a. Rate, type and degree of metabolism | |
| | b. Material balance | b. Mobility in environment |
| 3. Identification level | Residues comprising more than 10% of the initial application or 0·01 ppm, whichever is greater | |
| 4. Number of soils or of use areas | a. One aerobic soil<br>b. Three or more soils where the product has broad uses | a. Four for field and vegetable crop uses<br>b. Two for tree fruits and nut crop uses<br>c. Three for pasture land uses<br>d. Two for domestic outdoor, parks, ornamental and turf uses<br>e. Two for rights-of-way, shelterbelts and related uses |
| 5. Temperature range and moisture level | 18–30°C<br>≤75% of 0·33 bar moisture | —<br>— |
| 6. Depth of soil sampling | — | a. For 4a–c: 30 cm<br>b. For 4d–e: 15 cm |
| 7. Duration of experiments | a. Until 90% loss of the pesticide and until patterns of formation and decline of metabolic products are established | |
| | b. But not more than 1 year | b. Or to the maximum time specified:<br>For 4a: 18 months<br>For 4b–c: 12 months<br>For 4d: 4 months<br>For 4e: 2 months |

metabolism include aerobic, anaerobic and microbial studies in the laboratory as well as field dissipation experiments. The details of the requirements for aerobic laboratory and field soil dissipation studies are summarized in Table I. The experimental conditions for anaerobic metabolism are the same as for aerobic studies except that the duration of the experiments is restricted to 90 days (30 days of aerobic followed by 60 days of anaerobic incubation). In addition, soil surface photolysis experiments are also required to measure the influence of abiotic processes on breakdown of pesticides in soil.

The guideline of the German BBA (Biologische Bundesanstalt) demands rate studies of the parent compound in two standards soils under aerobic conditions and also information on the character of the major metabolites and an estimation of their persistence in the soil (Weinmann and Schinkel, 1976). The use of radioactive material is not mandatory but when $^{14}C$-labelled compounds are used in the studies the formation of $^{14}CO_2$ should be monitored. The soil samples should be incubated in Erlenmeyer flasks at a temperature of $22 \pm 2°C$ and a moisture level of 40% of the water holding capacity for a maximum period of 32 weeks.

The Dutch registration authorities require information on the pathway and the rapidity of degradation in several soil types. The experiments should be continued until 90% of the applied quantity has been broken down to "unimportant" degradation products. In addition, the percentage of soil bound residue shall be determined.

In contrast to these regulations, the British Pesticides Safety Precaution Scheme (Anon., 1971), is voluntary and includes the statement "the possible persistence of a pesticide in the soil and its subsequent effect on edible crops should not be overlooked" (Appendix C), allows most flexibility.

Pesticide regulations exist in more countries than those mentioned above and there is a clear need for a harmonization of registration requirements which is encouraged by several international organizations.

## III. Methodology

### A. Photochemical Transformation

Published information on experimental approaches for measuring photodegradation in soil or on soil surfaces is limited. Some studies have been carried out under natural sunlight and thus were affected by the meteorological variability (Nilles and Zabik, 1974, 1975; Liang and Lichtenstein, 1976; Smith *et al.*, 1978; Plimmer, 1978; Parochetti and Dec, 1978). Other workers have used germicidal lamps rich in radiation between 240 and

260 nm (Funderburk *et al.*, 1966; Liang and Lichtenstein, 1976). These studies are of limited value for estimating the photochemical stability of herbicides on soil surfaces since u.v. light below 290 nm does not reach the earth.

Parochetti and Hein (1973) used two Westinghouse FS 40T 12 lamps as the light source. These 40-W bulbs had a maximum energy output at 310 nm with 82% of the energy between 290 and 340 nm. The soil samples were kept in petri dishes which were covered with polished quartz and placed 33 cm directly below the lights which were mounted in an enclosed chamber.

A Rayonet Srinivasan-Griffin photoreactor employing eight RPR 3000 Å and eight RPR 3500 Å lamps in an alternate configuration was used by Nilles and Zabik (1974, 1975) for studying photolysis of fluchloralin and bentazon. Solid phase irradiations were done with a merry-go-round apparatus in order to ensure that all samples received uniform radiation.

Another system described by Hautala (1976, 1978) is shown in Fig. 1. It consists of a medium pressure mercury arc (450 W) immersed in a horizontal double-mantled Pyrex tube which simultaneously acts as cooler and filter for all radiation below 290 nm. The soil samples are photolysed on a custom-built glass tray with cooling water circulating on the under side to avoid evaporation losses and to maintain the initial moisture content. The trays are tightly covered with Pyrex glass on top and are placed about 14 cm below the light source.

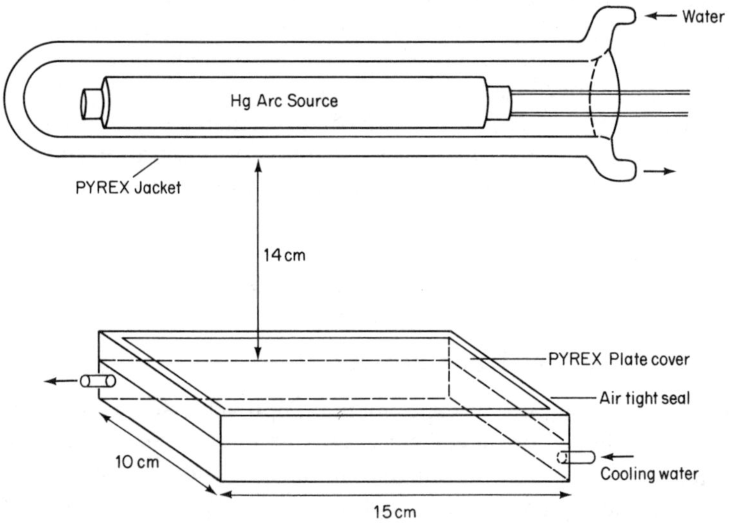


Fig. 1. Apparatus for photolysis on soil surfaces according to Hautala (1976, 1978). (Figure by courtesy of Professor R. R. Hautala, Athens, Georgia, U.S.A.).

Kennedy and Talbert (1977) used an environmental control chamber for their photolysis studies. The light source was a bank of eight Westinghouse fluorescent lamps. Two of the sun-lamps were 20-W FS 20 and the other six 30-W F 30T 8 BLB lights which have a peak energy output at 310 nm units with 82% of the energy between 290 and 340 nm. The samples prepared as thin layers of soil on glass plates were rotated under the lights on a revolving turntable to ensure even light radiation on all samples.

Burkhard and Guth (1979) used a commercially available Hanau Suntest apparatus which allowed experiments to be done at any time of the year independent of the meteorological variations of a particular environment.

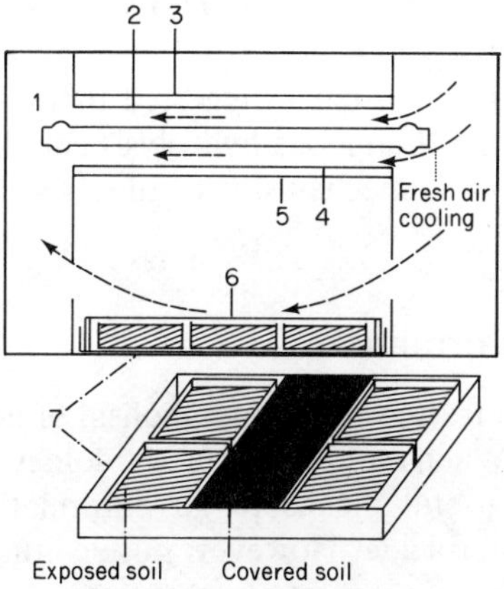


Fig. 2. The HANAU Suntest Apparatus (Burkhard and Guth, 1979). (Figure by courtesy of the Society of Chemical Industry, London, U.K.).

The system is schematically presented in Fig. 2. The radiation source was a xenon burner with a spectral energy distribution which very closely approaches that of sunlight if its infrared portion is filtered out. This can be done with a pane of quartz glass with an infrared reflecting coating and by infrared transmissive and ultraviolet light reflecting mirrors. All short u.v. light below 290 nm is eliminated by an additional pane of special glass. The total energy emitted by the xenon lamp was determined to be about 950 $J/m^2/s$ which is approximately double the energy of natural sunlight. The treated soil samples in open shallow metal boxes were placed 20 cm below the xenon burner.

## B. Chemical Transformation

Chemical transformations in soil have been classified by Goring *et al.* (1975) as hydrolytic nucleophilic transformations, oxidations and non-hydrolytic nucleophilic displacement reactions. Also reactions with free radicals in soil are possible. Considerable effort has been spent to distinguish between chemical and microbiological degradation in soil. The experimental approach commonly used is to compare the results of transformation studies in sterile and non-sterile soil. However, each sterilization method has certain shortcomings. Chemical sterilants may not eliminate the whole population of soil microorganisms, while heat and radiation sterilization also destroy soil enzymes and alter the chemical and physical properties of soil. The effects of biocidal treatments on metabolism in soil have been discussed by Powlsen and Jenkinson (1976) and Cawse (1975) and is also considered by Torstensson in Chapter 6.

Burkhard and Guth (1979) have suggested that a combination of photolysis on soil surfaces and aqueous hydrolysis could be used to evaluate the contribution of abiotic mechanisms to the overall breakdown of pesticides in soil.

## C. Microbial Transformation

Much work has been directed to the metabolism of herbicidal compounds in microbial cultures. When radio-tracer methodology was still in its infancy such studies were a useful means to get information on the inherent biodegradability of a herbicide. However, pure culture studies carried out with unrealistically high concentrations very often do not reveal the true fate of herbicides in natural soils and can never reproduce *in vivo* conditions (Cripps and Roberts, 1978). Both scientists and legislators should remember that microbial studies, although giving a very interesting insight into biodegradation mechanisms, have primarily only academic importance. The experimental aspects of soil microorganism studies are not included in this chapter.

## D. Transformation Studies in Soil

### 1. *Laboratory Methods*

A wide variety of experimental approaches have been designed to study herbicide transformations in soil. In principle, they can be classified into four categories: soil perfusion systems, soil biometers, gas flow-through systems, and integrated systems. The first three can be used for aerobic as well as anaerobic soil studies and have been extensively discussed by Kaufman (1977). Laboratory methods have recently been reviewed by Newby *et al.* (1979).

(*a*) *Soil perfusion devices* have been widely used to study soil microbiological processes. According to Kaufman (1977a) the perfusion system has the following advantages: it is adaptable for use with either positive or negative pressure systems; it can be maintained sterile if required; perfusate can easily be sampled; rate of perfusion can be controlled; gases of known composition can be used; infrequent adjustment required; selfsupporting and compact design; and relatively inexpensive parts. However, there are limitations. The data obtained are certainly not representative of what actually occurs in the environment. In particular, the perfusion system seems to favour the growth of bacteria. Other examples of perfusion apparatus have been described by Kimura and Yamaguchi (1978) and Fung and Uren (1977).

(*b*) *Biometer type systems* have been used to study degradation and metabolism of pesticides in soils. The "soil biometer" flask developed by Bartha and Pramer (1965) is shown in Fig. 3. The apparatus allows both $^{14}CO_2$ and total $CO_2$ production to be monitored and also all measurements to be made on the same soil sample. They suggested that the difference in amount of $CO_2$ produced by treated and untreated soil can be used as a measure of the extent of breakdown of a compound, but this would be misleading with compounds which stimulate microbial activity. The biometer flask has several limitations. One occurs if volatile degradation products are formed which cannot be absorbed in the KOH solution, or if they are

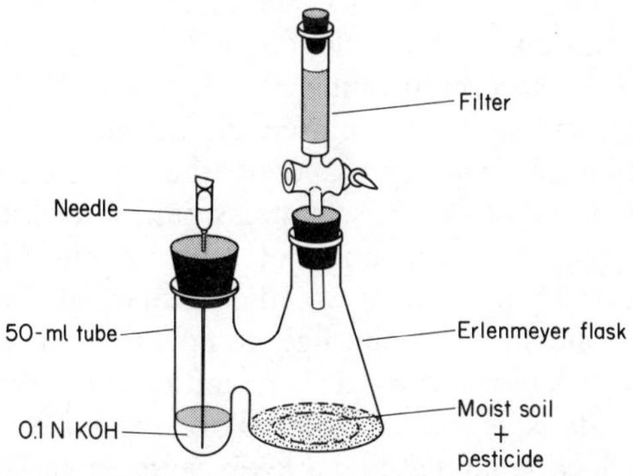


Fig. 3. Biometer flask according to Bartha and Pramer (1965) for measuring the persistence of pesticides in soil. (Figure by courtesy of The Williams and Wilkins Co., Baltimore, U.S.A.).

trapped, have to be separated by an additional analytical effort. A second is that during long-term experiments the microbial population of the soil consumes oxygen and produces carbon dioxide which is absorbed by the trapping solution so that without air exchange the nitrogen concentration steadily increases and finally might even produce anaerobic conditions. This problem can be circumvented by simply leaving the syringe needle open so that a negative pressure builds up in the flask and fresh air is drawn into the system but even this procedure will still lead to an increasing nitrogen atmosphere. Another alternative is to attach the open end of the syringe needle to an oxygen source. Thus, as the oxygen is utilized in the flask fresh $O_2$ is drawn in by the negative pressure.

An alternative to the Bartha–Pramer biometer flask has been developed by Anderson (1975). The system consists of an Erlenmeyer flask and a trapping tower containing glasswool treated with paraffin oil to absorb volatile compounds and an upper layer of soda lime to absorb $CO_2$. The advantage of the system over the Bartha–Pramer flask is that it allows exchange of air between the system and the surrounding atmosphere and in addition allows the trapping of volatile products. However, there is the question whether diffusion is fast enough to guarantee completely free air circulation.

The latter point has been examined by Marvel *et al.* (1976, 1978) over several years with a similar system modified to allow for separate analysis of evolved $^{14}CO_2$ and atmospheric carbon dioxide. They compared flasks with and without trapping towers and demonstrated that the growth of *E. coli*, a strict aerobic bacterium, was not significantly different and therefore concluded that the system maintained aerobic conditions.

(*c*) *Flow-through systems* have also been used in soil pesticide transformation investigations. They have varied from relatively simple systems such as those of Goswami and Koch (1976) and Kearney and Kontson (1976) to the complex manifold developed by Parr and Smith (1969). Flow-through systems have several advantages. They allow the conduct of soil respiration experiments under more precisely defined conditions, the use of different gaseous environments and operation on either positive or negative pressure. In addition, these systems permit measurement of volatilization losses and characterization of both volatilized products and $^{14}CO_2$ evolved. For absorption of volatile compounds ethylene glycol, ethanolamine, or solids, such as silica gel or polyurethane plugs (Kearney and Kontson, 1976), have been used. Parr and Reuszer (1962) found that the aeration flow rate is very important for organic matter decomposition since the $CO_2$ production increased markedly with increasing air flow rate. Also Kaufman (1977a) has observed a more rapid and extensive degradation of $^{14}C$-labelled pesticides in flow-through systems than in the biometer flask used in comparable experiments.

The major criticism of flow-through approaches is that they are complex in design and generally are expensive. However, once the equipment has been installed it need not be removed and can be used for further experiments.

The soil metabolism equipment used in the author's laboratories is graphically presented in Fig. 4. It consists of a gas washing bottle filled with distilled water to saturate the incoming air in aerobic experiments, a needle valve for adjustment of gas flow rate (air or nitrogen), a one-litre glass bottle (Steilbrust) containing the treated soil, several gas absorption bottles filled with ethylene glycol, sulphuric acid (0·1N) and sodium hydroxide (2N) to trap volatiles and radioactive carbon dioxide and finally a flow meter measuring the gas flow rate. The purge gas is delivered from a gas pressure cylinder. The system can be used for aerobic, anaerobic and sterile experiments and all of its parts are commercially available.

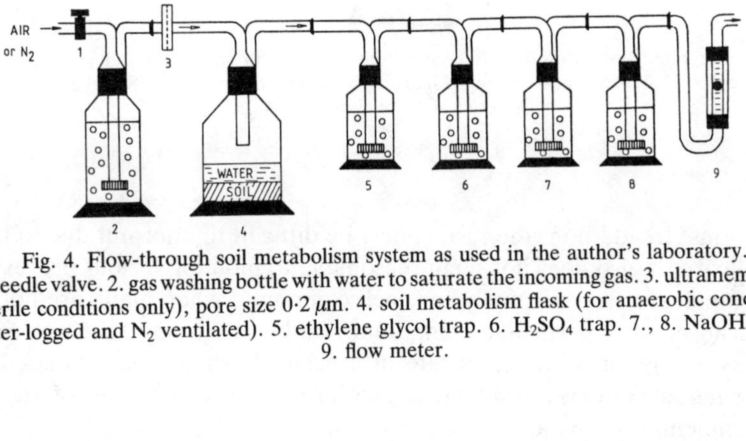


Fig. 4. Flow-through soil metabolism system as used in the author's laboratory. 1. needle valve. 2. gas washing bottle with water to saturate the incoming gas. 3. ultramembrane (sterile conditions only), pore size 0·2 μm. 4. soil metabolism flask (for anaerobic conditions water-logged and $N_2$ ventilated). 5. ethylene glycol trap. 6. $H_2SO_4$ trap. 7., 8. NaOH traps. 9. flow meter.

(*d*) *Integrated systems* have been developed by other laboratories. One example is that of Best and Weber (1974) shown in Fig. 5 which allows simultaneous monitoring of degradation, volatilization, respiration, plant uptake and leaching processes. A similar system has been used by Lichtenstein *et al.* (1974) when studying translocation and metabolism of phorate. Beall *et al.* (1976) have described the use of an enclosed "agro-ecosystem" in which a number of plants may be grown in soil and pesticide degradation examined under varying regimes of rainfall, wind velocity and light. All these integrated systems have tried to introduce the complexity of the environment but are not really able to simulate actual environmental

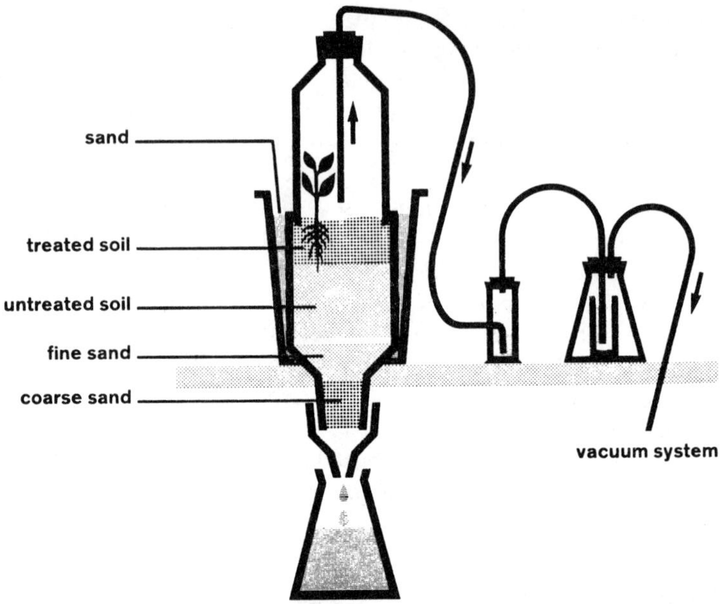


Fig. 5. Integrated system used by Best and Weber (1974) for studying the degradation and movement of 1,3,5-triazines in soil. (Figure by courtesy of the Weed Science Society of America, Champaign, U.S.A.).

conditions. In addition, the results may be difficult to interpret due to this complexity. It also does not seem reasonable to monitor various processes in a complex system routinely for each compound when many are not of significance for a particular compound. Rather the importance of each process for the environmental fate of a pesticide should be estimated in simple model systems by comparison with reference compounds of known environmental behaviour.

### 2. *Field Studies*

Field data are less easy to obtain than laboratory data and "balance studies" are difficult to establish. In spite of the difficulties a few systems have been successfully used. The approach used by Roberts (1976) is shown in Fig. 6. Plants are grown in boxes which are placed in a pit in such a way that the soil surface in the boxes is at the same level as the soil in the surrounding field. The floor of the pit slopes towards a central well containing a pipe linked to a drain. This pipe can be closed and the water can be periodically monitored for radioactivity. The plywood boxes (60 × 60 × 60 cm) are

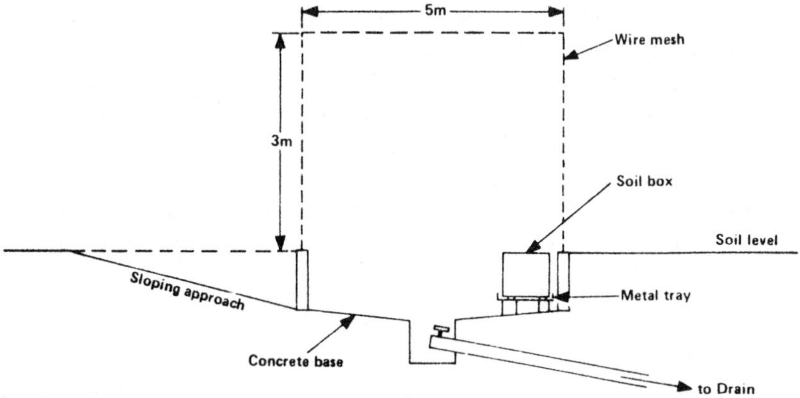


Fig. 6. Diagram of outdoor enclosure for radiochemical experiments according to Roberts (1976). (Figure by courtesy of the British Crop Protection Council, London, U.K.).

fitted with slats at the bottom to allow drainage. Flaps are fitted on one side so that soil samples may be taken horizontally at different depths. The boxes are filled with top soil and allowed to settle for 1–2 months. Plants are then grown in the boxes and they can be treated with the formulated pesticide at any time required.

Führ *et al.* (1976) have used standardized lysimeters as shown in Fig. 7 for pesticide balance and metabolism studies. The lysimeters are made of polyvinyl chloride walls which are 6 mm thick. They are packed with 40 cm of surface soil. A 5-cm layer of washed quartz gravel at the base of lysimeter prevents impoundment of the percolate. In the centre of the lysimeter, a separated control vessel is installed containing untreated soil and plants. Percolates from the treated portion and the control vessel can be collected separately.

However, these systems do not simulate actual environmental conditions in every aspect. The fact that artificial soil layers are used in these lysimeters instead of intact soil profiles allows no clear statement on the leaching behaviour of a pesticide under field conditions. The only way of simulating actual environmental conditions in every respect is to use monolith lysimeter systems such as the one recently described by Jarczyk (1978).

Transformation of pesticides in soil under field conditions can also be studied by applying radiolabelled compounds to small field plots and taking soil samples from various depths at regular intervals. This procedure does not allow measurement of the contribution of volatilization to the overall disappearance of the compounds and its breakdown products. The method described by Hill and Arnold (1978) attempts to overcome this deficiency.

The treated soil surface is covered by 60 cm high bell jars mounted on stainless steel rings embedded 5 cm into the soil. The air is sucked out of the bell jars and through trapping solutions for collection of volatile radiolabelled products. To reduce the heating effects inside the jar, volatiles are sampled only during a 24 hour period from midday to midday every 48 hours. However, the use of such an elaborate approach cannot be recommended for routine use for every compound. The importance of volatilization should first be assessed in a cheaper laboratory system.

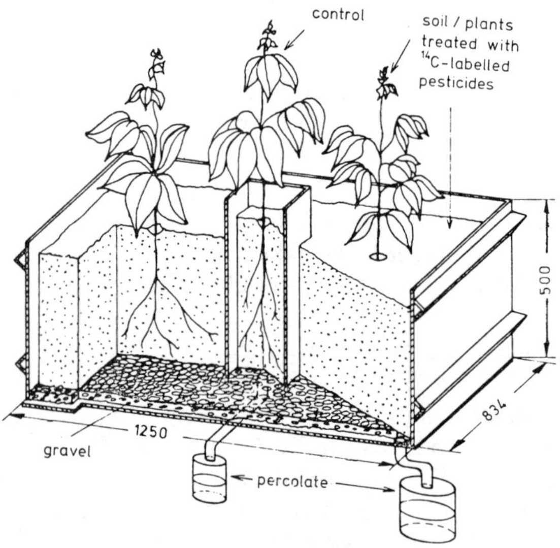


Fig. 7. Lysimeter system used by Führ *et al.* (1976). The dimensions are given in mm. (Figure by courtesy of Dr F. Führ, Jülich, and J. D. Sauerländer's Verlag, Frankfurt, FRG).

The use of radiolabelled pesticides is not always possible because of restrictions in their usage at field sites and at a time of steadily growing public concern about radioactivity the situation may become more difficult in the future. The limitations this would place on field studies might be overcome by the use of $^{13}$C-enriched pesticides as described recently by Roberts (1976).

At this point, the question may be raised whether field studies are really necessary to obtain information on the pathways of transformation. This will be discussed later in the section "interpretation of results".

# IV. Results and their Interpretation

## A. Pathways of Transformation

### 1. *Photochemical Pathways*

Published information on pathways of photodecomposition on soil is as limited as that on the experimental techniques available for studying the process. Nilles and Zabik (1974) studied the photodegradation of fluchloralin on sandy loam thin-layers in the Rayonet reactor. The major reactions observed were photodealkylation and benzimidazole formation.

The same photolytic *N*-dealkylation reaction was observed in our own studies with profluralin (Burkhard, 1978) on sandy and silty loam soil surfaces. In addition, oxidation of the trifluoromethyl group to the corresponding carboxy derivative and the formation of non-extractables was found but only in the slightly alkaline sandy soil (Fig. 8).

Bentazone photolysis on soil surfaces was investigated by Nilles and Zabik (1975) but no clear pathway was established. The major routes of photoalteration were found to be oxidative dimerization and loss of sulphur dioxide.

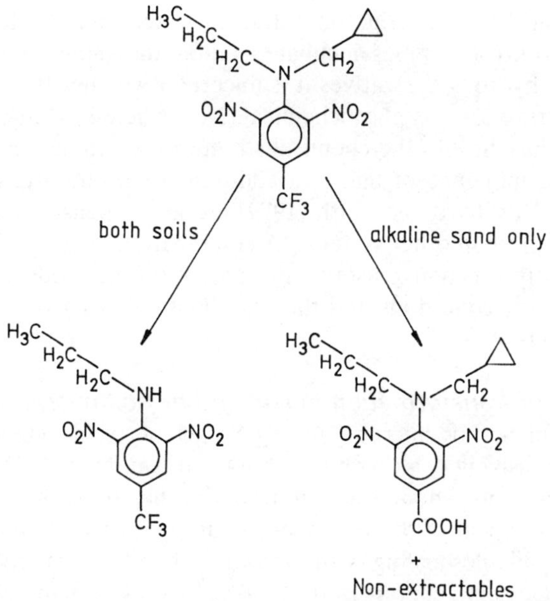


Fig. 8. Profluralin and its breakdown products formed under the influence of light on surfaces of a silty loam and a slightly alkaline sandy soil (Burkhard, 1978).

Table II. Degradation of the chlorotriazine herbicides simazine, atrazine, propazine and terbuthylazine on silty loam surfaces under the influence of light. The numbers give the percentages of compounds found after 24 h of irradiation. (Burkhard, 1978)

| | | Unchanged herbicide | Hydroxy parent | Non-extractables |
|---|---|---|---|---|
| Simazine | exposed | 62·0 | 4·7 | 22·2 |
| | covered | 71·9 | 3·5 | 17·4 |
| Atrazine | exposed | 71·3 | 8·7 | 17·6 |
| | covered | 82·1 | 5·4 | 11·5 |
| Propazine | exposed | 76·1 | 10·9 | 8·5 |
| | covered | 85·5 | 4·6 | 5·8 |
| Terbuthylazine | exposed | 68·8 | 10·6 | 11·4 |
| | covered | 83·8 | 3·3 | 6·8 |

In our laboratories, the photochemical degradation of simazine, atrazine, propazine and terbuthylazine on silty loam soil has also been studied (Burkhard, 1978). The only breakdown products detected were the corresponding hydroxytriazines but also the formation of non-extractable radioactivity was observed (Table II). None of the *N*-dealkylated products which are typical for *s*-triazine photolysis in water could be found. Since the covered control samples incubated under the same conditions also contained the hydroxy derivatives it is uncertain whether the degradation of the chlorotriazines was photochemical. Nevertheless, such experiments give an early insight into the chemical structures which may be formed in soil under the influence of light. In addition, such photolysis studies as conducted by Burkhard and Guth (1979) are less expensive and easier to perform than those in other systems. A combination of soil photolysis and aqueous hydrolysis can give useful information on the products formed in soil under abiotic conditions and thus can be an alternative to the sterile soil experiments.

### 2. *Pathways of transformation in soil and by soil microorganisms*

Pathways in soil and by microorganisms of each herbicidal class have been very comprehensively dealt with (Kearney and Kaufman, 1975, 1976) and in the meantime not enough new information has been published which would justify a repeat of this exercise. In addition, several other excellent reviews are available dealing with pathways of soil and microbial metabolism of pesticides in general (Bollag, 1972, 1974; Kaufman, 1974; Goring *et al.*, 1975; Williams, 1977; Hill, 1978) and herbicides in particular (Kaufman and Kearney, 1976; Cripps and Roberts, 1978). A more general treatment of the subject was published by Meikle (1972) on qualitative aspects of the decomposition of organic chemicals in the soil environment.

## B. Interpretation of Results

Degradation experiments in soil are carried out to obtain information on rates and pathways and on the nature of terminal residues. The rate of breakdown of the parent compound and the rate of formation of transformation products will depend on the properties of the soil types used as well as on the climatic conditions. A major question is whether this also applies to the pathways of pesticide transformation. In the following section, problems of the interpretation of metabolism results will therefore be discussed in order to answer two main questions: (i) Do the pathways of transformation differ from one soil to another? (ii) Are metabolism results obtained in the laboratory also representative of field situations?

### 1. *Pathways in Different Soils*

In recent years a considerable amount of published data has accumulated on the transformation of herbicides in more than one soil type. The available information is summarized in Table III which lists the herbicide, the number and type of soils used, the major transformation products, the differences observed and the corresponding reference. From this Table it can be easily seen that with the exception of one unrealistically high application rate, the pathways of transformation of the herbicides listed are the same in different soils.

Such results have been reported not only for herbicides but also for other chemical structures and it may be useful to mention additional examples. They include the insecticides isoxathion (Nakagawa *et al.*, 1975), *O*-ethyl *O*-[4-(methylthio) phenyl]-*S*-propyl phosphorodithioate (Bull *et al.*, 1976), endosulfan (Martens, 1977), lindane (Mathur and Saha, 1977), the pyrethroid WL 41706 (Roberts and Standen, 1977) and oxamyl (Harvey and Han, 1978), the fungicides edifenphos (Tomizawa *et al.*, 1976) and denmert (Ohkawa *et al.*, 1976) and finally, the antioxidant 2,6-di-*t*-butyl-4-methylphenol (Mikami *et al.*, 1979).

These data show that the pathways of a particular compound in several soils are practically identical when the incubation is carried out under similar experimental conditions. There is no doubt that changes in pathways may occur when an aerobic milieu is changed into an anaerobic or sterile one and vice versa. Although the overall qualitative picture in various soils seems to be the same, the amounts of products determined at identical sampling intervals obviously vary from soil to soil due to the differing reaction rates. It may well be that a product which is found in one soil is

Table III. Pattern of herbicide transformation in different soils

| Herbicide | Soils used: Number | Soils used: Types | Major transformation products observed | Differences in pattern between different soils | Reference |
|---|---|---|---|---|---|
| Atrazine | 3 | loam<br>clay<br>silt loam | 2-hydroxy-4-ethylamino-6-isopropylamino-*s*-triazine | None | Skipper and Volk (1972) |
| Benzoylprop-ethyl | 4 | sandy loam<br>medium loam<br>clay loam<br>peat | *N*-benzoyl-*N*-(3,4-dichlorophenyl) alanine,<br>non-extractables | None | Beynon *et al.* (1974) |
| Chlorthiamid | 5 | sand<br>clay<br>loam<br>peat<br>brick earth | dichlobenil,<br>2,6-dichlorobenzamide | None | Beynon and Wright (1972) |
| | 3[a] | sandy loam<br>medium loam<br>clay | | | |
| Cyanazine | 4 | sandy loam<br>medium loam<br>clay loam<br>peat | 2-chloro-4-(1-carbamoyl-1-methyl-ethylamino)-6-ethylamino-*s*-triazine,<br>2-chloro-4-(1-carboxy-1-methyl ethyl-amino)-6-ethylamino-*s*-triazine,<br>2-hydroxy-4-(1-carboxy-1-methyl ethyl-amino)-6-ethylamino-*s*-triazine,<br>non-extractables | None | Beynon, Stoydiu and Wright. (1972) |
| | 5[a] | sandy clay loam (2)<br>peat loam<br>[illegible] | 2-chloro-4-(1-carbamoyl-1-methyl-ethylamino)-6-ethyl,<br>[illegible] | None[b] | Beynon, Bosio and Elgar.<br>[illegible] |

| | | | | | |
|---|---|---|---|---|---|
| Dicamba | 3 | sandy loam<br>silty clay<br>heavy clay | 3,6-dichlorosalicylic acid,<br>carbon dioxide | None | Smith (1974) |
| Dichlobenil | 3[a] | sandy loam<br>medium loam<br>clay | 2,6-dichlorobenzamide | None | Beynon and Wright (1972) |
| Diclofop-methyl | 3 | sandy loam<br>silty clay<br>heavy clay | 2-[4-(2,4-dichlorophenoxy) phenoxy] propionic acid,<br>non-extractables | None | Smith (1974) |
| | 2 | sandy loam<br>loamy sand | 2-[4-(2,4-dichlorophenoxy)-phenoxy] propionic acid,<br>carbon dioxide[c] | None | Martens (1978) |
| Flamprop-isopropyl | 4 | sandy loam<br>medium loam<br>clay loam<br>peat | *N*-benzoyl-*N*-(3-chloro-4-fluorophenyl)alanine,<br>non-extractables | None | Hitchings and Roberts (1979) |
| Flamprop-methyl | 3 | sandy loam<br>medium loam<br>clay loam | *N*-benzoyl-*N*-(3-chloro-4-fluorophenyl) alanine,<br>non-extractables | None[d] | Roberts and Standen (1978) |
| Fosamine ammonium | 4 | sandy loam<br>silt loam (2)<br>fine sand | carbamoylphosphonic acid,<br>carbon dioxide,<br>non-extractables | None | Han (1979) |
| Glyphosate | 3 | sandy loam<br>silt loam<br>silty clay loam | aminomethylphosphonic acid,<br>carbon dioxide,<br>non-extractables | None[e] | Rueppel *et al.* (1977) |
| | 5 | soil texture not given | | None | Nomura and Hilton (1977) |
| Oxadiazon | 2 | sandy loam<br>loam | 3 products (Structures given in Fig. 9) | None[e] | Ambrosi *et al.* (1977) |

Table III—*continued*

| Herbicide | Soils used: Number | Soils used: Types | Major transformation products observed | Differences in pattern between different soils | Reference |
|---|---|---|---|---|---|
| Propanil | 5 | sandy loam<br>sandy clay loam<br>clay loam<br>clay<br>loam | 3,4-dichloroaniline,<br>3,3′,4,4′-tetrachloroazo-benzene | None[f] | Chisaka and Kearney (1970) |
| Trifluralin | 2 | silty clay loam<br>sandy loam | α,α,α-trifluoro-2,6-dinitro-*N*-propyl-*p*-toluidine,<br>α,α,α-trifluoro-2,6-dinitro-*p*-toluidine,<br>2-ethyl-7-nitro-1-propyl-5-(trifluoro-methyl)benzimidazole,<br>2-ethyl-7-nitro-5-(trifluoromethyl)benzimidazole,<br>non-extractables | None[g] | Wheeler *et al.* (1979) |

[a] Field studies.
[b] The other breakdown products of cyanazine were not monitored.
[c] $^{14}CO_2$ was not monitored by Smith (1977).
[d] Anaerobic experiments.
[e] Aerobic and anaerobic conditions.
[f] Only at the abnormal concentration of 850 ppm an additional metabolite was formed which was not observed at normal rates of application.
[g] No differences in pathways even at concentrations in soil of 1000 and 20000 mg/kg.

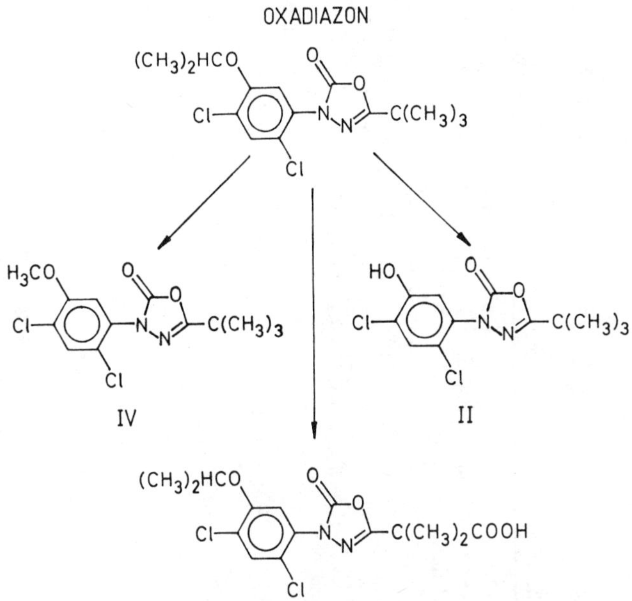


Fig. 9. Metabolism of oxadiazon in a sandy loam and a loam soil according to Ambrosi *et al.* (1977) see Table III. (Figure by courtesy of the American Chemical Society, Washington, D.C., U.S.A.).

not detected in another at the same interval since the compound has already been degraded or will appear later in the experiment. Every metabolism study should therefore be designed in such a way that sufficient and appropriate sampling intervals are included in the planning. It is not unlikely that studies carried out only with one or two sampling periods can produce different results which then would be misleading for the establishment of pathways of transformation.

## 2. *Laboratory Versus Field Results*

Table IV summarizes all data available to the author which contain information on the pathways of herbicide transformation in the laboratory or glasshouse and in the field. The few examples given demonstrate that the pathways in laboratory and field soils are also identical and that practically the same terminal residues are formed.

The information on this aspect is somewhat limited and there might be some doubts on the general validity of these findings as well as surprise that so little evidence is available. However, there exists certainly a large

Table IV. Differences in transformation pattern of herbicides between laboratory and field.

| Herbicide | Major transformation products observed | Differences in pattern between laboratory and field | Reference for laboratory data | Reference for field data |
|---|---|---|---|---|
| Atrazine | 2-Hydroxy-4-ethylamino-6-isopropylamino-*s*-triazine | None | Skipper and Volk (1972) | Cassidy and Marco (1975) Hermes *et al*. (1976, 1977) |
| Benzoylprop-ethyl | *N*-benzoyl-*N*-(3,4-dichlorophenyl) alanine, non-extractables | None | Beynon *et al*. (1974) | Smith (1979) |
| Chlorthiamid | dichlobenil 2,6-dichlorobenzamide | None | Beynon and Wright (1972) | Beynon and Wright (1972) |
| Chlortoluron | 3-(3-chloro-4-methylphenyl)-1-methylurea, 3-(3-chloro-4-carboxyphenyl)-1,1-dimethylurea | None | Gross *et al*. (1979) | Gross *et al*. (1979) |
| Cyanazine | 2-chloro-4-(1-carbamoyl-1-methylethylamino)-6-ethylamino-*s*-triazine | None | Beynon, Stoydin and Wright (1972) | Beynon, Bosio and Elgar (1972) |
| Dichlobenil | 2,6-dichlorobenzamide | None | Verloop and Nimmo (1970), Briggs and Dawson (1970) | Beynon and Wright (1972) |
| Dichloroaniline[a] | 3,4,3′,4′-tetrachloroazobenzene, non-extractables | None | Kearney *et al*. (1969) Kearney and Plimmer (1972) | Viswanathan (1978) |

| | | | | |
|---|---|---|---|---|
| Diclofop-methyl | 2-[4-(2,4-dichlorophenoxy)phen-oxy]propionic acid, non-extractables | None | Smith (1977) Martens (1978) | Smith (1979) |
| Flamprop-methyl | *N*-benzoyl-*N*-(3-chloro-4-fluoro-phenyl)alanine, non-extractables | None | Roberts and Standen (1978) | Roberts (1977) Smith (1979) |
| Fosamine ammonium | carbamoylphosphonic acid, non-extractables | None | Han (1979) | Han (1979) |
| Oxamyl[b] | methyl *N*-hydroxy-*N'*,*N'*-dimethyl-1-thioxamimidate, carbon dioxide, non-extractables | None | Harvey and Han (1978) | Harvey and Han (1978) |
| Trifluralin | $\alpha,\alpha,\alpha$-trifluoro-2,6-dinitro-N-propyl-p-toluidine, 2-ethyl-7-nitro-1-propyl-5-(trifluoromethyl)-benzimidazole, 2-ethyl-7-nitro-5-(trifluoromethyl)-benzimidazole, non-extractables | None | Kearney *et al.* (1976) Wheeler *et al.* (1979) | Golab *et al.* (1979) |

[a] 3,4-Dichloroaniline is not a herbicidal compound, but is part of the molecular structure of various widely used herbicides and fungicides such as linuron, diuron, neburon, propanil, imugan, etc.

[b] Insecticide-nematicide.

amount of unpublished work done for registration purposes which would confirm the above findings. Moreover, it is difficult to believe that the degradation of a chemical should follow different pathways in a field soil than in a subsample of the same soil incubated under controlled laboratory conditions.

## V. Potential Future Research Activities

### A. Biological Activity of Soils

Since microbial metabolism plays a significant role in the transformation of herbicides in soil, the microbiological activity of the test soils is of great importance, particularly for *in vitro* systems. Pramer and Bartha (1972) stated that soil sampled for use in metabolism studies should be treated as a living tissue and that sample handling is very important. Air-drying, prolonged storage, and freezing and thawing will drastically alter biochemical activity of soils by inactivating extracellular enzymes and by modifying the composition and population density of microorganisms in the soil. Similar observations were made by Lay and Ilnicki (1975) when measuring the effect of soil storage on propanil degradation. Based on their results the authors recommended that any soil biodegradation studies should be conducted with fresh soil samples.

Since the latter might not always be practicable, an attempt to circumvent changes in biological activity of soils during transport and processing was made by Chisaka and Kearney (1970). They reactivated Japanese rice soils by growing rice plants in the glasshouse before the degradation experiments were started. This principle has also been applied in our laboratories and several plants have been used for this purpose. When maize and cotton were used for reactivation, the evolution of $^{14}CO_2$ from chlortoluron-treated soils was drastically increased compared to non-activated soil (Laanio, 1974), particularly after the use of cotton as shown in Table V. In other experiments, the half-life of a thiadiazolyldimethylurea herbicide in a German standard soil was 45 weeks, whereas in soil reactivated with rape for 15 weeks the half-life period was reduced to 7 weeks (Keller, 1976 and 1977). Also enhanced rates of degradation of ethirimol and a thiadiazolidin-oxime carbamate occurred in soil when barley and maize, respectively, were present (Hill, 1978). Clearly the extent of reactivation of soils is influenced by the plant species used, presumably because of effects on the nature of the microbial population. On the other hand, Ellgehausen (1978) measured microbial functions (basic respiration, nitrification) of such plant-reactivated soils and observed they were depressed in comparison to soils

Table V. Reactivation of soil degradative capacity by plant roots measured by the rate of mineralization of $^{14}C$-labelled chlortoluron (Laanio, 1974).

| Pretreatment of soil[a] (4 weeks growth) | Amount (%) of $^{14}CO_2$ formed within 12 weeks |
|---|---|
| None | <1 |
| Maize | 8 |
| Cotton | 25 |

[a] Clay loam, pH = 6·2, 5·4% organic matter, cation exchange capacity = 25 meq/100 g, water holding capacity = 55·3%.

which were only rewetted with water. Thus, although this activation technique has been used in several laboratories, knowledge of the fundamental processes involved is still lacking and more research is needed on this aspect.

Another problem in this respect arises from proposals to use so-called standard soils for degradation studies. These soils were originally selected to measure the leaching behaviour of pesticides in different laboratories under standardized conditions in order to obtain reproducible and comparable results. Unfortunately, they were used later for soil degradation studies. After a large number of experiments on the rate of breakdown of various pesticides in soil it became obvious that in some cases relatively low degradation rates were observed in the standard soils when compared with soils of similar properties which were taken freshly from agricultural use areas. An example is shown in Fig. 10 which demonstrates that the herbicide di-allate degraded 2·5–5 times faster in the fresh soils than in standard soil 2·2 (Guth, 1978). Studies just under way with the phosphorus insecticide parathion show a similar tendency. These differences were found to be mainly due to different microbial activities. The microbial biomass of the Speyer standard soil 2·2 used in the studies was only about one fourth to one fifth of the two fresh soils (Domsch, 1978). Thus the concept of using standard soils seems to be very questionable. Besides the difficulties with the biochemical activity of soils probably caused by the processing, another critical point is to guarantee the same quality of soil over a period of years since the original soil source will sooner or later be exhausted.

The biological activity of soils is critical not only at the beginning of soil degradation experiments but also for the whole duration of a metabolism experiment. Some registration authorities require studies to be carried out for a period of one year without considering that soils can scarcely retain

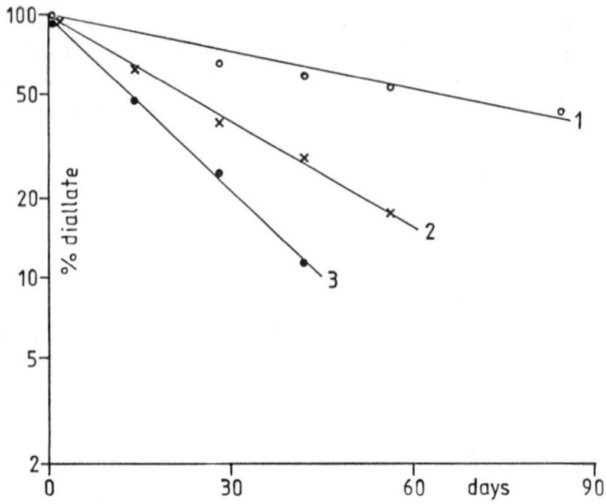


Fig. 10. Rate of degradation of di-allate in various soils with different microbial activities under aerobic conditions at 22 ± 2°C (Guth, 1978).
1. loamy sand, pH = 5·6, 2·2% organic carbon, 8·8% clay, 6·4% silt, 84·8% sand, Speyer standard soil 2·2. 2. loamy sand, pH = 8·0, 1·5% organic carbon, 3·0% clay, 14·6% silt, 82·4% sand. 3. sandy loam, pH = 7·8, 1·8% organic carbon, 6·3% clay, 24·4% silt, 69·3% sand.

their biological activity under *in vitro* conditions for such a long time. It is known from microbial studies that both $CO_2$ production and $O_2$ consumption, one measure of soil microbiological activity, steadily decrease when soils are incubated in laboratory systems (Marsh *et al.*, 1977; Davies and Marsh, 1977; Marsh and Davies, 1978). Observations have also been made in our laboratories that $^{14}CO_2$ evolution from radiolabelled pesticide markedly decreased between 3 and 6 months after starting the experiment but could be stimulated by stirring the soil sample. This effect will happen earlier and more drastically when soils with a lower initial bioactivity are used. Therefore it is very important to start degradation experiments with fresh soils which have their full biological potential. In addition, the biological activity of soils in laboratory systems should be monitored at regular intervals. This could be done, for example, as in our laboratories by measuring at least one parameter, $CO_2$ production from basic soil respiration.

Although the information reported in this section is mainly derived from kinetic studies these aspects are also of importance for studies on pathways of transformation. Low biological activity of soils causes slow breakdown of the pesticide and hence slow formation of metabolites. This can create problems in identifying the major metabolites as well as the terminal pesticide residues.

## B. Soil Bound Residues

During soil degradation experiments, a certain amount of radioactivity always remains unextractable and is referred to as "bound residue". A possible definition of soil bound residues is

> that unextractable and chemically unidentifiable pesticide residue remaining in fulvic acid, humic acid, and humin fractions after exhaustive sequential extraction with non-polar organic and polar solvents (Anon., 1975).

Non-extractable residues often represent a major metabolic fraction but despite this, relatively little progress has been made in understanding the processes involved. A more complete understanding of the significance of soil bound residues requires major advances in methodology. Knowledge of the distribution of non-extractable residues in the fulvic acid, humic acid and humin is of little value because the results cannot be interpreted.

In some cases, bound soil residues may be the only terminal residues or may become part of the terminal residue fraction. The degradation of propazine in aerobic soil is shown in Fig. 11 as an example (Keller, 1979). Besides the hydroxypropazine, the non-extractable residues represent the

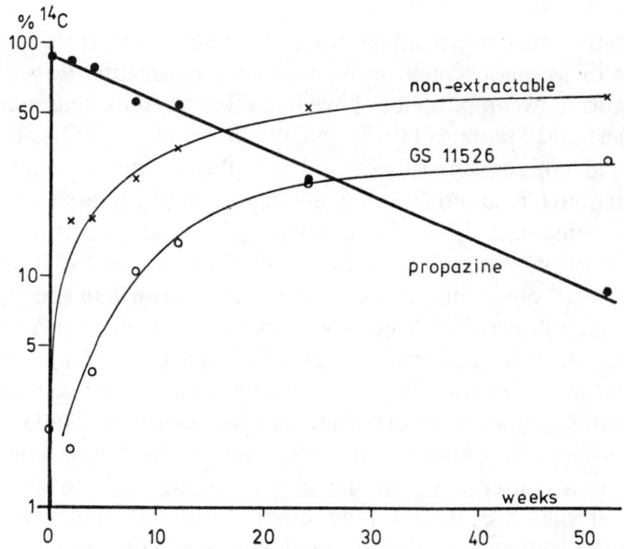


Fig. 11. Rate of $^{14}$C-propazine degradation and of formation of hydroxy-propazine (GS 11526) and non-extractable radioactivity in a loamy sand under aerobic conditions at 25°C. The soil properties were as follows: pH = 5·6, 2·2% organic carbon, cation exchange capacity = 8·4 meq per 100 g, 8·8% clay, 6·4% silt, 84·8% sand (Keller, 1979).

major metabolic fraction. Under such circumstances, the bioavailability to both plants and exposed aqueous or soil fauna would need to be examined. This concept is followed by the US Environmental Protection Agency when requiring studies on the uptake of weathered pesticide residues by rotational crops and on fish bioaccumulation in a static aquatic system containing aged, pesticide-treated soil. There is no legislative problem as long as such residues show no tendency to accumulate in significant amounts. However, from a scientific standpoint it is unsatisfactory not to understand properly the nature of soil bound residues. This is undoubtedly a large area for more academic research and it is to be hoped that research activities will also go more in this direction. At least a few successful steps have already been made. For example, bound glyphosate residues could be extracted with 0·5N ammonium hydroxide and identified as parent compound and one major metabolite (Nomura and Hilton, 1977; Rueppel *et al.*, 1977). Aqueous solutions of anionic surfactants and cold base released larger amounts of soil bound residues of dinitramine (Smith *et al.*, 1973). Hitchings and Roberts (1979) used 5% aqueous sodium hydroxide for extraction of bound residues and were able to identify at least a few percent of the bound material.

## C. Influence of Other Chemicals on Herbicide Transformation

Combinations of pesticide residues in soil can result from direct application of mixtures or as a consequence of successive treatments. Possible types of interactions between pesticides have been described by Kaufman (1972, 1977b), Priest and Stephens (1975) and Wallnöfer *et al.* (1977). Increased persistence, decreased persistence and complex residue formation were effects which have resulted from the application and degradation of some pesticide combinations. Potential reasons can be direct effects on enzymes by inhibition of or decomposition by soil enzymes or, on the other hand, indirect effects through inhibition or stimulation of growth of soil organisms which will again directly influence soil enzymes (Cervelli *et al.*, 1978).

These temporary effects were observed in soil perfusion systems and microbial cultures, but not always in soil under laboratory or greenhouse conditions. Unfortunately, no experiments have been carried out to confirm whether the above laboratory results also apply to field situations.

Possibly as a consequence of the above findings, laboratory or field dissipation studies are required by the US Environmental Protection Agency for all products intended for combination and tank mixtures. These studies will compare the rates of disappearance of each active ingredient of the mixture when applied to the soil as a mixture with the separate dissipation rates of each component when applied individually.

With regard to the interactions between herbicides, the information is rather limited. The author is aware of only three published studies dealing with this subject. The rate of dalapon degradation in soil was significantly reduced when applied in combination with amitrole than when dalapon was applied alone (Kaufman, 1966). The influence of monuron and di-allate on the rate of mineralization of monolinuron and linuron was reported by Wallnöfer *et al.* (1977). Hurle (1977) examined the influence of DNOC and 2,4-D on each other as well as that of diuron on 2,4-D and DNOC on simazine. DNOC decreased the rate of degradation of 2,4-D but the effect of 2,4-D on DNOC breakdown was less pronounced. Increasing diuron concentration caused only slight inhibition of 2,4-D degradation in the initial period, whereas the persistence of simazine was independent of the DNOC concentration.

Unpublished field studies of a large number of herbicide combinations at the recommended use rates did not show any significant effects on the rate of breakdown of the herbicides in mixtures compared with individual applications (Ballantine *et al.*, 1976–1979). Metolachlor was tested in two soils in combination with atrazine, linuron, metribuzin, dicamba, propazine, terbutryn, naptalam + dinoseb, cyanazine, paraquat and glyphosate. Profluralin was mixed with linuron and naptalam + dinoseb, and terbutryn was combined with atrazine and propazine.

Besides pesticides other chemicals such as fertilizers, surfactants, organic solvents, etc. may reach the soil environment but there is little knowledge of their influence on the degradation of herbicides. Moreover, all information available deals only with the influence on persistence and it is not known whether effects on the pathways of transformation also occur. This seems unlikely from our experience with several herbicide combinations tested in the field where not even the rates of degradation were significantly changed. Nevertheless, these interaction problems could still be a large area for academic research. It is important that any *in vitro* effects should be confirmed under practical field conditions, otherwise artificial problems will be created.

## D. Influence of Rhizosphere on Herbicide Transformation

Transformation by soil microorganisms is one of the most important metabolism mechanisms for many pesticides. The microbial population of soil can be altered both qualitatively and quantitatively by the presence of plant roots (Rovira and McDougall, 1967) and the regions around root systems are more heavily populated by soil microbes than soil without plant roots. Therefore, the transformation of a chemical might be also changed quantitatively and possibly qualitatively in the presence of plant roots. The great influence of growing plants in a laboratory soil on the rate of breakdown has been discussed earlier, but little is known of pathway differences caused by the rhizosphere. Although examples exist where soil metabolism experiments have been carried out in the presence and absence of plant roots (i.e. Roberts, 1977; Gross *et al.*, 1979), the author is not aware of any significant differences.

There is certainly a lack of studies designed to measure the influence of the rhizosphere on the pathways of transformation of herbicides in soil. Although structures not usually formed in plant-free soil will probably be detected in plant metabolism experiments, it is desirable for scientific reasons to obtain more specific information on this subject.

Studies on the influence of the rhizosphere on herbicide degradation are likely to use a volume ratio of root-containing versus root-free soil in the laboratory and the greenhouse that is far away from that in the field. Also the method of herbicide application may differ. Therefore, once again if significant effects occur in indoor experiments, they must be confirmed in the field soil environment before any significance can be given to them.

## VI. Conclusions

This chapter has tried to review the methodology available for studying the pathways of transformation of herbicides in soil, has given some thought on the interpretation of the results obtained with these methods and finally has proposed potential future research activities under consideration of registration requirements and of scientific interests.

Registration requirements on pathways of transformation exist in several countries and there are pressures to harmonize these requirements on an international level. However, it is important that any harmonization of requirements does not result in a strict standardization of methods so that an internationally approved procedure would have to be used by each research laboratory in the world. Such an approach would not stimulate research activities and certainly would restrict scientific freedom.

Moreover, it is absolutely unnecessary to use a mandatory standard method when pathways of transformation have to be established. Although the so-called flow-through systems are preferable for soil studies in the laboratory for several reasons the use of biometer flasks will also yield representative results if fundamental critical points are considered during the experiments. Soil perfusion systems might be useful for microbial degradation studies but do not reproduce practical conditions. Also the complex, integrated systems have several shortcomings and do not simulate the real environmental situation.

Although field studies with radiolabelled compounds have been carried out in the past it might become more difficult in the future due to the increasing public concern on the use of radioisotopes. Data have been presented which indicate that pathways of transformation in field soils can be predicted from studies with soils incubated under controlled laboratory conditions. Moreover, there is considerable information available on the identity of transformation pathways in different soils in the field and in the laboratory. It seems that the experiments with one fresh biologically active field soil incubated in laboratory systems will provide sufficient and representative data on the pathway of transformation of any pesticidal compound in soil since the differences observed between various soils are only of quantitative nature.

Future research activities should be directed to the biological activity of test soils, to soil bound residues and to the influence of other chemicals and of the rhizosphere on herbicide transformation.

Biological activity of the test soils is of great importance for the transformation of herbicide in soil but has certainly been overlooked in the standard soils concept. In this context, the duration of experiments will also adversely affect the bioactivity of soils in laboratory systems. It is essential that the microbial activity of soils is measured before and during transformation experiments and that fast and simple methods are available for such measurements.

Soil bound residues although found with many pesticides as a major metabolic fraction should not present problems for the legislator as long as they are not taken up by non-target organisms in significant amounts. For scientific reasons, more information on the nature of the non-extractable residues would be desirable but for this purpose appropriate methods have to be first developed.

Interactions between herbicides and other pesticides or pesticide related compounds have been reported but mainly dealing with the influence on the persistence of products in soil. Experimental data obtained under practical conditions indicate that combinations of herbicides may not be a problem in this respect. Little is known of the influence of other chemicals such as fertilizers, detergents, organic solvents, etc. on the degradation of herbicides in soil. In addition, information is lacking as to whether the pathways of transformation can also be affected.

Reactivation of soil microbiology by various plant species and thus an acceleration of the rate of mineralization indicated the potential influence of the rhizosphere. However, little knowledge exists whether and to what extent the rhizosphere can affect the mechanism of herbicide transformation in soils and this should induce further research.

## References

Ambrosi, D., Kearney, P. C. and Macchia, J. A. (1977). Persistence and metabolism of oxadiazon in soils. *J. agric. Fd Chem.*, **25**, 868–872.

Anderson, J. P. E. (1975). Einfluss von Temperatur und Feuchte auf Verdampfung, Abbau und Festlegung von Diallat im Boden. *Z. Pflkrankh Pflschutz, Sonderheft VII*, 141–146.

Anon. (1971). British Pesticides Safety Precaution Scheme agreed between Government Departments and Industry (PSPS), Appendix C. Ministry of Agriculture, Fisheries and Food.

Anon. (1975). Registration of pesticides in the United States—Proposed guidelines. *Federal Register*, **40** (123), June 25.

Anon. (1978). Registration of pesticides in the United States—Proposed guidelines. *Federal Register*, **43** (132), July 10.

Ballantine, L. G., Herman, M. M., Newby, L. C. and Camp, H. B. (1976–1979). Unpublished data. CIBA-GEIGY Ltd., Agricultural Division.

Bartha, R. and Pramer, D. (1965). Features of a flask and method for measuring the persistence and biological effects of pesticides in soil. *Soil Sci.*, **100**, 68–70.

Beall, M. L., Nash, R. G. and Kearney, P. C. (1976). Agroecosystem. A laboratory model ecosystem to simulate agricultural field conditions for monitoring pesticides. *Proc. EPA Conference on Modeling and Simulation*, Environ. Protect. Agency, 790–793.

Best, J. A. and Weber, J. B. (1974). Disappearance of *s*-triazines as affected by soil pH using a balance-sheet approach. *Weed Sci.*, **22**, 364–373.

Beynon, K. I. and Wright, A. N. (1972). The fates of the herbicides chlorthiamid and dichlobenil in relation to residues in crops, soils, and animals. *Residue Rev.*, **43**, 23–53.

Beynon, K. I., Bosio, P. and Elgar, K. E. (1972). The analysis of crops and soils for the triazine herbicide cyanazine and some of its degradation products. *Pestic. Sci.*, **3**, 401–408.

Beynon, K. I., Roberts, T. R. and Wright, A. N. (1974). Degradation of the herbicide benzoylprop-ethyl in soil. *Pestic Sci.*, **5**, 451–463.

Beynon, K. I., Stoydin, G. and Wright, A. N. (1972). The breakdown of the triazine herbicide cyanazine in soils and maize. *Pestic. Sci.*, **3**, 293–305.

Bollag, J. M. (1972). *Biochemical transformations of pesticides by soil fungi*. CRC Critical Reviews in Microbiology, pp. 35–58.

Bollag, J. M. (1974). Microbial transformation of pesticides. *Advances in Applied Microbiology*, **18**, 75–130.

Briggs, G. G. and Dawson, J. E. (1970). Hydrolysis of 2,6-dichlorobenzonitrile in soils. *J. agric. Fd Chem.*, **18**, 97–99.

Bull, D. L., Whitten, C. J. and Ivie, G. W. (1976). Fate of *O*-Ethyl *O*-[4-(methylthio)phenyl]*S*-propyl phosphorodithioate (BAY NTN 9306) in cotton plants and soil. *J. agric. Fd Chem.*, **24**, 601–605.

Burkhard, N. (1978). Unpublished data. CIBA-GEIGY Ltd., Agricultural Division.

Burkhard, N. and Guth, J. A. (1979). Photolysis of organophosphorus insecticides on soil surfaces. *Pestic. Sci.*, **10**, 313–319.

Cassidy, J. E. and Marco, G. J. (1975). Unpublished data. CIBA-GEIGY Ltd., Agricultural Division.

Cawse, P. A. (1975). Microbiology and biochemistry of irradiated soils. In *Soil Biochemistry* (E. A. Paul and A. D. McLaren, Eds.) Vol. III, 213–267.

Cervelli, S., Nannipieri, P. and Sequi, P. (1978). Interactions between agrochemicals and soil enzymes. In *Soil Enzymes* (R. G. Burns, Ed.), 251–293. Academic Press, London and New York.

Chisaka, H. and Kearney, P. C. (1970). Metabolism of propanil in soils. *J. agric. Fd Chem.*, **18**, 854–858.

Cripps, R. E. and Roberts, T. R. (1978). Microbial degradation of herbicides. In *Pesticide Microbiology*. (I. R. Hill and S. J. L. Wright, Eds.), 669–730. Academic Press, London and New York.

Davies, H. A. and Marsh, J. A. P. (1977). The effect of herbicides on respiration and transformation of nitrogen in two soils. II. Dalapon, pyrazone and trifluralin. *Weed Res.*, **17**, 373–378.

Domsch, K. H. (1978). Personal communication.

Ellgehausen, H. (1978). Unpublished data. CIBA-GEIGY Ltd., Agricultural Division.

Führ, F., Cheng, H. H. and Mittelstaedt, W. (1976). Pesticide balance and metabolism studies with standardized lysimeters. *Landwirtschaftl. Forsch., Sonderheft* **32**, 272–278.

Funderburk, H. H., Negi, N. S. and Lawrence, J. M. (1966). Photochemical decomposition of diquat and paraquat. *Weeds*, **14**, 240–243.

Fung, K. K. H. and Uren, N. C. (1977). Microbial transformation of *S*-methyl *N*-[(methylcarbamoyl)oxy]thioacetimidate (methomyl) in soils. *J. agric. Fd Chem.*, **25**, 966–969.

Golab, T., Althaus, W. A. and Wooten, H. L. (1979). Fate of [$^{14}$C] trifluralin in soil. *J. agric. Fd Chem.*, **27**, 163–179.

Goring, C. A. I., Laskowski, D. A., Hamaker, J. W. and Meikle, R. W. (1975). Principles of pesticide degradation in soil. In *Environmental Dynamics of Pesticides* (R. Haque and V. H. Freed, Eds.), 135–172. Plenum Press

Goswami, K. P. and Koch, B. L. (1976). A simple apparatus for measuring degradation of $^{14}$C-labelled pesticides in soil. *Soil Biol. Biochem.*, **8**, 527–528.

Gross, D., Laanio, T., Dupuis, G. and Esser, H. O. (1979). The metabolic behaviour of chlorotoluron in wheat and soil. *Pestic. Biochem. Physiol.*, **10**, 49–59.

Guth, J. A. (1978). Unpublished data. CIBA-GEIGY Ltd., Agricultural Division.

Han, J. C.-Y. (1979). Stability of [$^{14}$C] fosamine ammonium in water and soils. *J. agric. Fd Chem.*, **27**, 564–571.

Harvey, J. jr. and Han, J. C.-Y. (1978). Decomposition of oxamyl in soil and water. *J. agric. Fd Chem.*, **26**, 536–541.

Hautala, R. R. (1976). Photolysis of pesticides on soil surfaces. *Symposium on Nonbiological Transport and Transformation of Pollutants on Land and Water: Processes and Critical Data Required for Predictive Description*. Gaithersburg, Maryland (Abstract Volume).

Hautala, R. R. (1978). Surfactant effects on pesticide photochemistry in water and soil. *EPA-Report 600/3-78-060*. U.S. EPA, Athens, Georgia.

Hermes, P. A., Ryskiewich, D. P. and Marco, G. J. (1976, 1977). Unpublished data. CIBA-GEIGY Ltd., Agricultural Division.

Hill, I. R. (1978). Microbial transformation of pesticides. In *Pesticide Microbiology*. (I. R. Hill and S. J. L. Wright, Eds.), 137–202. Academic Press, London and New York.

Hill, I. R. and Arnold, D. J. (1978). Transformations of pesticides in the environment—The experimental approach. In *Pesticide Microbiology*. (I. R. Hill and S. J. L. Wright, Eds.), 203–245. Academic Press, London and New York.

Hitchings, E. J. and Roberts, T. R. (1979). Degradation of the herbicide flamprop-isopropyl in soil under laboratory conditions. *Pestic Sci.*, **10**, 1–13.

Hurle, K. (1977). Wechselseitiger Einfluss von Herbiziden auf ihre Abbaugeschwindigkeit im Boden. *Mitt. Biol. Bundesanstalt Land-Forstwirtsch.*, **178**, 234.

Jarczyk, H. J. (1978). Behaviour of pesticides in soil as determined by undisturbed soil cores and lysimeter systems. *4th International Congress of Pesticide Chemistry (IUPAC), Zürich, July 24–28*, Abstract Volume V, 35.

Kaufman, D. D. (1966). Microbial degradation of herbicide combinations: amitrole and dalapon. *Weeds*, **14**, 130–134.

Kaufman, D. D. (1972). Degradation of pesticide combinations. *Pest. Chem.*, **6**, 175–205.

Kaufman, D. D. (1974). Degradation of pesticides by soil microorganisms. *In* "Pesticides in Soil and Water" (W. D. Guenzi, Ed.), 133–202.

Kaufman, D. D. (1977a). Approaches to investigating soil degradation and dissipation of pesticides. In *Terrestrial Microcosms and Environmental Chemistry*. (J. M. Witt and J. W. Gillett, Eds.). 41–53. National Science Foundation/RA 79-0026, Washington D.C.

Kaufman, D. D. (1977b). Biodegradation and persistence of several acetamide, acylanilide, azide, carbamate, and organophosphate pesticide combinations. *Soil Biol. Biochem.*, **9**, 49–57.

Kaufman, D. D. and Kearney, P. C. (1976). Microbical transformations in the soil. In *Herbicides. Physiology, Biochemistry*, Ecology" (L. J. Audus, Ed.), Vol. 2, 29–64. Academic Press, London and New York.

Kearney, P. C. and Kaufman, D. D. (1975) *Herbicides. Chemistry, Degradation, and Mode of Action, Vol. 1.* Marcel Dekker Inc. New York and Basel.

Kearney, P. C. and Kaufman, D. D. (1976). *Herbicides. Chemistry, Degradation, and Mode of Action, Vol. 2.* Marcel Dekker Inc. New York and Basel.

Kearney, P. C. and Kontson, A. (1976). A simple system to simultaneously measure volatilization and metabolism of pesticides from soils. *J. agric. Fd Chem.*, **24**, 424–426.

Kearney, P. C., Plimmer, J. R. and Guardia, F. B. (1969). Mixed chloroazobenzene formation in soil. *J. agric. Fd Chem.*, **17**, 1418–1419.

Kearney, P. C. and Plimmer, J. R. (1972). Metabolism of 3,4-dichloroaniline in soils. *J. agric. Fd Chem.*, **20**, 584–585.

Kearney, P. C., Plimmer, J. R., Wheeler, W. B. and Kontson, A. (1976). Persistence and metabolism of dinitroaniline herbicides in soil. *Pestic. Biochem. Physiol.*, **6**, 229–238.

Keller, A. (1976, 1977, 1978, 1979) Unpublished data. CIBA-GEIGY Ltd., Agricultural Division.

Kennedy, J. M. and Talbert, R. E. (1977). Comparative persistence of dinitroaniline type herbicides on the soil surface. *Weed Sci.*, **25**, 373–381.

Kimura, R. and Yamaguchi, M. (1978). Microbial decomposition of 2-oxo-4-methyl-6-ureido-hexahydropyrimidine in soil. *Soil Biol. Biochem.*, **10**, 503–508.

Laanio, T. L. (1974). Unpublished data. CIBA-GEIGY Ltd., Agricultural Division.

Lay, M. M. and Ilnicki, R. D. (1975). Effect of soil storage on propanil degradation. *Weed Res.* **15**, 63–66.

Liang, T. T. and Lichtenstein, E. P. (1976). Effects of soils and leaf surfaces on the photodecomposition of [$^{14}$C] azinphosmethyl. *J. agric. Fd Chem.*, **24**, 1205–1210.

Lichtenstein, E. P., Fuhremann, T. W. and Schulz, K. R. (1974). Translocation and metabolism of [$^{14}$C]-phorate as affected by percolating water in a model soil-plant ecosystem. *J. agric. Fd Chem.*, **22**, 991–996.

Marsh, J. A. P. and Davies, H. A. (1978). The effect of herbicides on respiration and transformation of nitrogen in two soils. III. Lenacil, terbacil, chlorthiamid and 2,4,5-T. *Weed Res.* **18**, 57–62.

Marsh, J. A. P., Davies, H. A. and Grossbard, E. (1977). The effect of herbicides on respiration and transformation of nitrogen in two soils. I. Metribuzin and glyphosate. *Weed Res.* **17**, 77–82.

Marvel, J. T., Sutherland, M. L., Brightwell, B. B., Malik, J. M. and Rueppel, M. L. (1976). A simple apparatus and quantitative method for determining the persistence of pesticides in soil and soil/water mixtures. *172nd National Meeting of the American Chemical Society*, San Francisco, August 1976, PEST 02.

Marvel, J. T., Brightwell, B. B., Malik, J. M., Sutherland, M. L. and Rueppel, M. L. (1978). A simple apparatus and quantitative method for determining the persistence of pesticides in soil. *J. agric. Fd Chem.*, **26**, 1116–1120.

Martens, R. (1977). Degradation of endosulfan-8,9-$^{14}$C in soil under different conditions. *Bull. environ. Contam. Toxic.*, **17**, 438–446.

Martens, R. (1978). Degradation of the herbicide [$^{14}$C]-diclofop-methyl in soil under different conditions. *Pestic. Sci.*, **9**, 127–134.

Mathur, S. P. and Saha, J. G. (1977). Degradation of lindane-$^{14}$C in a mineral soil and in an organic soil. *Bull. environ. Contam. Toxic.*, **17**, 424–430.

Meikle, R. W. (1972). Decomposition: qualitative relationships. *In* "Organic Chemicals in the Soil Environment" (C. A. I. Goring and J. W. Hamaker, Eds.), Vol. I, 145–251.

Mikami, N., Gomi, H. and Miyamoto, J. (1979). Studies on degradation of 2,6-di-tert-butyl-4-methylphenol (BHT) in the environment. Part.-I. Degradation of $^{14}$C-BHT in soil. *Chemosphere*, **5**, 305–310.

Nakagawa, M., Ando, M. and Obata, Y. (1975). Fate of isoxathion [*O*,*O*-diethyl *O*-(5-phenyl-3-isoxazolyl)-phosphorothioate] in soils. *Agr. biol. Chem.*, **39**, 1763–1773.

Newby, L. C., Ballantine, L. G., Ellgehausen, H. and Guth, J. A. (1978). Herbicide soil metabolism. Laboratory methods. *Weed Sci. Soc. Amer. Meeting, San Francisco*, Abs. No. 247.

Nilles, G. P. and Zabik, M. J. (1974). Photochemistry of bioactive compounds. Multiphase photodegradation of basalin. *J. agric. Fd Chem.*, **22**, 684–688.

Nilles, G. P. and Zabik, M. J. (1975). Photochemistry of bioactive compounds. Multiphase photodegradation and mass spectral analysis of basagran. *J. agric. Fd Chem.*, **23**, 410–415.

Nomura, N. S. and Hilton, H. W. (1977). The adsorption and degradation of glyphosate in five Hawaiian sugarcane soils. *Weed Res.* **17**, 113–121.

Ohkawa, H., Shibaike, R., Okihara, Y., Morikawa, M. and Miyamoto, J. (1976). Degradation of the fungicide Denmert® (*S*-*n*-butyl *S*′-*p*-tert-butylbenzyl *N*-3-pyridyldithiocarbonimidate, S-1358) by plants, soils and light. *Agr. biol. Chem.*, **40**, 943–951.

Parochetti, J. V. and Dec, G. W. jr. (1978). Photodecomposition of eleven dinitroaniline herbicides. *Weed Sci.*, **26**, 153–156.

Parochetti, J. V. and Hein, E. R. (1973). Volatility and photodecomposition of trifluralin, benefin, and nitralin. *Weed Sci.*, **21**, 469–473.

Parr, J. F. and Reuszer, H. W. (1962). Organic matter decomposition as influenced by oxygen level and flow rate of gases in the constant aeration method. *Soil Sci. Soc. Amer. Proc.*, **26**, 552–556.

Parr, J. F. and Smith, S. (1969). A multipurpose manifold assembly: use in evaluating microbiological effects of pesticides. *Soil Sci.*, **107**, 271–276.

Plimmer, J. R. (1978). Photolysis of TCDD and trifluralin on silica and soil. *Bull. environm. Contam. Toxic.*, **20**, 87–92.

Powlson, D. S. and Jenkinson, D. S. (1976). The effects of biocidal treatments on metabolism in soil. II. Gamma irradiation, autoclaving, air-drying and fumigation. *Soil Biol. Biochem.*, **8**, 179–188.

Pramer, D. and Bartha, R. (1972). Preparation and processing of soil samples for biodegradation studies. *Environ. Letters*, **2**, 217–224.

Priest, B. and Stephens, R. J. (1975). Studies on the breakdown of *p*-chlorophenyl methylcarbamate I. In soil. *Pestic. Sci.*, **6**, 53–59.

Roberts, T. R. (1976). Experimental models for studying the fate of pesticides in plants. *British Crop Protection Council Monograph 17, Persistence of Insecticides and Herbicides* (K. I. Beynon, Ed.), 159–168.

Roberts, T. R. (1977). The metabolism of the herbicide flampropmethyl in wheat. *Pestic. Sci.*, **8**, 463–472.

Roberts, T. R. and Standen, M. E. (1977). Degradation of the pyrethroid insecticide WL 41706, (±)-α-cyano-3-phenoxybenzyl 2,2,3,3-tetramethylcyclopropane carboxylate, in soils *Pestic. Sci.*, **8**, 600–610.

Roberts, T. R. and Standen, M. E. (1978). Degradation of the herbicide flampropmethyl in soil under anaerobic conditions. *Pestic. Biochem. Physiol.*, **9**, 322–333.

Rovira, A. D. and McDougall, B. M. (1967). Microbiological and biochemical aspects of the rhizosphere. In *Soil Biochemistry* (A. D. McLaren and G. H. Peterson, Eds.), Vol. I, 417–463. Edward Arnold, London.

Rueppel, M. L., Brightwell, B. B., Schaefer, J. and Marvel, J. T. (1977). Metabolism and degradation of glyphosate in soil and water. *J. agric. Fd Chem.*, **25**, 517–528.

Skipper, H. D. and Volk, V. V. (1972). Biological and chemical degradation of atrazine in three Oregon soils. *Weed Sci.*, **20**, 344–347.

Smith, A. E. (1974). Breakdown of the herbicide dicamba and its degradation product 3,6-dichlorosalicylic acid in prairie soils. *J. agric. Fd Chem.*, **22**, 601–605.

Smith, A. E. (1977). Degradation of the herbicide diclofop-methyl in prairie soils. *J. agric. Fd Chem.*, **25**, 893–898.

Smith, A. E. (1979). Extraction of free and bound carboxylic acid residues from field soils treated with the herbicides benzoylprop-ethyl, diclofop-methyl, and flamprop-methyl. *J. agric. Fd Chem.*, **27**, 428–432.

Smith, C. A., Iwata, Y. and Gunther, F. A. (1978). Conversion and disappearance of methidathion on thin layers of dry soil. *J. agric. Fd Chem.*, **26**, 959–962.

Smith, R. A., Belles, W. S., Shen, K.-W. and Woods, W. G. (1973). The degradation of dinitramine ($N^3$,$N^3$-diethyl 2,4-dinitro-6-trifluoromethyl-*m*-phenylenediamine) in soil. *Pestic. Biochem. Physiol.*, **3**, 278–288.

Tomizawa, C., Uesugi, Y., Ueyama, I. and Yamamoto, H. (1976). Movement and metabolism of *S*-benzyl *O,O*-diisopropyl phosphorothiolate (KITAZIN P®) and *O*-ethyl *S,S*-diphenyl phosphoro-dithiolate (edifenphos) in various types of soils. *J. Environ. Sci. Health*, **B 11**, 231–251.

Verloop, A. and Nimmo, W. B. (1970). Metabolism of dichlobenil in sandy soil. *Weed Res.* **10**, 65–70.

Viswanathan, R., Scheunert, I., Kohli, J., Klein, W. and Korte, F. (1978). Long-term studies on the rate of 3,4-dichloroaniline-$^{14}C$ in a plant-soil-system under outdoor conditions. *J. Environ. Sci. Health*, **B 13**, 243–259.

Wallnöfer, P., Poschenrieder, G. and Engelhardt, G. (1977). Verhalten von Pestizidkombinationen im Boden. *Z. Pflkrankh. Pflschutz, Sonderheft VIII*, 199–207.

Weinmann, W. and Schinkel, K. (1976). *Unterlagen zum Verhalten von Pflanzenbehandlungsmitteln in Boden im Rahmen des Zulassungsverfahrens*. Merkblatt Nr. 36 der Biologischen Bundesanstalt für Land- und Forstwirtschaft.

Wheeler, W. B., Stratton, G. D., Twilley, R. R., Ou, L.-T., Carlson, D. A. and Davidson, J. M. (1979). Trifluralin degradation and binding in soil. *J. agric. Fd Chem.*, **27**, 702–706.

Williams, P. P. (1977). Metabolism of synthetic organic pesticides by anaerobic microorganisms. *Residue Rev.*, **66**, 63–135.

# 6 Role of Microorganisms in Decomposition

*Lennart Torstensson*
Swedish University of Agricultural Sciences
Department of Microbiology
S-750 07 Uppsala 7, Sweden



## I. Introduction

Microorganisms usually occupy a volume of less than 0·1% of the soil, but are responsible for numerous transformations that cycle elements and energy in nature. Population densities of bacteria may be high, $10^9$ per g soil is not uncommon and the length of fungal hyphae may be some hundreds of thousands of metres per g soil. The biomass of microorganisms per hectare may reach several tonnes.

The diversity of species of microorganisms in soil is great. The microbial population exists in a dynamic equilibrium formed by interactions of abiotic and biotic factors that can be altered by modifying environmental conditions. For example, in the region of plant roots, the rhizosphere, the numbers, types and metabolic activity of microorganisms and the physicochemical soil conditions are different from those in soil outside the rhizosphere.

Microorganisms are able to degrade a wide variety of chemicals, from simple polysaccharides, amino acids, proteins, lipids, etc. to more complex materials such as plant residues, waxes and rubbers. Without the activity of microorganisms such organic compounds would accumulate and pollute our environment.

Microorganisms are also able to degrade chemical compounds synthesized by man. A variety of mechanisms underly these activities. In this chapter the role of microorganisms in the decomposition of herbicides, the relevant underlying mechanisms, and factors influencing rates and routes of transformations are discussed.

## II. Mechanisms of Decomposition

Many microbial species are capable of growth on a wide variety of organic compounds, and the necessary catabolic enzymes may be subject to several types of control. Enzyme induction is a phenotypic response of a culture to the presence of the appropriate chemical signal in the environment (Clarke, 1972), resulting in an increase in the synthesis of enzymes so controlled. The synthesis of an induced enzyme is under the control of a regulator gene. The inducer need not be the enzyme substrate.

During growth on compounds which are converted by enzymes into intermediates, the enzymes may be related to a number of separately controlled regulatory units. There may be a sequential induction, in which the first enzyme is induced by the substrate and the later enzymes are induced by intermediate metabolites. This allows growth on intermediates occurring on the pathway.

Table I. Classification of microbial activities in connection with decomposition of herbicides.

A. Decomposition to smaller molecules
   1. Enzymatic reactions
      (a) Direct degradation of herbicides in central metabolism of microorganisms in which they serve as energy sources to supply growth (catabolism) and where adaptation phenomena appear.
      (b) Incidental transformation of herbicides by microorganisms, via peripherical metabolic processes in the absence of the perfect coordination of the process which is characteristic in central metabolism (co-metabolism).
      (c) Incidental transformation of herbicides by extracellular enzymes
   2. Non-enzymatic reactions
      (a) Contribution through pH changes
      (b) Production of substances that interact with herbicides in photochemical and chemical reactions

B. Formation of new substances in which herbicides or metabolites of herbicides are parts

For several of the enzymes involved in the metabolism of aromatic compounds the inducer is the product of the enzyme and not its substrate (Clark, 1972). These enzymes are also subject to repression by products of the metabolism, so that the rates of synthesis are balanced according to the metabolic demands of the bacteria and the nutrients available in the environment.

As well as by enzymatic reactions, microorganisms may also contribute to the transformation of herbicides via non-enzymatic reactions. A classification of different microbial activities in connection with decomposition of herbicides is given in Table I.

The classification is not made on the basis of microbial species or any individual herbicide. Indeed, the nature of metabolism can be influenced largely by environmental factors, which in turn affect the physiological conditions of the microbes. It is therefore possible that the metabolism of a herbicide by certain microorganisms might differ depending on the environmental conditions.

## A. Decomposition to Smaller Molecules

### 1. *Enzymatic reactions*

Microbial metabolic activities need energy. Most organic materials can serve as a source of energy to at least some microorganisms. Another attribute of microbial metabolism is the adaptation of microorganisms through mutation and induction, particularly towards chemicals that are initially toxic to them.

An important criterion used in classifying the enzymatic reactions involved in microbial transformations of herbicides (Table I) is whether the microorganisms derive energy from the process or not. This consideration is significant from a practical view, since it is of importance for predictions of the herbicide's persistence. Thus catabolic metabolism requires a favourable chemical structure of the herbicide which allows it to be utilized as a carbon and energy source. It often becomes more obvious the higher the concentration.

Horvath and Alexander (1970) pointed out that in studies of chlorinated aromatic pesticides, selection of a microbial strain capable of degrading the pesticide might be possible by enriching the medium with a non-chlorinated analogue of the pesticide (co-metabolism). Such an approach might make even very stable and usually non-degradable pesticides susceptible to microbial attack. In such cases, however, the metabolic pathways do not go as far as complete mineralization to $CO_2$, $H_2O$ and $Cl^-$ and the processes themselves are seldom energy-yielding.

A way of distinguishing these classes of metabolic activities is to induce herbicide-metabolizing microbes by the herbicides themselves. These induction experiments should be done both in the presence and absence of other carbon sources. In the absence of other carbon sources only the microbes performing catabolic metabolism would be able to multiply by using the herbicide as a sole carbon and energy source.

In co-metabolism the microorganisms depend for their survival upon other more accessible carbon sources, which increase their general metabolic activities. Thus, the rate of microbial metabolism could be controlled by changing either the amount of herbicide or other carbon sources added, depending upon the type of microbial degradation activity. In general, incidental transformations (Table I) are the prevalent form of microbial metabolism when the amount of herbicide is low in comparison with other carbon sources (Meikle, 1972; Matsumura and Benezet, 1978).

Thus the two phenomena are of particular interest with regard to microbial decomposition of herbicides: (a) The mechanism by which a soil microbial population develops the capacity to degrade some herbicides (adaptation); (b) The nature of incidental microbial transformations by peripheral metabolic processes (co-metabolism). These two phenomena are examined more closely below.

(*a*) *Adaptation.* Several mechanisms have been proposed to explain how a soil microbial population develops the capacity to degrade a herbicide (Audus, 1960; Hamaker, 1972). The mechanism of initial enrichment could be due to the synthesis of inducible enzymes by responsible microorganisms, to chance mutation by certain strains, or to the presence of organisms with constitutive enzymes. Development of the capability through mutation has been considered unlikely, but cannot be excluded (Senior *et al.*, 1976).

A small initial dose of herbicide is often sufficient to adapt the soil microflora. A second application, made at a higher concentration, will then rapidly be broken down. The adaptation of soil microorganisms to decompose herbicides has been demonstrated with 2,4-D, MCPA, chlorpropham, 2-chloroethyl-*N*-(3-chlorophenyl)carbamate, endothal, TCA, dalapon and chloridazon (Waid, 1972).

The theory of inducible enzymes has been criticized because it cannot explain why the effect of adaptation persists long after the disappearance of the inducer. In natural environments it would be expected that organisms would lose their adaptation as they can change to alternative substrates. But in several different investigations (Fryer and Kirkland, 1970; Hurle and Rademacher, 1970; Kirkland and Fryer, 1972), a faster degradation of 2,4-D and MCPA is observed at a second application several months after the first and for MCPA even several years after the first (Torstensson and Rosswall, 1977). The same phenomenon has also been reported for TCA (Torstensson, 1976).

When the herbicide has been metabolized, the active organism will have to change to other substrates in order to survive. However, Gray (1970) suggested that as most soil organisms live under starvation conditions most of them are dormant. As a result these microbes might retain their adaptation for longer periods. Another suggestion is that the genetic capacity of the microorganisms to degrade certain herbicides is conserved through the action of the infectious transfer from donor cells to recipient cells of genetic factors which control the formation of enzymes decomposing herbicides (Waid, 1972).

It is well known that the resistence of a bacterial cell towards antibiotics can be transmitted to a drug-sensitive strain when the cells of the resistant-donor strain, which possesses a transfer factor, and a suitable recipient strain are in contact. The possibility for this depends on the occurrence in the donor bacteria of small pieces of DNA, which are known as plasmids, encoded for drug-resistence. Plasmids are transmissible from one bacterial cell to another by conjugation. The donor cell makes two identical copies of the plasmid. One copy is retained by the donor while the second copy is transferred to the recipient cell. In this way the plasmid and the genetic information it carries can be transferred to a population of bacterial cells which originally did not possess the plasmid.

The gene coding for many hydrocarbon degradations occur naturally in the form of transmissible or nontransmissible plasmids (Chakrabarty, 1978). There is now additional evidence that degradation of herbicides such as 2,4-D and MCPA may also be specified by plasmid-borne genes (Pemberton and Fisher, 1977). Fisher *et al.* (1978) have shown that the ability to degrade these herbicides by a strain of *Alcaligenes paradoxus* is encoded by a conjugal plasmid.

The findings of Senior *et al.* (1976) that a *Pseudomonas putida* acquired the ability to grow on dalapon through the evolution of an extant dehalogenase is another indication of plasmid transfer in connection with herbicide degradation. Investigation of the role of pesticide-plasmids seems to be of great importance. Hitherto, plasmid studies have been restricted nearly exclusively to bacterial strains. However, Grimaldi *et al.* (1978) found a fungus, *Fusarium oxysporum*, to contain a plasmid-like DNA molecule which can be introduced by transformation into *E. coli* K-12 strains, where it can be maintained as a bacterial plasmid.

The use of herbicides may already have caused a spread of pesticide-plasmids within the soil microbial populations. However, in the event of its occurrence this could be viewed as a "positive" event, in contrast to the spread of drug-resistance plasmids that causes so much trouble in the medical world. More detailed knowledge of the evolution of pesticide-plasmids might open interesting possibilities. Plasmids, containing genetic information for degradation of a specific herbicide, might be transferred in the laboratory to a suitable soil bacterium and then re-inoculated into soil at the time when the particular herbicide is to be degraded.

(*b*) *Co-metabolism.* The term co-metabolism has no strict definition and it has been used in several different ways. According to Horvath (1972), co-metabolism implies the concomitant oxidation of a non-growth substrate during the growth of a microorganism on a utilizable carbon and energy source; it also describes oxidation of non-utilizable substrates by resting cells. The energy derived from the oxidation does not support microbial growth. Horvath and Alexander (1970) have given as an example of co-metabolism the transformation of *m*-chlorobenzoate to 4-chlorocatechol by an *Arthrobacter* sp. Further examples are reviewed by Bollag (1974).

Golovleva and Skryabin (1978) consider co-metabolism to be the process of fermentative conversion of organic compounds carried out by microorganisms only in connection with the use of additional substrates. In their opinion, close interrelations between metabolism of the two substrates is the characteristic and distinctive feature of co-metabolism. As examples they list a number of reactions carried out co-metabolically, including: oxidation of methyl groups of n-alkanes; oxidation of methyl groups of aromatic and heterocyclic compounds; oxidation of cycloalkanes; hydroxylation of aromatic rings; hydroxylation of steroids; acetylation of amino-groups; co-reduction; and cyclization of aliphatic radicals.

Hulbert and Krawiec (1977) examined the descriptive features of co-metabolism. They concluded that none of them differed from those of ordinary microbial catabolism or anabolism and recommend, therefore, that only long-established categories of metabolism should be used in naming the transformations observed.

In studies of pesticide decomposition by microorganisms it seems, however, of practical value to distinguish between degradation by catabolism in connection with adaptation phenomena and that caused by other metabolic transformations. As Skryabin *et al.* (1978) point out, the use of the term co-metabolism for these transformations, though criticized by Hulbert and Krawiec (1977), has focused many investigators on the microbial interactions with several substances and on the interrelation between their metabolism. Therefore, since the phenomena embraced in the term co-metabolism are of great importance for classification of the decomposition of pesticides, use of the term might be justified.

In studies of co-metabolism it is unlikely that new regulatory mechanisms will be found differing from those already known. No specific enzyme mechanisms underlie processes of co-metabolism. Skryabin *et al.* presume that the basis for co-metabolism is the supply of energy, cofactors or metabolites at various levels from the transformation of one substrate to processes such as substrate transport, enzyme biosynthesis or functioning involved in the transformation of a second substrate. The role of the induction–repression or activation–inhibition of enzymes responsible for transformations by intermediates of the co-substrates cannot be excluded. Co-metabolism is connected with the absence of the perfect co-ordination of peripheral metabolic processes which is characteristic of central metabolism.

(*c*) *Extracellular enzymes in soil.* In some instances, enzymes have been isolated from soil microorganisms and found to catalyse the decomposition of specific herbicides (Lanzilotta and Pramer, 1970; Engelhardt *et al.*, 1973). This suggests that enzymes free in the soil may contribute to degradation of pesticides (Cervelli *et al.*, 1978).

Soil enzymes can be divided into cellular enzymes (living microorganisms and soil animals) and extracellular enzymes (released by proliferating microorganisms, dying microorganisms, dying plants, roots of growing plants and soil animals). There is only limited information available on the role of extracellular enzymes, how they are distributed and localized on soil constituents, and their availability for biochemical catalysis (Esser, 1970).

A clear separation of co-metabolic reactions from incidental transformations of herbicides by extracellular enzyme is not always possible, since the corresponding aim-substrate for co-metabolism is often unknown. The resultant metabolites may also be the same.

### 2. *Non-enzymatic Processes*

The processes by which microbial activities contribute to the overall alteration of herbicidal molecules by non-enzymatic mechanisms are not well known. The effect of pH is often neglected in the field of pesticide metabolism despite numerous reports of pH-dependent reactions of relatively labile molecules, both in soil and *in vitro*. Large pH changes (1 to 2 pH units or more) are often associated with microbial activities together with changes in nutritional sources, particularly in aqueous media. Degradation of proteins causes increased alkalinity while with carbohydrate metabolism the pH gets lower.

While the actual occurrence of pH effects resulting from microbial action in nature might be difficult to document, it is certainly easy to demonstrate the phenomenon *in vitro*, where the pH of spent culture media often becomes very low during the decay period. There are ways in which microbial products can promote photochemical reactions (Matsumura and

Benezet, 1978). First, microbial products can act as photosensitizers by absorbing energy from light and transmitting it to the herbicidal molecule (Crosby, 1976). Secondly, microbial products facilitate such photochemical reactions by serving as donors or acceptors, for example of hydrogen and $OH^-$, which are often needed for photochemical reactions (Crosby, 1979).

Microbial production of co-factors that are used in both enzymatic and non-enzymatic reactions is also an aspect that should not be overlooked (Matsumura and Benezet, 1978).

### B. Formation of New Substances

Little attention has been paid so far to the importance of the microbial formation of organic products capable of reacting with pesticides (Bollag, 1974; Matsumura and Benezet, 1978). Such reagents of microbial origin can be postulated to include amino acids, peptides, alkylating agents such as acetyl CoA, methylcobalamine, *S*-adenosylmethionine and organic acids. Tweedy *et al.* (1970) demonstrated an acetylation product, *p*-bromacetanilide, in the products of incubation of *Talaromyces wortmanii* and *Fusarium oxysporum* with metobromuron. Ross and Tweedy (1973) report that the aniline formed from the microbial degradation of chlordimeform is converted to its malonanilic acid derivative. Torstensson (1977) found a conjugate of 3,4-dichloroaniline with acetic acid when identifying metabolites during linuron degradation studies. Some aromatic herbicides are metabolized to structures in which the ring systems are substituted with functional groups such as $NH_2$, OH and COOH. These structures may then be incorporated by polymerization, oxidation, and reduction reactions into humic substances (Martin, 1972).

However, in none of these cases was any proof of an actual non-enzymatic conjugation reaction offered. These types of compound are the ones that have been demonstrated capable of conjugating with reactive groups of chemicals. Production of such compounds by microorganisms could conceivably facilitate the overall rate of microbial alteration of herbicidal chemicals, whether the subsequent conjugation mechanism is actually enzymatic or not.

## III. Methodology

It is difficult to make direct determinations of microbial transformations of herbicides by microorganisms in the environment. Chemical, microbial and other biotic reactions cannot be distinguished readily. Only the broad biological contribution can be assessed *in situ*, using reaction kinetics or sterilized and non-sterilized soils, while the specific processes of metabolism must be studied *in vitro* with isolated species of the natural microbial population.

Initial research in microbial herbicide metabolism was characterized solely by isolating organisms capable of using a compound as their only source of both carbon and energy, but latter it was realized that other processes, such as co-metabolic transformation, conjugation reactions, or the accumulation of a pesticide within a microbe, are important factors of microbial interference.

Investigations of microbial degradation of herbicides may be carried out at different levels of complexity with respect to the total microorganism population in soil, as illustrated in Table II. All levels from single pure cultures to intact soil in the field give useful information on the degradation processes and the influence of different factors on it. Investigations to estimate the role of microorganisms in decomposition of herbicides may proceed stepwise, as is illustrated in Table III. Comments on different points in the schedule are given below.

## A. Enzymatic Versus Chemical Reactions

It is not always easy to distinguish whether herbicide transformation is caused by enzymatic reactions or by purely chemical reactions (Table III, A). In practice, the products of biological, chemical and photochemical activity on herbicides are often the same or similar.

Microbial transformation of herbicides has frequently been inferred from observation of the kinetics of the transformation processes. Where the initial disappearance of herbicide is slow or absent (lag phase) but increases progressively with time and reaches a steady state (usually a logarithmic rate of change) for a period of time, it is deemed likely that the herbicide is used as a microbial substrate (Table III, B). Further applications of the pesticide may be transformed more rapidly, either with or without a reduced lag phase.

It is more difficult to distinguish between co-metabolic, extracellular and chemical transformations. The simplest procedure is to compare persistence of the test substance in sterilized and unsterilized portions of a soil sample, but it is difficult to sterilize soil without modifying its physical structure, chemical composition or catalytic properties.

The most commonly used methods of controlling the biological reactivity of soil organisms are steam sterilization (autoclaving) and irradiation. High energy ionizing radiation may be provided by an electron beam, X-rays or $\gamma$-rays. Of the three latter, the $\gamma$-ray method is the one generally used. An alternative approach has been to inhibit the microorganisms in the soil by

Table II. Investigations of microbial degradation of herbicides at different complexity levels and possibilities for the study of specific phenomena and factors in connection with degradation.

| Complexity level | | Specific phenomena | | | | Different factors | | | | |
|---|---|---|---|---|---|---|---|---|---|---|
| | | | | | | Soil and climate factors | | | | |
| | | Detailed studies of degradation mechanism[a] | Synergistic phenomena | Adaptation | Metabolites appearing | Soil type, $H_2O$ | pH, Temp., Aerob/ anaerob | Aim substrates and other nutrients | Influence of plants and rhizosphere | Influence of Chemical form[b]; Formulation components[c]; Pesticide combinations |
| Laboratory | Pure cultures of microorganisms | * | | * | * | | * | * | | * |
| | Mixtures of microorganisms: | | | | | | | | | |
| | a) two or more known organisms | * | * | * | * | | * | * | | * |
| | b) suspensions of soil or other microorganism-containing matters | | | * | * | | * | * | * | * |
| | Soil incubations in flasks, bags, pots or percolators | | | * | * | * | * | * | * | * |
| Field | Investigations in natural environments | | | * | (*) | (*) | (*) | (*) | (*) | (*) |

[a] e.g. catabolism, enzymes, plasmids
[b] e.g. salt or ester forms
[c] e.g. solvents, emulsifiers, inert fillers, wetters, activators, adsorbents, buffers and foaming agents
(*) only confirmative tests

Table III. Proposed steps in investigation of the role of microorganisms in decomposition of herbicides.

A. Is the transformation biological, chemical or photochemical?
B. Which mechanism underlies a biological transformation?
   - Direct degradation with the herbicide utilized as a source of carbon and energy
   - Co-metabolic degradation
   - Degradation by extracellular enzymes

C.1. Which metabolites could appear?
   - Pure cultures, mixed cultures, in soil
   - Influence of different factors
     - Aerobic/anaerobic conditions
     - Different pH levels
     - Different substrates

C.2. Which factors influence rate of transformation?
   - Soil and climate factors (e.g. temp., pH, $O_2$-levels, $CO_2$-levels)
   - Substrates
   - Chemical form, formulation components, pesticide combinations
   - General biological activity level in soil

D. Are metabolites that appear degradable by microorganisms?
   - Intermediate metabolites (from microorganisms, plants and animals)
   - Conjugation products (from microorganisms and plants)

chemical means. Reviews of the methods used have been given by Pramer and Bartha (1972) and by Hill and Arnold (1978). However, a few points of special interest will be discussed here.

Heat is an effective sterilant, but samples of soil should be autoclaved intermittently on two or three successive days, to allow heat persistent spores to germinate. The duration and need for repetition of treatment will depend on the quantity of soil to be sterilized. Autoclaving is quite a drastic process which may well do more than simply kill the microorganisms, and it is unfortunate that in many cases this is the only direct evidence offered that microbiological degradation has taken place. The high temperatures and pressures of steam sterilization may alter soil surface adsorption characteristics and cause some organic matter degradation.

$\gamma$-Irradiation does not seem to alter the cation exchange capacity, particle size distribution or moisture desorption properties of soil in the same way as autoclaving, but both methods result in an increased extractability of nitrogen, phosphorus, sulphur and manganese. Both the organic matter content of the soil and its moisture content may affect the relative merits of the different sterilization methods.

$\gamma$-Irradiation does not remove all biological activity since enzymic activity derived from microbes and/or plants is retained by the sterilized soil, often for prolonged periods. Comparisons of pesticide degradation in autoclaved and $\gamma$-irradiated soils have shown more rapid degradation in the latter, indicating the presence of enzyme activity in the irradiated soils.

Antibiotics and many organic biocides such as phenol are degraded and of limited use as soil sterilants, but high levels of $HgCl_2$ or $NaN_3$ will effectively suppress respiration and other manifestations of biological activity in soil. Concentrations of inhibitors must be excessive (0·1% or greater) because distribution in soil is never uniform and because compounds are adsorbed and retained on the surfaces of soil colloids. The addition of such quantities may, however, affect the chemical reactivity and adsorptive properties of the soil. It has been suggested that the use of the gaseous sterilant propylene oxide is preferable to both sodium azide and autoclaving for preparing sterile soils for studies of herbicidal transformations. However, propylene oxide-treated soils may be slightly phytotoxic.

Unfortunately, many investigators fail to report whether tests to check sterility have been performed. The slow degradation rate sometimes observed in the sterilized soil may be caused either by incomplete sterilization or unintended reinoculation with microorganisms.

## B. Pure Cultures of Microorganisms

Laboratory studies of the microbial metabolism of herbicides have utilized organisms either obtained from culture collections or isolated from soil, aquatic environments, sewage, rumen content, etc. (Table III,C, 1 and 2). Methods for isolation have been reviewed by Hill and Arnold (1978). The isolation of organisms able to degrade organic compounds has often followed enrichment of the source with the product under study or with a structurally related chemical. Isolation can be carried out under field or laboratory-simulated field conditions, or in the laboratory by using soil suspensions or perfusion techniques. Isolation of cultures of these organisms is generally carried out by dilution and plating techniques on appropriate synthetic media. The methods used rarely isolate more than a small proportion of the viable microbial population and in particular are likely to miss auxotrophic organisms, many non-sporing fungi and organisms with highly specific nutrient or physical requirements. Furthermore, enrichment techniques tend to favour organisms which can metabolize the pesticide as a carbon and energy source and may exclude those capable of cometabolic transformation. Another disadvantage of enrichment procedures is that the chemical is often applied at high concentrations, which may inhibit organisms capable of metabolizing lower concentrations.

The maintenance of laboratory cultures of microorganisms able to metabolize pesticides has not always proved easy. The progressive loss of metabolic ability of many organisms when maintained by regular subculture and growth on or in nutrient media has been frequently noted, even when the medium contains the pesticide. In order to avoid or reduce these losses of activity and to prevent the fall in viability which commonly occurs with some soil organisms maintained on laboratory nutrient media, microbial cultures can be stored either lyophilized (freeze-dried) or at low temperatures (e.g. under liquid nitrogen; at −196°C).

The use of microorganisms for laboratory culture studies of herbicide metabolism, both *per se* and as an aid to the interpretation of field data, is well established. They may also be used as a tool for the production of transformation products useful in identification of unknown products extracted from soil. In this case growth in fermenters under controlled environmental conditions is a valuable technique.

Observations of herbicide degradation in pure cultures of microorganisms, or in other laboratory studies can, however, rarely be extrapolated to field circumstances. Some herbicides which are degraded *in vitro* may not be degraded in the field, because the organisms responsible for their degradation lack competitive ability and are consequently uncommon in that environment. The opposite situation may also occur. Single microorganisms may be unable to attack a herbicide because of the lack of a specific growth factor. Thus, in natural environments it is possible that microbes may act synergistically to effect the degradation of the herbicide. Bordeleau and Bartha (1971) showed that propanil was transformed by the synergistic interaction of two common soil fungi, *Penicillium piscarium* and *Geotrichum candidum.* In certain cases a herbicide can be decomposed only in the presence of two different microbial species (Gunner and Zuckerman, 1968). In a natural ecosystem like the soil, with an abundance of various microorganisms, there is an even greater possibility for combined transformations.

Laboratory studies with pure and mixed microbial cultures have often been used to determine or predict the fate of herbicides in the environment. Such simplified systems are unlikely to reproduce accurately the complex soil environment, where microhabitats do not necessarily reflect the gross natural soil conditions.

## C. Factors Influencing Microbial Activity

The routes and/or rates of microbial transformations of herbicides are influenced by environmental factors, agricultural techniques and the properties of the herbicide. The different factors may themselves be interrelated, causing difficulty in determining their individual effect. Some comments to investigations of the influence of factors mentioned in Table III, C, 1 and 2 follow below.

Studies of the fate of a herbicide in soil must be carried out using soil in conditions that approach those in the field as closely as possible. Soil collected for use in the studies should be handled as living tissue rather than as a geological specimen. Practices such as air-drying, prolonged storage, and freezing and thawing will drastically alter the biochemical capability by inactivating extracellular enzymes and by modifying the composition and density of populations of microorganisms indigenous to soils. Sampling, processing, storing and incubation of soil samples for biodegradation studies is described exhaustively by Pramar and Bartha (1972).

In the temperate zone, biochemical activity of soil is a function of season. It is greatest during late spring to early autumn. This influences the composition of the soil microflora. Likewise, earlier pesticide applications may have caused changes in the microorganism population. Soil samples for herbicide studies must then be taken from localities that have not been treated with pesticides during at least the previous three years.

The temperatures generally accepted as optimal for the activity of microorganisms indigenous to soil range between 20–30°C. In the temperate zone such temperatures are seldom reached in the soil. In studies of herbicides intended for use in these areas, incubations at 10–15°C are more realistic.

When the soil environment becomes anaerobic, herbicide degradation will be a reductive rather than an oxidative process. Aerobic and anaerobic metabolism may have different rates and result in different metabolites (Williams, 1977). Many investigations of degradation in soil may have unwillingly introduced this factor by incubating the moist soil for long periods in sealed jars. If the air space is limited, the soil respiration may quickly exhaust the oxygen. According to Pramer and Bartha (1972) 1 g of an "average" soil could exhaust the oxygen in 1 $cm^3$ of air in 24 h, and they therefore recommend daily aeration of closed vessels where the ratios of vessel volume (ml) to sample weight (g) are 10 or less.

A herbicide subject to co-metabolism does not support the proliferation of the microorganisms and, should the initial number of organisms be small, the reaction will proceed at a slow rate. It is also possible that a slow rate of co-metabolism may be due to the lack of a suitable aim substrate. In the field, with growing crops, some herbicides applied to soil enter and either remain in, or pass through, rhizosphere zones. A herbicide in the rhizosphere may be transformed at different rates or conceivably even by different routes from that in non-rhizosphere soil as a result of the altered soil conditions (e.g. nutrients, oxygen, carbon dioxide, water), the presence of exuded root enzymes, and the metabolic activity of the highly specialized rhizosphere microflora (Rovira and Davey, 1974).

The general biological activity level of soil samples used for studies of microbial herbicide degradation (measured, for example, as $O_2$-consumption or $CO_2$-evolution) should be determined. It provides an indication of the biological activity and thereby the conditions that determine the level of extracellular enzymes in the soil. Alternatively the soil biomass could be estimated (Jenkinson and Powlson, 1976). It should indicate the possible rate of microbial co-metabolism transformation. Direct microbial degradation of herbicides is not dependent on the general biological activity or the total biomass in the soil to the same extent as co-metabolic degradation.

The use of radio-labelled herbicides allows the greatest degree of methodological flexibility (Hill and Arnold, 1978) and in closed systems (either sealed or with a gas flow) it enables the detection of both non-volatile and volatile herbicides and metabolites.

## IV. Research Priorities

### A. Mechanism of Decomposition

There are chemical and physical factors that influence the fate of a herbicide in the environment, but the least predictable factor is decomposition by microorganisms. Because of this there is a need to investigate further the actual biochemical reactions involved in microbial herbicide metabolism. This will provide a basis for the understanding of their short or long persistence in a natural environment. It will also contribute to clarifying the relationship between chemical structure of herbicides and their possible microbial transformations.

### B. Methods

To assess the contribution of the soil microbial population to herbicide decomposition, additional methods of eliminating, controlling or reducing microbiological activity in soil are necessary. Methods such as fumigation, $\gamma$-irradiation, the use of specific microbial inhibitors, enrichment or isolation of effective microorganisms should be used to augment the more traditional methods.

In studies of details in microbial herbicide degradation mechanisms, we still have to rely upon work with single isolated microorganism species. These species should be isolated from real application situations and not from laboratory enrichment experiments under special conditions and high herbicide doses. By not doing so, we take the risk of working with species that turn out to be of no importance in the real herbicide application situation.

Microorganisms do not function as individual species in the environment, but as members of a community with great diversity and where a multitude of interrelations between members exist. This has to be considered when designing experiments to evaluate the role of microorganisms in decomposition of herbicides.

Laboratory incubations should be designed to produce, as closely as possible, the conditions existing in the proposed field application situation. This applies particularly to factors like temperature, pH, $O_2$-level, moisture and available nutrients. Temperature in soil in the temperate zones is seldom as high as the incubation temperatures used in laboratory studies today. In soil pH is normally estimated as a mean value of many microsites. Under field conditions it is well buffered during the period of herbicide degradation. However, in many reported laboratory incubation systems there is no provision for compensation for pH changes in the medium. Another aspect that must be considered is that of aerobic-anaerobic conditions. In a well aerated soil there are also sites with anaerobic conditions. Unintentionally introduced anaerobic conditions in the laboratory by incubation for long periods in sealed jars should be avoided. The moisture level in soil varies during the growing season and is seldom at the optimal level that is generally used in laboratory incubations. Also available nutrients in soil vary with time of sampling.

Obviously there seem to be many reasons for evolving methods in which environmental factors could be closely controlled. because microorganisms respond to changes of environmental factors, the temperature, pH, $O_2$-level, moisture, nutrients and other factors in investigations should then be adjusted to conditions predominant in field soil during the actual application situation. Failure to give realistic values of these factors to laboratory incubations, create results useless for prediction of rates and/or routes of herbicide decomposition.

The soil factors that influence rates of degradation are themselves interrelated, making it difficult to determine the effect of each individually. However, more work is clearly needed before the effects of different soil factors on microbiological activity of the degradation of herbicides can be expressed on a quantitative basis.

There may, moreover, also be soil properties that are better related to microbial degradation of herbicides than those normally considered. Hamaker (1972) mentioned titratable acidity, zymogenic respiration and adsorption properties. Soil respiration could be a useful index for microbiological potential for degrading herbicides. It might be necessary to distinguish between basal respiration, $CO_2$-evolution or $O_2$ consumption of the moist soil as it is, and zymogenic respiration (after addition of excess nutrients). Estimation of soil biomass may also be of value.

Another problem that has so far met with little interest is the possibility that soil organisms other than bacteria and fungi are responsible for herbicide decomposition. Such organisms include algae, protozoa and members of the soil fauna. The possibility that herbicides are incorporated into soil organisms by an active or passive accumulation mechanism provokes special concern, since this, as in conjugation reactions, only temporarily removes a toxic compound. All these problems need the development of new methods, which is also the case for research on the importance of extracellular enzyme activity for herbicide decomposition in soil.

Conjugation reactions are syntheses by which a herbicide, or any of its metabolites, is combined with naturally occurring compounds. Conjugations of herbicides are common and frequent reactions in higher organisms, but they have not been found to a similar extent in microorganisms. One explanation is that they may not be frequent here. Another is that methods used for extraction of herbicide metabolites seldom take into consideration conjugates, which have other chemical properties. Nevertheless, this mode of transformation requires further exploration, since a herbicide, or an intermediate of it, is only bound to another molecule. A conjugated compound can easily be cleaved again, and the released chemicals can subsequently exert a toxic influence.

## V. Conclusions

The role of microorganisms in decomposition of herbicides is well documented. Microbial transformations with different underlying biochemical mechanisms are the main route of disappearance of these chemicals from soil. Rate and/or route of degradation is largely controlled by the environmental factors that normally influence microbial activity.

To improve understanding of the role of microorganisms in decomposition, future research activities on mechanisms of microbial degradation of herbicides should be concentrated on:

The way a microbial population develops a capacity to degrade herbicides (induction of enzymes, mutation, plasmids, etc.) and the period this capacity is retained under field conditions.

The process of co-metabolism in microbial transformations with reference to its underlying mechanisms (aim substrates, induction-repression of enzymes) and transformation capabilities (capacity under different environmental conditions).

Future research activities on methods for assessment of the role of microorganisms in decomposition of herbicides should concentrate on:

Methods of eliminating or controlling microbial activity in soil.
Procedures and substrates for isolation of active herbicide-degrading microorganisms in the soil environment.
Methods for incubation experiments in which environmental factors of importance for microbial activity (e.g. temperature, pH, $O_2$-level, moisture, nutrients) can be closely controlled and given values that are typical of the proposed application situation of the herbicide. Rhizosphere conditions should be especially considered.
Methods to evaluate the influence of interrelated factors (e.g. pH and organic matter) on microbial herbicide degradation.
Methods to evaluate the possible usefulness of other factors (e.g. titratable acidity, respiration, biomass, adsorption properties of the herbicide) than those normally referred to for prediction of microbial capacity to degrade herbicides.
Methods to evaluate the possible importance of other soil organisms than bacteria and fungi (e.g. soil algae, protozoa, members of the soil fauna) and extracellular enzymes on decomposition of herbicides. Methods to evaluate incorporation (active or passive) of herbicides in soil organisms are also desirable.
Methods for study of and evaluation of the importance of microorganisms in formation of conjugates between herbicides and naturally occurring compounds, and also for study of decomposition of this and other conjugates formed by other organisms (e.g. formed by and exuded via roots of higher plants).

For registration, data are needed that define clearly the importance of microorganisms as decomposers of the substances compared with other possible transformation reactions (non-sterile conditions and sterile conditions created by different agencies). In the case of herbicides mainly decomposed by microbial reactions, incubation of soil samples should take into consideration those factors influencing the activity of microorganisms. This means that conditions in the different soil types to which the herbicide is likely to be applied, should be reproduced in the incubated soil. The environmental factors that could be studied include temperature, pH, $O_2$-level (aerobic/anaerobic conditions), moisture and various nutrient levels. The role of microbial decomposition of intermediate metabolites and conjugation products from microorganisms and plants may need to be evaluated. Existing registration requirements generally cover these points if the data is evaluated sensibly.

## References

Audus, L. J. (1960). Microbial breakdown of herbicides in soils. In *Herbicides and the Soil.* (E. K. Woodford, and G. R. Sager, Eds.) pp. 1–19. Blackwell Scientific Publications, Oxford.

Bollag, J.-M. (1974). Microbial transformations of pesticides. In *Advances in Applied Microbiology*. (D. Perlman, Ed.) **18**. pp. 75–130. Academic Press, New York and London.

Bordeleau, L. M. and Bartha, R. (1971). Ecology of a herbicide transformation: synergism of two soil fungi. *Soil Biol. Biochem.* **3**, 281–284.

Cervelli, S., Nannipieri, P. and Sequi, P. (1978). Interactions between agrochemicals and soil enzymes. In *Soil Enzymes*. (R. G. Burns, Ed.) pp. 251–295. Academic Press, London and New York.

Clarke, P. H. (1972). Induction of bacterial enzymes. *Biochem. J.* **130**, 47–48.

Chakrabarty, A. M. (1978). Molecular mechanism in the biodegradation of environmental pollutants. *ASM News*, **44**, 687–690.

Crosby, D. G. (1976). Nonbiological degradation of herbicides in the soil. In *Herbicides: Physiology, Biochemistry, Ecology*, (L. J. Audus, Ed.) 65–97. 2nd Edition, Academic Press, London.

Crosby, D. G. (1979). The significance of light induced pesticide transformations. In *Advances in Pesticide Science*, Symp. Pap. Int. Congr. Pestic. Chem. (H. Geissbühler, Ed.) 568–576. 4th part 3. Pergamon Press, Oxford.

Engelhardt, G., Wallnöfer, P. R. and Plapp, R. (1973). Purification and properties of an aryl acylamidase of *Bacillus sphaericus*, catalyzing the hydrolysis of various phenylamide herbicides and fungicides. *Appl. Microbiol.* **26**, 709–718.

Esser, H. O. (1970). The biodegradation of pesticides in the soil. *Meded. Rijksfac. Landbouww. Gent*, **35**, 753–758.

Fisher, P. R., Appleton, J. and Pemberton, J. M. (1978). Isolation and characterization of the pesticide-degrading plasmid pJP1 from *Alcaligenes paradoxus*. *J. Bacteriol.* **135**, 798–804.

Fryer, J. D. and Kirkland, K. (1970). Field experiments to investigate long-term effects of repeated applications of MCPA, tri-allate, simazine and linuron: Report after six years. *Weed Res.* **10**, 133–158.

Golovleva, L. A. and Skryabin, G. K. (1978). Cometabolism of foreign compounds. "Symp. Environmental Transport and Transformation of Pesticides." EPA-600/9-78-003. 73–85.

Gray, T. R. G. (1970). Microbial growth in soil. *Proc. Int. Symp. Pesticides in the Soil*, East Lansing. 36–41.

Grimaldi, G., Guardiola, J. and Martini, G. (1978). Fungal extrachromosomal DNA and its maintenance and expression in *E. coli* K-12. *TIBS*, Nov. 248–249.

Gunner, H. B. and Zuckerman, B. M. (1968). Degradation of "Diazinon" by synergistic microbial action. *Nature,* **217**, 1183.

Hamaker, J. W. (1972). Decomposition: quantitative aspects. In *Organic Chemicals in the Soil Environment*. (C. A. I. Goring and J. W. Hamaker, Eds.) Vol. I. 253–340. Marcel Dekker, New York.

Hill, I. R. and Arnold, D. J. (1978). Transformations of pesticides in the environment—The experimental approach. In *Pesticide Microbiology*. (I. R. Hill and S. J. L. Wright, Eds.) pp. 203–245. Academic Press, London and New York.

Horvath, R. S. (1972). Microbial co-metabolism and the degradation of organic compounds in nature. *Bact. Rev.* **36**, 146–155.

Horvath, R. S. and Alexander, M. (1970). Cometabolism of *m*-chlorobenzoate by an *Arthrobacter*. *Appl. Microbiol.* **20**, 254–258.

Hulbert, M. H. and Krawiec, S. (1977). Cometabolism: A critique. *J. theor. Biol.* **69**, 287–291.

Hurle, K. and Rademacher, B. (1970). Untersuchungen über den Einfluss langjährig wiederholten Anwendung von DNOC und 2,4-D auf ihren Abbau im Boden. *Weed Res.* **10**, 159-164.

Jenkinson, D. S. and Powlson, D. S. (1976). The effect of biocidal treatments on metabolism in soil—V. A method for measuring soil biomass. *Soil Biol. Biochem.* **8**, 209–213.

Kirkland, K. and Fryer, J. D. (1972). Degradation of several herbicides in a soil previously treated with MCPA. *Weed Res.*, **12**, 90–95.

Lanzilotta, R. P. & Pramer, D. (1970). Herbicide transformation II. Studies with an acylamidase of *Fusarium solani. Appl. Microbiol.*, **19**, 307–313.

Martin, J. P. (1972). Side effects of organic chemicals on soil properties and plant growth. In: *Organic Chemicals in the Soil Environment.* (Goring, C. A. I. and Hamaker, J. W., (Eds.) 733–792. Marcel Dekker, New York.

Matsumura, F. and Benezet, H. J. (1978). Microbial degradation of insecticides. In: *Pesticide Microbiology.* (Hill, I. R. and Wright, S. J. L. Eds.), pp.623–667. Academic Press, London and New York.

Meikle, R. W. (1972). Decomposition: Qualitative relationships. In: *Organic Chemicals in the Soil Environment.* (Goring, C. A. I. and Hamaker, J. W. Eds.) pp. 145–251. Marcel Dekker, New York. 145–251.

Pemberton, J. M. and Fisher, P. R. (1977). 2,4-D plasmids and persistence. *Nature*, **268**, 732–733.

Pramer, D. & Bartha, R. (1972). Preparation and processing of soil samples for biodegradation studies. *Environmental Letters*, **2**, 217–224.

Ross, J. A. and Tweedy, B. G. (1973). Malonic acid conjugation by soil microorganism of a pesticide—derived aniline moiety. *Bull. Environ. Contam. Toxicol.*, **10**, 234–236.

Rovira, A. D. and Davey, L. B. (1974). Biology of the rhizosphere. In: *The Plant Root and Its Environment.* (Carson, E. W. Ed.) pp. 153–204. University Press of Virginia, Charlottesville.

Senior, E., Bull, A. T. and Slater, J. H. (1976). Enzyme evolution in a microbial community growing on the herbicide dalapon. *Nature*, **263**, 476–479.

Skryabin, G. K., Golovleva, L. A. and Golovlev, E. L. (1978). *Co-metabolism: Biological sense of the phenomenon.* Scientific Centre of Biological Research, Academy of Sciences of the USSR in Pushchino. 3–17.

Torstensson, L. (1976). Effects of some factors on persistence of TCA in arable soil. *Weeds and Weed Control, 17th Swedish Weed Conf.*, Uppsala. K8–13.

Torstensson, L. (1977). Microbial degradation of linuron. *Weeds and Weed Control, 18th Swedish Weed Conf.*, Uppsala. 16–10.

Torstensson, N. T. L. and Rosswall, T. (1977). The effect of 20 year's application of 2,4-D and MCPA on the soil-flora. *Int. Symp. Interaction Soil Microflora Environmental Pollut.* Vol. I. Pulawy. 170–176.

Tweedy, B. G., Loepphy, C. and Ross, J. A. (1970). Metobromuron: Acetylation of the aniline moiety as a detoxification mechanism. *Science*, **168**, 482–483.

Waid, J. S. (1972). The possible importance of transfer factors in the bacterial degradation of herbicides in natural ecosystems. *Residue Rev.*, **44**, 65–71.

Williams, P. P. (1977). Metabolism of synthetic organic pesticides by anaerobic microorganisms. *Residue Rev.*, **66**, 63–135.

# 7 Plant Availability and Uptake of Herbicides from Soil

*R. R. Schmidt*
Bayer AG, Pflanzenschutz,
Biol. Forschung
D-5090 Leverkusen

*W. Pestemer*
Biologische Bundesanstalt
für Land und Forstwirtschaft,
Institut für Unkrautforschung
D-3300 Braunschweig

---



## I. Introduction

The availability of herbicides in the soil to plants depends on many factors including the physico-chemical properties of the compound, climatic and soil conditions, the activities of microflora and -fauna in the soil and the method of application all of which influence the fate of the active compound. They all exist simultaneously and influence each other, resulting in a complex dynamic behaviour.

Details are dealt with in further chapters of this book and in several reviews (e.g. Kearney and Kaufman, 1976; Audus, 1976; Hurle, 1975).

Herbicides must first be taken up into the plant before becoming active. Depending on the type of application different sites of uptake are possible. From post emergence application the compound penetrates into the plant via the shoot and the leaves, but root uptake might also occur under these conditions. After pre-sowing or pre-emergence application, i.e. after application of the herbicide onto or into the soil, the active compound may be taken up via the seed, the embryo, the seedling, or the adult plant. The herbicide may be taken up actively or may passively penetrate into the plant. This article will describe the availability, sites of uptake and mechanisms of uptake of soil applied herbicides.

## II. Availability for Uptake of Soil-applied Herbicides

### A. Influence of Pore Space Distribution and Water Content of the Soil on the Availability of Herbicides to Plants

The pore size distribution and the related water content available to the plant are of special importance for the availability to plants of herbicides in different soils.

The soil water available to the plant is usually considered to be that held between wilting point and field or water holding capacity. Primarily only the herbicide dissolved in the available water may be taken up by plants. This water occupying the middle pores (pF 2·5–4·2) has been termed "effective water content" by Green and Obien (1969) when considering the action of soil applied herbicides. However, molecular diffusion in the soil water and soil air may, depending on temperature and vapour pressure, change the fraction available to the plant by drawing on the fine and coarse pores. The latter pores might partially be filled with air. The dependence of the herbicide equilibrium and the related herbicidal amount available

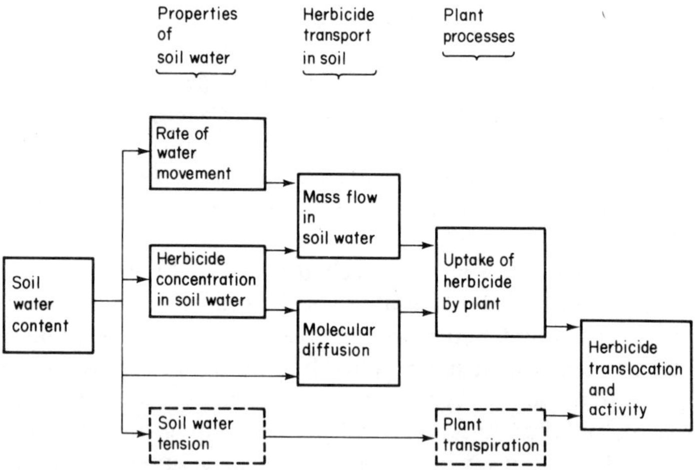


Fig. 1. Schematic diagram showing probable mechanisms by which soil water content influences the activity of a soil-applied herbicide at water contents above the wilting point (after Green and Obien, 1969).

to the plant upon the water content and upon the movement of water in the soil is presented in Fig. 1.

Water and air conditions of a soil depend on the climate and on the landscape, but primarily on the soil type and soil structure and texture. Water movement by mass-flow in soil may occur in the liquid phase following a potential gradient from places with high to places with low water potential (Rowse *et al.*, 1978; Ehlers, 1978). The movement may even oppose the direction of gravity. In addition transport may occur as a gas following gradients of decreasing vapour pressure.

The distribution of a herbicide among the solid, liquid and gaseous phase can be calculated according to Riley (1978). Three equilibrium constants are required (Fig. 2).

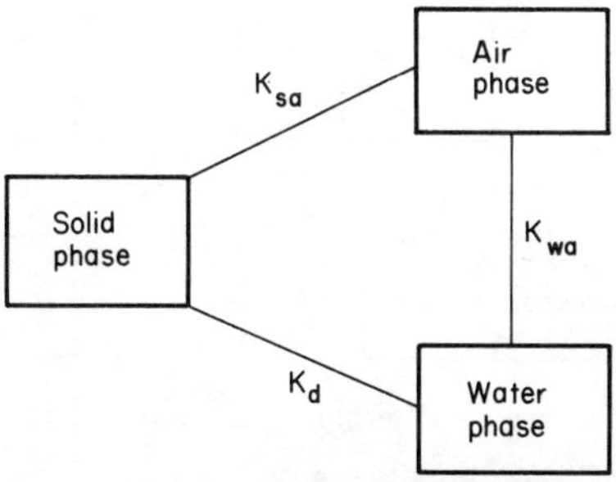


$K_d$ is normally expressed as $\frac{\mu g \text{ pesticide/g soil}}{\mu g \text{ pesticide/ml water}}$

$K_{wa}$ is normally expressed as $\frac{\mu g \text{ pesticide/ml water}}{\mu g \text{ pesticide/ml air}}$

$K_{sa}$ is normally expressed as $\frac{\mu g \text{ pesticide/g soil}}{\mu g \text{ pesticide/ml air}}$

Fig. 2. Sorption equilibria in soil (Riley, 1978).

Herbicides with low $K_{wa}$-values (e.g. trifluralin) move primarily by diffusion via the gaseous phase while compounds with high $K_{wa}$-values (e.g. diuron, atrazine) move by mass transport in the water phase. When water mass-flow does not occur, transport by molecular diffusion can be significant, at least over short distances.

## B. Influence of Adsorption, Desorption and Degradation on the Availability to Plants

The key processes adsorption, desorption and degradation determine the availability of chemicals to plants in the soil (Kerpen and Schleser, 1977).

These processes are time dependent as shown in Fig. 3. This scheme illustrates that only a part of the amount of herbicide originally applied is available for the plant.

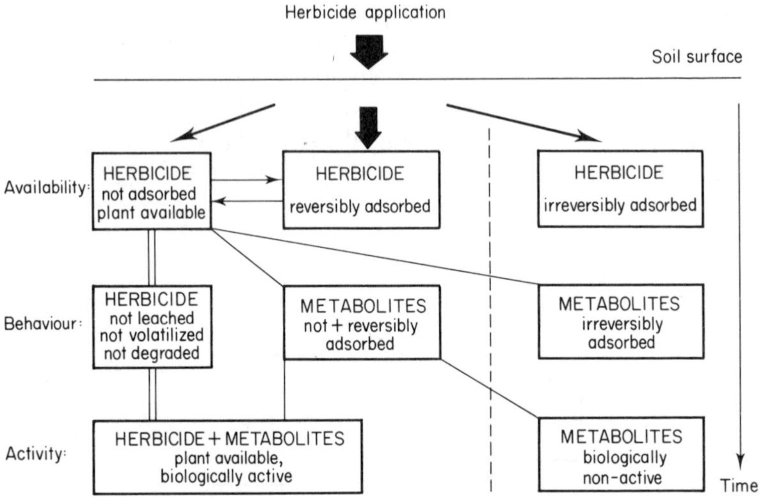


Fig. 3. Main factors determining plant availability.

## III. Methods for Measuring Availability and Uptake of Herbicides

A number of methods are available to measure both the total quantity of herbicide in the soil and the quantity available to plants. The methods in the first group are the standard physico-chemical methods used in residue analysis whereas for the second group biological tests are preferred.

### A. Physico-chemical Methods (Instrumental Methods)

These have been surveyed e.g. by Hance and McKone (1976) and by Jarczyk (1977). For the measurement of herbicides in availability and uptake studies chromatographic and radiochemical methods are the most important. Physico-chemical methods should preferably be employed when bioassays are impossible or impracticable, e.g. when a herbicide is bio-available, but the concentration is too low for quantifiable biological action.

Other sources of information on chromatographic methods include Blinn (1971), Yip (1971), Byast *et al.* (1977), Ebing (1970–1978) and Anon. (1969–1979).

Radiochemical assays are particularly useful for studies of the uptake and distribution of the herbicide, and possibly also its metabolites, in the plant. They may be qualitative by autoradiography, or quantitative by liquid scintillation counting (e.g. Crafts and Yamaguchi, 1964; Lavy *et al.*, 1972).

### B. Bioassays

Bioassays may be conducted *in situ*, *in vivo* or *in vitro*. Reviews have been published for the detection of herbicides in the soil by Hance and McKone (1976) and Hurle (1977), for quantitative bioassays of photosynthetic inhibitors in soils by Pestemer (1976), for bioassays for herbicide development by Gerber (1977), and for the application of bioassay techniques to herbicide investigations by Horowitz (1976).

The following methods may be used:

1. *Bioassays in water culture* (nutrient solutions or soil extracts)

(a) rapid tests (e.g. isolated chloroplasts, isolated mitochondria, Chlorella algae, leaf disc sinking test; $CO_2$- or $O_2$-measurements either manometrically or polarographically, $CO_2$-measurements by infrared gas analysis)

(b) growth tests—visual evaluation only (no special equipment required)
—quantitative measurements

2. *Bioassays in hydroponic culture* (nutrient solutions or soil extracts) using growth test as above (1.b)

3. *Bioassays in soils* which may be

(a) rapid tests (employing soil extracts for the tests as in 1.a) or

(b) growth tests as above (1.b).

Comparative studies of instrumental and bioassay methods are listed by Hurle (1977). They showed in some cases satisfactory agreement, but residues determined chemically were sometimes higher or lower. Various factors can explain these discrepancies (Eberle and Gerber, 1976).

## IV. Methods for Studying Availability and Uptake of Soil-applied Herbicides

The majority of studies conducted so far have involved application of the herbicides to particular plant parts followed by measurement of effects in these plant parts or the whole plant. However, the site of herbicide uptake does not necessarily coincide with the site of action. Uptake studies with intact plants may be conducted in several ways.

## A. Treatment via Water Culture

The great advantage of water culture systems is that the compound is fully available to the plant roots as far as the water solubility allows.

To obtain homogeneous plant material the seeds of the test plant can be pregerminated in either aqueous sand, vermiculite, perlite or on filter paper. After the cotyledons have unfolded and before the first true leaf is differentiated, seedlings uniform in size and development are transferred to the herbicide containing nutrient solution. The reserves of the endosperm will usually be depleted at this stage and an immediate herbicidal response may be obtained. The nutrient solution should be aerated; if not, the upper root parts should then not be submerged.

Several nutrient solutions for uptake studies are suitable (Hoagland and Arnon, 1950; Hewitt, 1952; Penningsfeld and Kurzmann, 1966).

## B. Treatment via Hydroponic Culture (non-adsorbing media)

The media usually employed include vermiculite, perlite, cellulose or agar-agar. Quartz sand is not ideal because of difficulties in maintaining both aeration and an adequate water supply.

The test plants may either be seeded directly or transferred as seedlings after pregermination. The herbicides may either be mixed with the medium or applied with the nutrient solution.

Herbicidal treatments via water culture or hydroponic culture are generally used when herbicide uptake and distribution in the plant are to be measured.

## C. Treatment in Soil (sorptive substrate)

### 1. *Surface treatment and uniform incorporation of herbicides*

This application is similar to the treatment via hydroponic culture. When working with soils low in nutrients, fertilizers may be required. A constant water content in the soil during the experiment may be obtained by daily weighing, by water supply via a peristaltic pump or by capillarity via glassfibre or cotton wicks. This is of special importance when working with small pots.

The herbicide may be applied by surface treatment or may be incorporated into the soil. In the latter case the amount may be expressed per unit weight (mg/kg soil) or preferably per unit volume (mg/l soil). The test plants may be seeded or may be transplanted.

Several methods have been described for studying herbicide uptake separately by the seed, the root and the shoot.

### 2. *Treatment of layers with the herbicide*

A herbicide treated layer of soil of the thickness corresponding with the root, seed or shoot zone is used while the remaining soil layers are untreated. Movement of the herbicide out of the treated layer may be prevented by plastic foil (Appleby and Furtick, 1965), by the dual-pot-method (Eshel and Prendeville, 1967) or by layers of activated carbon (charcoal) (Gray and Weierich, 1969). These measures should ensure that uptake of the herbicide occurs only by the plant parts present in the treated layer.

### 3. *Split root technique*

This is the method of choice when root inhibition is of interest: The primary roots of seedlings (e.g. oats) are removed, the secondary roots are divided by a plastic shield and one half is grown in herbicide treated soil and the other half is kept in untreated soil and serves as control (Deli and Warren, 1970). However, when systemic herbicides are considered the results may not be reliable.

## D. Estimation of the Fraction of Herbicides in the Soil Available to Plants

### 1. *Sampling soil water with ceramic filters*

Water held between field capacity and wilting point may be sampled by applying suction to ceramic filters. The experiments by Green and Obien (1969) have shown that herbicide concentrations in such samples agree reasonably well with values calculated on the assumption that water held below wilting point (~pF 4·2) does not participate in equilibrium with the applied herbicide.

### 2. *Extraction of available herbicide from soil with water*

Available herbicide may be extracted from soil by shaking a sample of 100 g with 100 ml water for 1 h. The concentration of herbicide in this extract is proportional to the quantity available to the plant (Stalder and Pestemer, in press).

The available fraction of a compound in the soil may also be calculated by comparing the dose-response-curve in the soil with the curve obtained in a non-adsorbing substrate (Pestemer, 1975 and 1977; Stalder and Pestemer, in press).

## V. Sites of Uptake

Since the majority of the economically important weeds belong to the Spermatophyta, this review will concentrate on the uptake of herbicides by these plants. However some species from the taxonomic group of mosses (Bryophyta), ferns (Pteridophyta) and algae (Phycophyta) also have some importance as weed plants and uptake by these groups will be discussed briefly as well.

### A. Seed

#### 1. *General*

Herbicide uptake by the seed is very much influenced by the physiological state of the seed, i.e. whether the seed is dry or imbibed, dormant or germinating.

#### 2. *Seed coat*

The first barrier to be met by the herbicide is the seed coat. This is crossed by a physical process independent of the physiological state of the seed (Ebert and Dumford, 1976; Rieder *et al.*, 1970). In contrast to herbicide uptake by the seed or the embryo, which is correlated with the uptake of water and varies with temperature, the uptake of herbicides by the seed coat is nearly independent of the water uptake (Hocombe, 1968).

The uptake into the seed coat mainly occurs by diffusion (Scott and Phillips, 1971 and 1973). However, it is not known whether this diffusion occurs through special preformed spaces in the seed coat or whether the permeability of the seed coat as a whole changes during its residence in the soil.

#### 3. *Seeds as "sinks"*

Many seeds act as "sinks" for certain herbicides in the soil in that they accumulate some compounds to concentrations very much in excess of those existing in the soil (Scott and Phillips, 1971 and 1973). Detailed investigations of which seeds are best as "sinks" for which herbicides, have not yet been reported. However, Scott and Phillips (1971) found that the concentration of herbicides in the seed increased as seed size and radii decreased.

### 4. *Lipid content*

The lipid fraction may play an important role in herbicide uptake by seeds, as shown for trifluralin by Hilton and Christiansen (1972). The idea is that trifluralin is trapped in the lipid compartments and is therefore no longer available for herbicidal action. Related data have been obtained by Phillips *et al.* (1972) for the absorption of several herbicides by soya beans where the herbicide accumulation was correlated with the seed oil content.

## B. Seedlings and Adult Plants

In addition to uptake by seeds the herbicide may be taken up later by the seedling after cracking of the seed coat or by the adult plant either via the root or via the shoot.

### 1. *Shoot*

(*a*) *General.* Shoot uptake of soil herbicides is possible via the primary shoot or via secondary parts such as shoot-borne roots, rhizomes, stolons and tubers. Whether the uptake occurs via primary or via secondary shoot organs mainly depends on the age of the seedling and on the morphological, anatomical and physiological stage of the plant.

Leaf uptake generally may be neglected with soil-applied herbicides. However, volatile compounds may cause crop damage by uptake through aerial parts.

The following sections describe the organs mainly considered to be primary sites of uptake during the germination phase.

(*b*) *Dicotyledoneae.* The seedling of dicotyledoneous plants consists of hypocotyl, epicotyl and cotyledons, all of which may take up herbicides.

(i) Hypocotyl. The transport of herbicides to the hypocotyl and subsequent entry occur primarily by diffusion, as has been shown for *s*-triazines with *Raphanus* (Shone and Wood, 1976).

An interesting observation has been reported where a pretreatment with DCPA increased the subsequent uptake of $^{14}$C-DCPA, presumably because permeability is increased by this pretreatment (Nishimoto and Warren, 1971a).

An especially important site of uptake appears to be the hypocotyl hook (Barrentine and Warren, 1971b).

(ii) Cotyledons. There are only a few reports of the uptake of herbicides by cotyledons, which include napropamide (Hopp, 1972), bifenox (Leather and Foy, 1978) and trifluralin and nitralin (Barrentine and Warren, 1971a and b).

(iii) Epicotyl. The uptake of herbicides by the epicotyl may be easier to measure than uptake by cotyledons. This is especially true for the Leguminosae as their epicotyl may elongate considerably. It appears that different parts of the epicotyl differ in the amount they can take up as well as in their sensitivity (Eshel *et al.*, 1975; O'Brien and Prendeville, 1972; O'Donovan and Prendeville, 1975). These investigations have also shown that different herbicides are taken up at different places on the epicotyl, which may explain differences in sensitivity and selectivity. Whether the epicotyls are still growing or not should also be important, since a differentiated tissue presumably takes up less herbicide than a meristematic one.

(*c*) *Monocotyledoneae.* Information on the sites of uptake by Monocotyledoneae is mostly confined to the Gramineae (Appleby *et al.*, 1965; Knake and Wax, 1968; Parker, 1966; Prendeville, 1968; Prendeville *et al.*, 1968; Gray and Weierich, 1969). Very little information exists for other monocotyledonous families (Liliacea, Iridaceae, Juncaceae, Typhaceae, etc.). However, some work has been done with Cyperaceae (see Tubers in the next section). The shoot parts important for herbicide uptake by the Gramineae are the coleoptile and the mesocotyl.

(i) Coleoptile. The coleoptile has been shown to be the site of uptake and/or action for dinitroanilines (Barrentine and Warren, 1971a), acetamides (Chandler *et al.*, 1974), thiocarbamates (Nalewaja, 1968; Prendeville *et al.*, 1968; Appleby *et al.*, 1965) and piperophos (Tylson, 1979). The amount taken up has been shown to differ between the coleoptile tip and base (Barrentine and Warren, 1971b).

(ii) Mesocotyl (= subcoleoptilar internode = first internode) Uptake via the mesocotyl has been described for the thiocarbamates (Nishimoto and Warren, 1971b; Prendeville, 1968). Some dinitroanilines may also be taken up via the mesocotyl, but herbicidal activity is less from this route than by root uptake (Barrentine and Warren, 1971b).

Knowledge of possible differences in the uptake of individual herbicides via the coleoptile or the mesocotyl in different plant species may be significant as a source of selectivity. The exact location of the growing point may be also important, as for example in the case of tri-allate in *Avena fatua* and in wheat (Fryer and Makepeace, 1977). In *Avena* the mesocotyl extends to place the sensitive zone in the herbicide-treated soil, whereas the extension of the mesocotyl of wheat is less so that the sensitive zone remains below the treated layer of soil.

(*d*) *Secondary sites of uptake.* These include all plant parts not directly belonging to the primary shoot, i.e. the stem axis and the leaves. They are (i) stem-borne roots and (ii) secondary shoot formations above or below the ground primarily functioning as depository or storage organs such as rhizomes, tubers, bulbs and stolons.

(i) Stem-borne roots. Some herbicides which normally are taken up by the root may also act after shoot-application. Detailed investigations have, however, shown that not the shoot but that stem-borne roots were responsible for the herbicide uptake after these applications. Nishimoto and Warren (1971b) for example were able to show that in the case of maize and *Sorghum* trifluralin inhibited the uptake of diuron and ametryne by suppressing the formation of stem-borne roots. Similarly O'Donovan and Prendeville (1975) showed that trifluralin and nitralin inhibited stem-borne roots of *Vicia sativa* and hence reduced damage from shoot-zone applications of atrazine, prometryne, simazine, diuron and linuron. The same explanation presumably applies to the reduction of metribuzin phytotoxicity by trifluralin in soya beans (Ramthun and Talbert, 1975; Ladlie *et al.*, 1977; Moomaw and Martin, 1978). The action of some antidotes may in part also be explained by a similar effect.

When investigating the shoot uptake of herbicides, therefore, care is needed to differentiate clearly between uptake by the shoot itself and uptake by secondary stem-borne roots. Some apparent contradictions (Knake and Wax, 1968; Eshel, 1969) may possibly be explained in that way.

(ii) Rhizomes. Rhizomes not only permanently bear roots, either all around or only on the lower side, but also colourless, membraneous bracts. Both organs may function as sites of herbicide uptake. Some important rhizome-building weed grasses are *Agropyron repens*, *Agrostis gigantea*, *Cynodon dactylon* and *Sorghum halepense*. While no information is available for the last three, detailed investigations have been published for the uptake of TCA (e.g. Fryer and Makepeace, 1978) and pronamide (Carlson *et al.*, 1975) by the rhizomes of *Agropyron repens*.

(iii) Stolons. No references have been found for the possible uptake by this organ. However, stolon-uptake of herbicides may play a role in the case of *Achillea millefolium*, *Cynodon dactylon*, *Poa trivialis* and *Ranunculus repens*. Also perennial plants with horizontal, creeping shoots, which differ morphologically from true stolons, may be mentioned here, e.g. *Agrostis stolonifera*, *Ajuga repens* and *Veronica filiformis*. The control of these plants with soil herbicides is difficult because of very limited uptake.

(iv) Tubers. Among the most important tuberous weeds the Cyperaceae should be considered first. Their tubers are formed by a dense net of rhizomes; the tubers closest to the soil surface may form new shoots (Holm *et al.*, 1977), while the deeper tubers may form rhizomes with "basal bulbs". These "basal bulbs" also may form shoots, rhizomes and roots, all of which may take up herbicides. In the case of alachlor, for example, good herbicidal action against *Cyperus* has been demonstrated after uptake via the shoot or via the parts of the rhizome above the tuber and below the "basal bulbs"

(Armstrong *et al.*, 1973). On the other hand, butylate and dichlobenil showed their greatest activity after application in the tuber zone (Ingle and Worsham, 1971).

In the case of tuber-bearing crop plants special attention should be paid to (a) the possibility of herbicidal uptake via the tuber with resulting phytotoxicity and (b) the possible accumulation of residues. This applies to the Brassicae (e.g. kohlrabi, radish), beets (fodder beet, beetroot, sugar beet in part) and potato tubers.

(v) Bulbs. These organs, representing extremely shortened shoots with thick, fleshy bracts, are mainly found among the Liliaceae and Iridaceae. They may be relevant as a site of uptake for cultivated as well as for wild *Allium* species. Herbicides applied for weed control in tulips, hyacinths, etc. should be considered in this respect.

## 2. *Root*

(*a*) *General.* Several reviews have been published concerning the uptake of soil-applied herbicides by roots (Bukovac, 1976; Klingman and Ashton, 1975; Kearney and Kaufman, 1976; Koch, 1970; Fryer and Makepeace, 1977).

It is well established that photosynthesis-inhibiting herbicides such as ureas, *s*-triazines, *as*-triazinones and uracils, are taken up from the soil by the roots (Kearney and Kaufman, 1976; Eue, 1971; Schmidt *et al.*, 1975). Other herbicides for which an uptake or absorption by the roots has been demonstrated include, aminotriazole (Steckó, 1975), benzamide (Sarjas and Perrot, 1969), benzoic acids and dalapon (Herbicide Handbook, 1974), bifenox (Leather and Foy, 1978), dichlobenil (Price and Putnam, 1969), diphenylethers (Kearney and Kaufman, 1976), ethofumesate (Eshel *et al.*, 1978), isocarbamid (Hack and Schmidt, 1972; Mittelstaedt and Führ, 1975), napropamide (Hopp, 1972), oxadiazon (Ishizuka *et al.*, 1975), perfluidone (Lamoureux and Stafford, 1977), phenoxyalkanoic acids (Kearney and Kaufman, 1976; Schmidt, 1972; Maas, 1971), chloridazone (Eshel and Sompolinsky, 1970), tetrafluoropropionate (Aelbers *et al.*, 1969), phenols, TCA, thio-carbamates (Herbicide Handbook, 1974) and inorganic herbicides. Also for dinitroanilines (Barrentine and Warren, 1971a) and the chloro-acetanilides, alachlor (Eshel, 1969; Chandler *et al.*, 1974) and metolachlor (Diner *et al.*, 1977; Pillai *et al.*, 1977) root uptake is possible.

Some dicotyledoneous plants and some Gymnospermae form tap roots which penetrate deep into the soil. These tap roots usually are not exposed to herbicides applied to the soil surface. Consequently, weeds with this type of root, e.g. *Cirsium*, *Rumex*, etc. are difficult to control.

(*b*) *Influence of symbionts on the uptake of herbicides.* The physiological activity of roots may be enhanced by an ectotrophic or endotrophic mycorrhiza. Plants with a mycorrhiza have an increased capability for the uptake of nutrients (Safir, *et al.*, 1972), hence they also may have an increased capacity for the uptake of herbicides. No investigations with plants growing either with or without a mycorrhiza seem to have been conducted to elucidate this point. It is also unknown whether or not root nodules contribute to the uptake of herbicides.

(*c*) *Influence of the rhizosphere*. The roots of higher plants excrete a variety of inorganic and organic compounds, e.g. bicarbonate and amino acids, into the soil. These materials strongly influence the growth of the bacteria and fungi inhabiting the rhizosphere. The pH-value, which is lower in the rhizosphere, affects the uptake of dissociating or proton-accepting herbicides either directly, by controlling the ionic state and hence penetration, or indirectly by affecting sorption processes and thus availability.

Another aspect that requires attention is the possible influence of the rhizosphere on the bioavailability of bound residues.

(*d*) *Influence of other agrochemicals*. The availability of herbicides may also be influenced by the presence of other agrochemicals, e.g. fungicides, insecticides, other herbicides, fertilizers and mineral salts (Wilkinson and Karunen, 1976; Köcher, 1972; Hamill *et al.*, 1972; Wolf and Hurle, 1977). The effects may be direct, for example decreased availability as a result of decreased solubility in the presence of mineral salts, or indirect for example urea increases the xylem stream and hence increases the uptake of water and thus any herbicide dissolved in water (Minshall *et al.*, 1977). An especially marked effect may be exerted by liming the soil which leads to a pH-shift (Schmidt and Hoyer, 1975) and may also affect microbial activity and therefore persistence. The action of antidotes may also involve the reduction of herbicide uptake.

The use of different formulations, for example wettable powders, emulsion concentrates, granules, capsules or controlled release formulations may also increase or decrease the availability of herbicides (Riley and Morrod, 1976).

## C. Uptake by Parasitic Weeds

Many parasitic weeds penetrate into the root of their host plants with the aid of haustoria. Through their haustoria these parasitic weeds are directly connected to the vascular system of their host plants.

It may be possible, that herbicides, which have been taken up by the host plant and move in their xylem, are transferred into the parasitic weed plant and may act there (Whitney, 1973). A direct uptake by parasitic plants from the soil is really only possible through the shoot, since only rudimentary roots and root hairs exist.

It is not clear whether or not some herbicides, primarily dinitroanilines, act on parasitic weed via their shoot, as suggested by Saghir *et al.* (1973).

The most effective control method presently available employs soil fumigants which act on the seeds. The state of imbibition is of special importance for the adsorption of fumigants as has been shown by Kasasian (1971) with *Orobanche.*

### D. Uptake by Phycophyta, Bryophyta and Pteridophyta

In places where they grow as weeds, algae (Phycophyta), mosses (Bryophyta, classes Hepaticae and Musci) and ferns (Pteridophyta) may also be controlled by herbicides.

Mosses may take up herbicide by the rhizoids which have the function of roots in these plants. Most of the green algae growing in the soil take up the herbicides directly, since most of them lack any root-like organs. The Pteridophyta, however, have true roots, all of which are shoot-borne. *Equisetum*, which as a very important pteridophyte weed only has rhizomes and rhizome-tubers which have been shown to take up herbicides. However, the translocation is very poor in the phloem and in the xylem (Köhler, 1971). The reasons for this poor translocation are unknown.

## VI. Mechanisms of Uptake

Herbicides may be taken up into roots by mechanisms which are basically similar to those responsible for the uptake of nutrients. The uptake may either be passive and apoplastic (via non-living plant parts) or active and symplastic (through living cells), or a combination of both.

### A. Passive Uptake

#### 1. *Apoplastic root uptake*

In this case the herbicides are transported in the cell walls (free space). However, only those compounds which are able to penetrate the endodermis and the Casparian strip may reach the xylem. This path of uptake is mainly in association with the uptake of water (Mengel and Kirkby, 1978). The xylem transport is basically controlled by the transpiration-rate. The apoplastic uptake and the transport in the xylem to the aerial parts has been documented for chlorinated fatty acids, urea compounds, phenoxyalkanoic acids and triazines (Koch, 1970). Also *as*-triazinones (Müller and Sanad, 1976; Herbicide Handbook, 1974) and uracils (Steckó, 1975) are acropetally transported in the apoplast/xylem.

(*a*) *Interception.* It may be possible, that herbicide molecules which have been immobilized in the soil are taken up by direct contact via the rhizodermis and the root hairs. However, no investigations seem to have been reported on this possibility.

(*b*) *Mass-flow.* Most of the soil applied herbicides are taken up by passive mass-flow, i.e. dissolved in the water taken up by the root. In these cases the amount of absorbed herbicide is usually proportional to the herbicidal concentration and to the amount of water which has been transpired by the plant (Bukovac, 1976; Walker, 1973a). For example Walker (1973b) found that 70% of the linuron supplied to the roots of wheat seedlings was translocated to the shoot. However, these correlations do not always apply, for example, in soya beans (Moody *et al.*, 1970). Further evidence that transpiration may not be directly involved in root uptake of herbicides comes from experiments with isolated roots and intact plants which both revealed similar uptake curves (Bukovac, 1976). However, in these cases the possibility of herbicide adsorption by the root should also be considered (Grover and Hance, 1968).

The distribution of the herbicide in the soil is not often the same as the distribution of the root system. Generally herbicides remain in the upper soil layer whereas the roots penetrate deeper into the soil and there take up water which is free of herbicides (Russell and Shone, 1972). Walker (1973a) reported that the uptake of atrazine and linuron paralleled the fraction of the root system present in the herbicide treated soil. Therefore a good correlation between the amount of herbicide taken up and the transpiration rate may only be obtained for plants growing in nutrient solution, where the active compound is taken up equally by all root parts.

(*c*) *Diffusion.* This is the term for the transport of a solute following a concentration gradient, in this context especially the diffusion of water soluble herbicides into the "free space" of roots. Herbicides which have entered the "free space" by diffusion are of course easily lost as soon as the roots are transferred into water which is free of herbicides; this has been demonstrated with radioactively labelled substances (Shone and Wood, 1977; Bukovac, 1976; Price and Putnam, 1969).

The efflux (exudation) out of the roots observed after application of dichlobenil may also be explained by diffusion (Price and Putnam, 1969). Root exudation may work as a herbicide-inactivation mechanism (Sanad and Müller, 1973), but may also lead to problems with phytotoxicity in other plants and in following crops.

### 2. *Shoot*

Below the soil surface the shoot does not take up appreciable amounts of water. Therefore herbicides must mainly enter the tissue by diffusion. This appears to be the case for most of the herbicides acting on the shoot, since they are easily volatile (Hartley, 1976).

### 3. *Seed*

In seeds the first part of the herbicide uptake, which occurs in the seed coat or testa, must also proceed by diffusion and consequently is purely physical and consumes no metabolic energy (Ebert and Dumford, 1976). This is the reason why no significant differences are obtained between the uptake into dry or into imbibed seeds respectively or into dead or into living seeds.

## B. Active Uptake

After the first phase of passive, apoplastic herbicide uptake the second, symplastic phase is an active process.

In this case the herbicide uptake and transport from cell to cell occurs through the plasma and plasmodesmata. A transfer into the apoplast is possible, especially into the xylem. The uptake into the basipetal stream of assimilates of the phloem is also possible. This second active uptake phase depends on metabolic energy and varies with temperature (Bukovac, 1976; Isensee *et al.*, 1971; Penner, 1971). Contrasting with the physicochemical mechanisms of uptake, the active uptake may be inhibited by metabolic inhibitors as has been demonstrated for example by Isensee *et al.* (1971) with reduced root uptake of picloram after application of 2,4-dinitrophenol, sodium azide and sodium arsenite.

# VII. Conclusions and Further Aspects

The availability of herbicides in the soil and hence their uptake by plants is affected by sorption, transport and degradation processes. As described elsewhere in this volume there are a variety of laboratory methods for studying these phenomena but they do not always give a reliable estimate of the likely field performance of a herbicide.

Ecological impact studies are usually concerned with the total residues in the soil but from the agricultural standpoint it is the quantity available to the plant that is of interest. This may depend on the plant species (or even cultivar) under consideration, the method of cultivation and on the presence of other agrochemicals. The relative importance of these factors in any particular circumstances cannot be quantitatively described in the current state of knowledge. The significance of bound or conjugated residues in this context is uncertain but at the moment is not thought to be great.

Laboratory experiments have shown that herbicides enter the plant more easily at some sites than at others but the actual site of uptake in the field is sometimes unknown. It may well vary according to the distribution of the herbicide in the soil, the growth stage of the plant and the environmental conditions under which it was grown. More information on these aspects is urgently needed if the reliability of soil applied herbicides is to be improved.

It is necessary to distinguish between site of entry and site of action. In some cases plants take up herbicides from the soil by their roots, without harm. This may be because the compound is not translocated into the shoot, which is the site of action as for example in the case of the diphenylethers (Shimotori and Kuwatsuka, 1978; Vanstone and Stobbe, 1978). In some cases selectivity may be explained by differences in root retention between species (Sargent, 1976; Rubin and Eshel, 1977), e.g. as for fluridone (Berard *et al.*, 1978), metobromuron and chlorbromuron (Hogue, 1978), metribuzin (Hargroder and Rogers, 1974; Smith and Wilkinson, 1974), or SAN 6706 and norflurazon (Strang and Rogers, 1974, 1975). The mechanisms of root retention are not very well known (Price, 1975). Uptake in such cases although not producing phytotoxicity or herbicidal activity does decrease the amount of herbicide in the soil, sometimes significantly (Alkämper, 1969). Hence further research on this subject could lead to better understanding of selectivity and perhaps persistence.

The passive apoplastic mechanisms of uptake (diffusion, mass-flow) are well understood, but the active symplastic and the symplastic-apoplastic pathways of uptake and translocation require more study for a proper understanding. Some herbicides are transported in the phloem, e.g. aminotriazole, dalapon and phenoxyalkanoic acids (Koch, 1970), but it is not known whether the uptake is also symplastic.

Another area of ignorance is the function of phytohormones in regulating the uptake and transport of nutrients and herbicides.

The future increase in our knowledge of the mechanisms of uptake and transport of herbicides into plant tissue largely depends on a close cooperation with plant physiologists since basic information is needed before herbicide behaviour can be adequately described.

## References


Aelbers, E., Bastiaansen, M. G. and Grijseels, A. J. (1969). 2,2,3,3,-Tetrafluorpropionate, a herbicide for the control of annual and perennial grasses in different crops. *3rd Symposium on new herbicides*, I, 17–24, Versailles.

Alkämper, J. (1969). Der Einfluß von Kulturpflanzen auf die Inaktivierung von Herbiziden. *Z. Acker- und Pflanzenbau*, **129**, 145–156.

Anon. (1969–1979). *Rückstandsanalytik von Pflanzenschutzmitteln* (Mitt. VI der Senatskommision für Pflanzenschutz-, Pflanzenbehandlungs- und Vorratsschutzmittel. Methodensammlung der Arbeitsgruppe "Analytik" der DFG). Verlag Chemie, Weinheim, New York.

Appleby, A. P. and Furtick, W. R. (1965). A technique for controlled exposure of emerging grass seedlings to soil-active herbicides. *Weeds* **13**, 172–173.

Appleby, A. P., Furtick, W. R. and Fang, S. C. (1965). Soil placement studies with EPTC and other carbamate herbicides on *Avena sativa*. *Weed Res.* **5**, 115–122.

Armstrong, A. P., Meggitt, W. F. and Penner, D. (1973). Absorption, translocation and metabolism of alachlor by Yellow Nutsedge. *Weed Sci.*, **21**, 357–360.

Audus, L. J. (1976). *Herbicides, Physiology, Biochemistry, Ecology*, 2nd Edition, Academic Press, London, New York, San Francisco, Vol. 1, 608 pp.; Vol. 2, 564 pp.

Barrentine, W. L. and Warren, G. F. (1971a). Differential phototoxicity of trifluralin and nitralin. *Weed Sci.*, **19**, 31–37.

Barrentine, W. L. and Warren, G. F. (1971b). Shoot zone activity of trifluralin and nitralin. *Weed Sci.* **19**, 37–41.

Berard, D. F., Rainey, D. P. and Lin, C. C. (1978). Absorption, translocation and metabolism of fluridone in crop species. *Weed Sci.* **26**, 252–254.

Blinn, R. C. (1971). Methodology of residue analysis for herbicides. In *Methods in Residue Analysis*, Vol. 4, pp. 493–523. Gordon and Breach, New York.

Bukovac, M. J. (1976). Herbicide entry into plants. In *Herbicides, Physiology, Biochemistry, Ecology*, Vol. 1, (L. J. Audus, Ed.) pp. 353–364. Academic Press, London and New York.

Byast, T. H., Cotterill, E. G. and Hance, R. J. (1977). *Methods of Analysis for Herbicides Residues*. Technical Report No. 15 (2nd edition), Agric. Res. Council Weed Res. Org., Begbroke Hill, Yarnton, Oxford.

Carlson, W. C., Lignowski, E. M. and Hopen, H. J. (1975). Uptake, translocation and adsorption of pronamide. *Weed Sci.*, **23**, 148–154.

Chandler, J. M., Basler, E. and Santelmann, P. W. (1974). Uptake and translocation of alachlor in soybean and wheat. *Weed Sci.* **22**, 253–258.

Crafts, A. S. and Yamaguchi, S. (1964). *The autoradiography of Plant Materials*. Univ. Calif. Agric. Exp. Sta. Ext. Ser. Manual 35, 143 p.

Deli, J. and Warren, G. F. (1970). Uptake, translocation, and herbicidal effect of diphenamid. *Weed Sci.* **18**, 692–696.

Diner, A. M., Davis, D. E. and Truelove, B. (1977). Absorption and translocation of root and foliar-applied $^{14}$C-metolachlor in soybean. In *Proc. 30th Ann. Meet. Southern Weed Sci. Soc.* p. 358.

Ebert, E. and Dumford, S. W. (1976). Effects of triazine herbicides on the physiology of plants. *Residue Rev.* **65**, 48–58.

Eberle, D. O. and Gerber, H. R. (1976). Comparative studies of instrumental and bioassay methods for the analysis of herbicide residues. *Arch. environ. Contam. Toxic.* **4**, 101–118.

Ebing, W. (1970–1978). *Gaschromatographie der Pflanzenschutzmittel.* Tabellarische Literaturreferate I-VII, Mitt. aus der Biologischen Bundesanstalt für Land- u. Forstwirtschaft, Berlin-Dahlem.

Ehlers, W. (1978). Wassergehalts- und Wasserspannungsmessungen im Felde zur Bilanzierung des Bodenwasserhaushalts. *Mitt. Deut. Bodenkundl. Ges.* **26**, 115–132.

Eshel, Y. (1969). Phytotoxicity, leachability and site of uptake of 2-chloro-2′,6′-diethyl-*N*-(methoxymethyl) acetanilide. *Weed Sci.* **17**, 441–444.

Eshel, Y. and Prendeville, G. N. (1967). A technique for studying root vs. shoot uptake of soil-applied herbicides. *Weed Res.* **7**, 242–245.

Eshel, Y., Kovacs, M. and Rubin, B. (1975). Effect of soil placement and shoot uptake of prometryne on peas. *Weed Res.* **15**, 369–372.

Eshel, Y. and Sompolinsky, D. (1970). Selectivity of pyrazon and benzthiazuron in sugar beet. *Weed Res.* **10**, 196–203.

Eshel, Y., Zimdahl, R. L. and Schweizer, E. E. (1978). Uptake and translocation of ethofumesate in sugar-beet plants. *Pestic. Sci.* **9**, 301–304.

Eue, L. (1971). 4-Amino-6-tert.-butyl-3-(methylthio)-1,2,4-triazin-5-on zur Unkrautbekämpfung im Kartoffelbau. *Med. Fac. Landbouww. Rijksuniv. Gent* **36**, 1233–1239.

Fryer, J. D. and Makepeace, R. J. (1977). *Weed Control Handbook. Principles.* Vol. 1, Blackwell Scientific Publ., Oxford, London, Edinburgh, Melbourne, 510 p.

Fryer, J. D. and Makepeace, R. J. (1978). *Weed Control Handbook. Recommendations.* Vol. 2, Blackwell Scientific Publ., Oxford, London, Edinburgh, Melbourne, 532 p.

Gerber, H. R. (1977). Biotests for herbicide development. In *Crop protection agents*, (N. R. McFarlane, Ed.) pp. 307–321. Academic Press, London, New York.

Gray, R. A. and Weierich, A. J. (1969). Importance of root, shoot, and seed exposure on the herbicidal activity of EPTC. *Weed Sci.* **17**, 223–229.

Green, R. E. and Obien, S. R. (1969). Herbicide equilibrium in soils in relation to soil water content. *Weed Sci.* **17**, 514–519.

Grover, R. and Hance, R. J. (1968). Adsorption of some herbicides by soil and roots. *Canad. J. Plant Sci.* **49**, 378–380.

Hack, H. and Schmidt, R. R. (1972). BAY 6339 H, a new herbicide for weed control in sugar and fodder beets. *Proc. 11th Br. Weed Control Conf.* 479–486.

Hamill, A. S., Smith, L. W. and Switzer, C. M. (1972). Influence of phenoxy herbicides on picloram uptake and phytotoxicity. *Weed Sci.* **20**, 226–229.

Hance, R. J. and McKone, C. E. (1976). The determination of herbicides. In *Herbicides, Physiology, Biochemistry, Ecology*, Vol. 2, (L. J. Audus, Ed.) pp. 393–445. Academic Press, London and New York.

Hargroder, T. G. and Rogers, R. L. (1974). Behaviour and fate of metribuzin in soybean and hemp sesbania. *Weed Sci.* **22**, 238–245.

Hartley, G. S. (1976). Physical behaviour in the soil. In *Herbicides, Physiology, Biochemistry, Ecology*, Vol. 2, (L. J. Audus, Ed.) pp. 1–28. Academic Press, London, New York, San Francisco.

Herbicide Handbook (1974). Weed Sci. Soc. of America, Champaign, 430 p.

Hewitt, E. J. (1952). *Sand and Water Culture Methods used in the Study of Plant Nutrition.* Commonwealth Agric. Bur., East Malling, Maidstone, Kent.

Hilton, J. L. and Christiansen, M. N. (1972). Lipid contribution to selective action of trifluralin. *Weed Sci.* **20**, 290–294.

Hoagland, D. R. and Arnon, D. I. (1950). *The water culture method for growing plants without soil.* Univ. Calif. Agric. Exp. Stn. Circ. 347, 32pp.

Hocombe, S. D. (1968). The uptake of atrazine by germinating seeds of turnip. *Weed Res.*, **8**, 68–71.

Hogue, E. J. (1978). Absorption and translocation of metobromuron and chlorbromuron. *J. Environ. Sci. Health*, **B 13**, 323–339.

Holm, L. G., Plucknett, D. L., Pancho, J. V. and Herberger, J. P. (1977). *The World's Worst Weeds. Distribution and Biology.* An East-West Center Book, University Press Hawaii, Honolulu, 609 p.

Hopp, H. (1972). Gewächshaus- und Freilandergebnisse mit dem neuen Rapsherbizid Devrinol. *Z. Pflanzenkrankh. Pflanzensch., Sonderheft VI*, 183–193.

Horowitz, M. (1976). Application of bioassay techniques to herbicide-investigations. *Weed Res.* **16**, 209–215.

Hurle, K. (1975). Verfügbarkeit von Herbiziden im Boden. *Z. Pflanzenkrankh. Pflanzensch., Sonderheft* **VII**, 101–108.

Hurle, K. (1977). Biotests for the detection of herbicides in the soil. In *Crop Protection Agents*, (N. R. McFarlane, Ed.), pp. 285–306. Academic Press, London and New York.

Ingle, E. L. and A. D. Worsham (1971). Effect of soil placement of several herbicides on yellow nutsedge (*Cyperus esculentus*) tubers. *Proc. 24th Ann. Meet. Southern Weed Sci. Soc.*, pp. 188–192.

Isensee, A. R., Jones, G. E. and Turner, B. C. (1971). Root absorption and translocation of picloram by oats and soybeans. *Weed Sci.* **19**, 727–731.

Ishizuka, K., Hirata, H. and Fukunaga, K. (1975). Fate of oxadiazon in rice plants. *Proc. 5th Asian Pacific Weed Sci. Soc. Conf.*, 204–208.

Jarczyk, H. J. (1977). Instrumental methods for the analysis of herbicides residues. *Weed Res.* **17**, 93–101.

Kasasian, L. (1971). Orobanche spp. *PANS* **17**, 35–41.

Kearney, P. C. and Kaufman, D. D. (1976). *Herbicides, Chemistry, Degradation and Mode of Action*, 2nd Edition, Vols 1 and 2, p. 1036, Marcel Dekker, New York and Basel.

Kerpen, W. and Schleser, G. (1977). Adsorption and Desorption von Methabenzthiazuron in verschiedenen Böden. *Z. Pfl. Ernähr. Bodenkunde* **140, b**, 697–705.

Klingman, G. C. and Ashton, F. M. (1975). *Weed Science, Principles and Practices.* J. Wiley & Sons, New York, London, Sydney, Toronto, p. 431.

Knake, E. L. and Wax, L. M. (1968). The importance of the shoot of giant foxtail for uptake of pre-emergence herbicides. *Weed Sci.* **16**, 393–395.

Koch, W. (1970). *Unkrautbekämpfung.* E. Ulmer, Stuttgart, 342 p.

Köcher, H. (1972). Möglichkeiten der Interaktionen von Herbiziden untereinander und mit anderen Agrochemikalien. *Z. Pflanzenkrankh., Pflanzensch., Sonderheft VI*, 163–170.

Köhler, J. (1971). *Verbreitung, Biologie und Bekämpfung des Sumpfschachtelhalmes.* Aktuelle Fragen des Landbaues, Heft 6., P. Parey, Berlin und Hamburg, 111 p.

Lamoureux, G. L. and Stafford, L. E. (1977). Translocation and metabolism of perfluidone (1,1,1-trifluoro-*N*-2-methyl-4-(phenylsulfonyl)phenyl methanesulfonamide) in peanuts. *J. agric. Fd. Chem.*, **25**, 512–517.

Ladlie, J. S., Meggitt, W. F. and Penner, D. (1977). Effect of trifluralin and metribuzin combinations on soybean tolerance to metribuzin. *Weed Sci.* **25**, 88–93.

Lavy, T. L., Messersmith, C. G. and Knocke, H. W. (1972). Direct liquid scintillation radioassay of $^{14}$C-labelled herbicides in soil. *Weed Sci.* **20**, 215–219.

Leather, G. R. and Foy, C. L. (1978). Differential absorption and distribution as a basis for the selectivity of bifonex. *Weed Sci.* **26**, 76–81.

Maas, G. (1971). Der Einfluß der Sorptionskapazität des Bodens auf die Wirksamkeit von Mecoprop gegen Galium aparine L. *Mitt. Biol. Bundesanstalt.* **146**, 165.

Mengel, K. and Kirkby, E. A. (1978). *Principles of Plant Nutrition.* Intern. Potash Inst. Worblaufen-Bern, 593 p.

Minshall, W. H., Sample, K. C. and Robinson, J. R. (1977). The effect of urea on atrazine uptake from soil. *Weed Sci.* **25**, 460–464.

Mittelstaedt, W. and Führ, F. (1975). Mineralization and uptake of the herbicidal compound isocarbamid. *Pflanzenschutz-Nachrichten Bayer* **28**, 3, 353–358 (engl. edition).

Moody, K., Kust, C. A. and Buchholtz, K. P. (1970). Uptake of herbicides by soybean roots in culture solutions. *Weed Sci.* **18**, 642–647.

Moomaw, R. S. and Martın, A. R. (1978). Interaction of metribuzin and trifluralin with soil type on soybean (*Glycine max*) growth. *Weed Sci.* **26**, 327–331.

Müller, F. and Sanad, A. (1976). Verteilung von $^{14}$C-Metamitron in jungen Zuckerrübenpflanzen. *Med. Fac. Landbouww. Rijksuniv. Gent*, **41**, **2**, 1211–1227.

Nalewaja, J. D. (1968). Uptake and translocation of diallate in wheat, barley, flax and wild oat. *Weed Sci.* **16**, 309–312.

Nishimoto, R. K. and Warren, G. F. (1971a). Site of uptake, movement and activity of DCPA. *Weed Sci.* **19**, 152–155.

Nishimoto, R. K. and Warren, G. F. (1971b). Shoot zone uptake and translocation of soil-applied herbicides. *Weed Sci.* **19**, 156–161.

O'Brien, L. P. and Prendeville, G. N. (1972). Shoot zone uptake of soil-applied herbicides in *Pisum sativum L.* Weed Res. **12**, 248–253.

O'Donovan, J. T. and Prendeville, G. N. (1975). Shoot zone uptake of soil-applied herbicides in some legume species. *Weed Res.* **15**, 413–417.

Parker, C. (1966). The importance of shoot entry in the action of herbicides applied to the soil. *Weed Sci.* **14**, 117–121.

Penner, D. (1971). Effect of temperature on phytotoxicity and root uptake of several herbicides. *Weed Sci.* **19**, 571–576.

Penningsfeld, F. and Kurzmann, P. (1966). *Hydrokultur und Torfkultur.* Handbuch des Erwerbsgärtners, Band 7, E. Ulmer, Stuttgart.

Pestemer, W. (1975). Verhalten gemüsebaulich genutzter Triazine in sorptionsstarken Böden. *Z. Pflanzenkrankh. Pflanzensch., Sonderheft VII*, 125–133.

Pestemer, W. (1976). Quantitativer Biotest zur Bestimmung von Photosynthesehemmern in Boden. *Weed Res.* **16**, 357–363.

Pestemer, W. (1977). Verhalten gemüsebaulich genutzter Harnstoff-Herbizide in sorptionsstarken Böden. *Z. Pflanzenkrankh. Pflanzensch., Sonderheft VIII*, 259–270.

Phillips, R. E., Egli, D. B. and Thompson, L. Jr. (1972). Adsorption of herbicides by soybean seeds and their influence on emergence and seedling growth. *Weed Sci.* **20**, 506–510.

Pillai, C. G. P., Davis, D. E. and Truelove, B. (1977). Site of uptake and mode of action of metolachlor. *Proc. 30th Ann. Meet. Southern Weed Sci. Soc.* 367.

Prendeville, G. N. (1968). Shoot zone uptake of soil-applied herbicides. *Weed Res.* **8**, 106–114.

Prendeville, G. N., Oliver, L. R. and Schreiber, M. M. (1968). Species differences in site of shoot uptake and tolerance to EPTC. *Weed Sci.*, **16**, 538–540.

Price, C. E. (1975). Penetration and movement within plants of pesticides and other solutes: Uptake mechanisms. In *Reports on the Progress of Applied Chemistry During 1974*, 59, Blackwell Sci. Publ. Oxford, London, Edinburgh, Melbourne, 310–326.

Price, H. C. and Putnam, A. R. (1969). Efflux of dichlobenil from shoots and roots of corn. *J. Agric. Fd. Chem.*, **17**, 135–137.

Ramthun, L. E. and Talbert, R. E. (1975). Factors affecting metribuzin rates in soybean. *Proc. 28th Ann. Meet. Southern Weed Sci. Soc.*, 25.

Rieder, G., Bucholtz, K. P. and Kust, C. A. (1970). Uptake of herbicides by soybean seed. *Weed Sci.*, **18**, 101–105.

Riley, D. (1978). Physical loss and redistribution of pesticides in the liquid phase. *British Crop Protection Council Monograph, No. 17, The Persistence of Insecticides and Herbicides* 109–116.

Riley, D. and Morrod, R. S. (1976). Relative importance of factors influencing the activity of herbicides in soil. *Proc. Br. Crop Prot. Conf. Weeds*, 971–980.

Rowse, H. R., Stone, D. A. and Gerwitz, A. (1978). Simulation of the water distribution in soil, II. The model for cropped soil and its comparison with experiment. *Pl. Soil* **49**, 533–550.

Rubin, B. and Eshel, Y. (1977). Absorption and translocation of terbutryne and fluometuron in cotton (*Gossypium hirsutum*) and snapbeans (*Phaseolus vulgaris*). *Weed Sci.* **25**, 499–505.

Russell, R. S. and Shone, M. G. T. (1972). Root function and the soil. *Proc. 11th Br. Weed Control Conf.*, 1183–1192.

Safir, G. R., Boyer, J. S. and Gerdemann, J. W. (1972). Nutrient status and mycorrhizal enhancement of water transport in soybean. *Plant Physiol.*, **49**, 700–703.

Saghir, A. R., Foy, C. L., Hameed, K. M., Drake, C. R. and Tolin, S. A. (1973). Studies on the biology and control of *Orobanche ramosa* L., *Proc. Eur. Weed. Res. Counc. Symp. Parasitic Weeds*, 106–116.

Sanad, A. and Müller, F. (1973). Untersuchungen über die Wurzelausscheidung von $^{14}C$-markiertem 2,4-D bei *Datura stramonium* L. und *Galium aparine* L. *Z.PflKrankh, Pflsch.* **80**, 74–80.

Sargent, J. A. (1976). Relationship of selectivity to uptake and movement. *In* “Herbicides, Physiology, Biochemistry, Ecology,” Vol. 2, (L. J. Audus, Ed.), Academic Press, London and New York pp. 303–312.

Sarjas, P. de and Perrot, A. J. (1969). A new benzamid herbicide: *N*-(1,1-dimethylpropynyl)-3,5-dichlorobenzamid. *3rd symposium on New Herbicides*, Vol. 1, Versailles, 249–259.

Schmidt, R. R. (1972). Bekämpfung von *Galium aparine* mit Mecoprop in Abhängigkeit von verschiedenen Bodenarten. *Weed Res.* **12**, 174–181.

Schmidt, R. R., Eue, L. and Draber, W. (1975). Metamitron—ein neues

Rübenherbizid auf der Basis eines Triazinons. *3rd International Meeting on Selective Weed Control in Beet Crops*, Paris, pp. 713–728.

Schmidt, R. R. and Hoyer, H. (1975). Untersuchungen zur biologischen Verfügbarkeit von Metribuzin. *Z. Pflanzenkrankh. Pflanzensch., Sonderheft VII*, 109–115.

Scott, H. D. and Phillips, R. E. (1971). Diffusion of herbicides to seed. *Weed Sci.* **19**, 128–132.

Scott, H. D. and Phillips, R. E. (1973). Absorption of herbicides by soybean seed. *Weed Sci.* **21**, 71–76.

Shimotori, H. and Kuwatsuka, S. (1978). Absorption, translocation and metabolism of CNP, a diphenylether herbicide and its amino derivate in rice plants. *J. Pesticide Sci.* **3**, 267–275.

Shone, M. G. T. and Wood, A. V. (1976). Uptake and translocation of some pesticides by hypocotyls of radish seedlings. *Weed Res.* **16**, 229–238.

Shone, M. G. T. and Wood, A. V. (1977), Longitudinal movement and loss of nutrients, pesticides and water in barley roots. *J. exp. Bot.*, **28**, 872–885.

Smith, A. E. and Wilkinson, R. E. (1974). Differential absorption, translocation and metabolism of metribuzin (4-amino-6-tert-butyl-3-(methylthio)-astriazine-5(4 H)one) by soybean cultivars. *Physiol. Plantarum*, **32**, 253–257.

Stalder, L. and Pestemer, W. (1980). Availability to plants of herbicide residues. Part 1: A rapid method for estimating potentially available residues of photosynthesis inhibitors in soils. *Weed Res.* (in press).

Steckó, V. (1975). Experiments with simazine, monuron and bromacil—three soil-applied herbicides. Reports and dissertations 29, Dep. Plant Husbandry, Agric. Coll. Sweden, Uppsala, p. 85.

Strang, R. H. and Rogers, R. L. (1974). Behaviour and fate of two phenylpyridazinone herbicides in cotton, corn and soybean. *J. agric. Fd Chem.*, **22**, 1119–1125.

Strang, R. H. and Rogers, R. L. (1975). Translocation on $C^{14}$-SAN 6706 in cotton, soybean and corn. *Weed Sci.* **23**, 26–31.

Tylson, E. C. (1979). Characterization and application of piperofos against *Echinochloa* in different rice varieties. *Nachrichtenbl. Deut. Pflanzenschutzd.* **31**, 116–117.

Vanstone, D. E. and Stobbe, E. H. (1978). Root uptake, translocation and metabolism of nitrofluorfen and oxyfluorfen by fababeans (*Vicia faba*) and green foxtail (*Setaria viridis*). *Weed Sci.* **26**, 389–392.

Walker, A. (1973a). Vertical distribution of herbicides in soil and their availability to plants: treatment of different proportions of the total root system. *Weed Res.* **13**, 416–421.

Walker, A. (1973b). Availability of linuron to plants in different soils. *Pestic. Sci.* **4**, 665–675.

Whitney, P. J. (1973). Transport across the region of fusion between bean (*Vicia faba*) and broomrape (*Orobanche crenata*). *Proc. Eur. Weed Res. Counc. Symp. Parasitic Weeds*, 154–166.

Wilkinson, R. E. and Karunen, P. (1976). Influence of *S*-ethyl dipropylthiocarbamate on atrazine absorption by wheat. *Ann. Bot.* **40**, 1043–1046.

Wolf, E. and Hurle, K. (1977). Einfluß von Mineralstoffen auf die Phytotoxizität von Herbiziden. *Z.Pflkrankh. Pflssch., Sonderheft VIII*, 301–310.

Yip, G. (1971). Herbicides and their methods for residue analysis. In *Methods in residue analysis*. Vol. 4, pp. 537–552. Gordon and Breach, New York.

# 8 Activity and Selectivity in the Field

*Allan Walker*
National Vegetable
Research Station,
Wellesbourne,
Warwick CV35 9EF, England



## I. Introduction

The essential requirement for a successful soil-applied herbicide is that it should accumulate in weed seedlings in toxic concentrations from soil treated at an economic rate without risk of damage to the crop (Graham-Bryce, 1972). The amounts taken up by seedlings over a period of time will be determined by the ability of the soil to supply the chemical and the ability of the plant to absorb it. To a large extent, the variations in plant response which occur when a herbicide is applied to different soils or on different occasions to the same soil will reflect differences in the amount of herbicide which becomes available to the plant. The main processes which control this availability—adsorption by soil, herbicide transport in the liquid or gas phases of soil and factors influencing plant uptake have been considered in detail in other chapters. The purpose here is to discuss the ways in which soil and environmental factors can interact to control activity and selective toxicity under practical field conditions.

## II. Activity

### A. Soil Organic Matter

Some herbicides may be available to plants as a vapour in the gas-phase

Table I. Correlations between the activities (ED50) of some soil-applied herbicides and soil properties

| Herbicide | Number of Soils | Simple correlation ($r$) with: Organic matter | Clay | Exchange capacity | Reference |
|---|---|---|---|---|---|
| CDEC | 6 | 0·90 | 0·55 | 0·65 | Upchurch and Mason (1962) |
| Chlorpropham | 32 | 0·95 | 0·24 | 0·88 | Harris and Sheets (1965) |
| | 6 | 0·77 | 0·30 | 0·62 | Upchurch and Mason (1962) |
| 2,4-D | 6 | 0·91 | 0·18 | 0·95 | Upchurch and Mason (1962) |
| Diuron | 6 | 1·00 | 0·13 | 0·95 | Upchurch and Mason (1962) |
| | 32 | 0·80 | 0·06 | 0·43 | Harris and Sheets (1965) |
| Diphenamid | 11 | 0·91 | 0·16 | 0·60 | Dubey and Freeman (1964) |
| EPTC | 6 | 0·92 | 0·22 | 0·95 | Upchurch and Mason (1962) |
| | 9 | 0·91 | 0·20 | — | Jordan and Day (1962) |
| Linuron | 11 | 0·90 | 0·06 | 0·57 | Dubey and Freeman (1964) |
| Methazole | 5 | 0·84 | 0·66 | — | Brockman (1975) |
| Metribuzin | 10 | 0·62 | 0·03 | 0·55 | Schmidt (1973) |
| Picloram | 7 | 0·90 | 0·55 | 0·65 | Grover (1968) |
| Simazine | 6 | 0·98 | 0·14 | 0·93 | Upchurch and Mason (1962) |
| | 32 | 0·36 | 0·06 | 0·25 | Harris and Sheets (1965) |
| | 16 | 0·88 | 0·42 | 0·62 | Sheets *et al.* (1962) |
| | 65 | 0·72 | 0·12 | 0·52 | Day *et al.* (1968) |

of the soil but most must be present in the soil solution before they can be absorbed appreciably by plant roots or shoots or germinating weed seeds. One of the main factors controlling soil solution concentration is adsorption and since soils vary considerably in their adsorptive properties, a particular rate of herbicide cannot be expected to be equally toxic in all soil types. In practice, the recommended rate is often varied according to soil type and is lower for soils with low clay or low organic matter compared with heavy or organic soils. Results from numerous glasshouse experiments have confirmed this inverse correlation between activity and soil organic matter or clay content and have usually shown organic matter to be the most important. Some examples of the correlations obtained are shown in Table I which is not a complete list but a selection presented to illustrate the general type of result obtained.

All the data in Table I were obtained under glasshouse or growing room conditions when herbicides were uniformly mixed throughout the rooting zone of test species. They reflect a direct effect of adsorption in reducing soil solution concentrations. In the field, herbicides are sprayed on the soil surface and usually not incorporated. Weather conditions, particularly rainfall will interact with adsorption to control the vertical distribution of herbicides in soil and hence their availability. The degree of correlation between activity and soil organic matter will be reduced and variability will increase. This increased variability becomes apparent during evaluation of new compounds with the change from glasshouse to field testing and was illustrated diagrammatically by Riley and Morrod (1976) (Fig. 1). In glasshouse trials, a compound having these properties shows good selectivity and a dose at any point between A and B would give acceptable weed control without crop damage. In the field, because of increased variability the margin of selectivity is effectively reduced and the rate for acceptable results is close to point C—much higher than A and much lower than B. This diagram can be extended to illustrate the effects with different soils (Fig. 2). In glasshouse experiments, increasing organic matter from soil 1 to soil 4 will probably indicate separate rates of application to give acceptable results (R1 to R4). In the field, increased variability leads to marked overlapping of the response curves and it becomes more realistic to group soils so that only two rates, R5 and R6, are required to allow for the differences in soil type. This of course is the situation in practice. Eagle (1976) presented evidence that although soils are grouped in this way for practical recommendations, recommended rates can still be related to soil adsorption of the herbicide. Eagle took a representative soil from each of 11 texture grades and measured simazine adsorption. The soils were then grouped into the texture grades: sands, very light soils, light soils, medium soils, heavy soils. When adsorption values were averaged within each

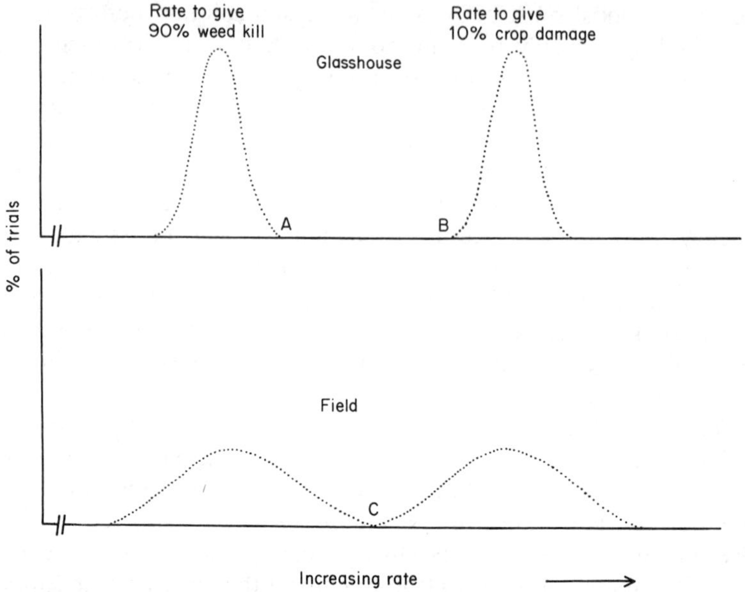


Fig. 1. Comparison of the activity of a herbicide against crop and weed under glasshouse and field conditions (modified from Riley and Morrod, 1976).

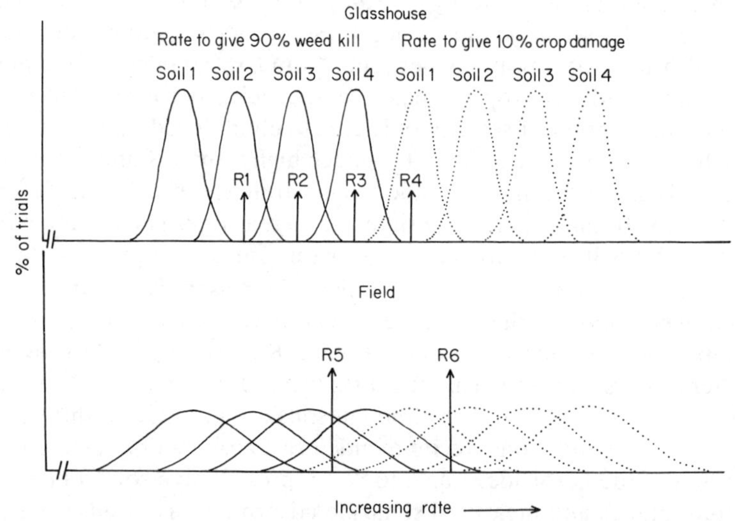


Fig. 2. Comparison of the activity of a herbicide in different soils in the glasshouse and the field. Soil organic matter increases from soil 1 through soil 4.

texture group, the relative mean values agreed well with current label recommendations for simazine use in field beans (Fryer and Makepeace, 1978).

Decreased correlation between activity of soil-applied herbicides and soil organic matter, and increased variability in the field compared with the glasshouse can be illustrated by data of Hance *et al.* (1968). In greenhouse tests with lenacil, simazine, linuron and prometryne, phytotoxicity to two test species was, in general, best related with the organic matter content of 11 different soils. In the field, the correlation between activity and organic matter was much poorer. The field experiments were carried out on two occasions, in autumn and in spring, and autumn phytotoxicities were not significantly correlated with spring phytotoxicities and on neither occasion were field phytotoxicities correlated with those obtained in the glasshouse. The conclusion was that weather conditions in the field were more important in determining activity within this group of soils than was the effect of soil type. A similar conclusion was reached by Upchurch *et al.* (1966) who examined the activity of five herbicides at 17 locations in North Carolina. Even though the organic matter content of the soils varied from 0·7 to 49% and proved to be the most important single soil property with which activity could be related, less than 66% of the total variation could be accounted for, again suggesting an important contribution from other soil and environmental factors.

## B. Rainfall and Soil Moisture

Rainfall is probably the most important climatic factor which affects the activity of soil-applied herbicides and the requirement for rainfall activation of herbicides applied to the soil surface has been recognized for many years. This is reflected in several current recommendations (Fryer and Makepeace, 1978) although the advice is usually of a general nature. Some examples are "Sufficient rainfall after application to carry the chemical into the root zone of the germinating weeds is necessary for maximum efficiency" (lenacil in sugar beet); "Good weed control depends on sufficient rain falling after treatment to at least maintain soil moisture" (linuron in carrot and parsnip). Walker and Roberts (1975) suggested that rainfall activation reflected the need for sufficient moisture to bring the herbicide into solution from its formulation, the need for some redistribution in surface soil layers and the need for adequate soil moisture to maximize herbicide availability to weed seedlings.

Results from several glasshouse tests have shown a marked effect of soil moisture level on the phytotoxicity of a number of herbicides including diuron (Lambert, 1966), atrazine, chloramben, EPTC, propachlor (Stickler *et al.*, 1969), linuron, simazine, lenacil, aziprotryne (Walker, 1971) and terbacil (Kratky and Warren, 1973), with increased toxicity in wetter soil. Although the effects of rainfall on soil moisture levels are undoubtedly of some significance in the field, effects on vertical redistribution of herbicide are probably more important and rainfall shortly after application is important for optimum weed control (Splittstoesser and Derscheid, 1962; Neururer, 1972). It is difficult to specify the actual amount of rain required for optimum activity. LeBaron (1970) suggested that this will depend on the solubility of the chemical, on soil texture and organic matter and on the particular weed species and their depth of germination.

There have been some attempts to devise suitable parameters for rainfall with which to compare the activity of herbicides. Upchurch *et al.* (1966) suggested a rain index in which rain received 0 to 5 days after spraying was rated five times more important than rain received 21 to 25 days after spraying. Since Upchurch *et al.* examined a wide range of soils, this rain index alone did not give significant correlations with the activity of any of the five herbicides examined. However, in multiple linear regressions it did improve the degree of correlation between activity of some herbicides and organic matter level suggesting, as expected, some interaction between adsorption and rainfall in determining activity. Mindreboe (1970) related the activity of pre-emergence herbicides to the days from spraying to first rain and to the accumulated rainfall in 1 to 2 weeks after spraying. The activity of linuron and atrazine was negatively correlated with days from spraying to first rain and the activity of linuron was also correlated with total rainfall in the first week after application. Walker and Roberts (1975) attempted to relate the activities of prometryne, aziprotryne, propyzamide and propachlor in 12 separate field experiments on the same site to a series of rainfall parameters. These were the delay to first rain, the delay to accumulation of 5, 10 or 15 mm rainfall, the delay to a rain of 5 or 10 mm, the accumulated rainfall in 7, 14 or 21 days after application and the rain index of Upchurch *et al.* (1966). A number of significant correlations were obtained for prometryne, aziprotryne and propyzamide but the particular rainfall parameter which gave the best correlation varied with the herbicide. No correlations were obtained between the activity of propachlor and any of the parameters and it was suggested that variations in the composition of the weed flora against which activity was assessed dominated over any possible rainfall effects. This highlights the primary problem of field experiments to examine the effects of one variable on activity—there are usually many variables which interact so that it becomes difficult to evaluate the individual contribution of each.

If it is assumed that the main effect of rainfall is to redistribute herbicide in the surface soil layers and move the compound into the zone where most weed seeds germinate and have their roots, the activity of surface-applied herbicides should be made more reliable following their incorporation into soil. It has been shown that the activity of propachlor (Menges and Tamez, 1973), prometryne, fluometuron, diuron (Wiese and Smith, 1970), simazine, atrazine (Splittstoesser and Derscheid, 1962), aziprotryne, prometryne, chlorpropham and propyzamide (Walker and Roberts, 1975) can be improved by incorporation into the surface 2 to 4 cm of soil, particularly when dry weather follows application. Wiese and Smith (1970) suggested that weed control by prometryne was better when surface applied if rain fell within two weeks after application but incorporated treatments gave better results if no rain fell in this time. This was confirmed by Walker and Roberts (1975) who found that differences between weed kill given by surface and incorporated treatments of prometryne, aziprotryne, propyzamide and propachlor could be significantly correlated with the delay between application and the accumulation of 10 mm rain.

There is little doubt that rainfall is essential for effective weed control by pre-emergence herbicides but it has proved difficult to quantify the effects. Such effects will be determined by the distribution and availability of the herbicide in soil, which will be controlled by the soil's adsorptive properties and moisture content, in relation to the position of the roots and other organs taking up the compound and hence plant age. There is a need for more work on how interactions between root and herbicide distribution in the soil, concentrations of herbicide available, time of exposure and plant age determine biological response.

## C. Soil Factors other than Organic Matter

Although organic matter content is the most important property of soil which affects herbicide activity, other properties do influence behaviour. Corbin *et al.* (1971) studied the effects of soil pH on the phytotoxicities of several herbicides. The phytotoxicity of the weak acids dicamba and 2,4-D and the weak bases prometone and aminotriazole increased as pH increased; the phytotoxicity of dalapon decreased as pH increased and there was little effect on the activities of picloram, chloramben, diuron and nitralin. The *s*-triazine herbicides are weak bases which become protonated in acid soil systems and their adsorption by soils is increased (McGlamery and Slife, 1966; Nearpass, 1965; Weber, 1970). This increase in adsorption should reduce plant availability and hence phytotoxicity and this has been confirmed for prometryne (Weber *et al.*, 1968) and simazine (Leefe, 1968). In modified field plots, Best *et al.* (1975) observed increased activity of both prometryne and atrazine following liming an acid soil from pH 5·5 to pH 7·5. Similar results in amended field plots were reported by

Ladlie *et al.* (1976) for metribuzin and protonation of the herbicide leading to increased adsorption was again suggested as the reason for lower activity at lower pH. An interesting practical consequence of a pH change in the field was reported by Schnappinger *et al.* (1977). It is common practice in no-till maize production in parts of the US to add a 33% nitrogen solution to the soil immediately over the seeded row. On some occasions, reduced activity of simazine and atrazine in the rows was observed. It was found that the nitrogen treatment gave reduced pH values in localized areas with which the reduced activity was associated.

In several glasshouse experiments, the nutrient status of the soil has also been shown to influence herbicide activity. Selman and Upchurch (1961) showed increased phytotoxicity of aminotriazole to cotton, soya bean, ryegrass and wheat as the soil phosphorus level was increased and this was subsequently confirmed in further tests (Selman and Upchurch, 1970). High levels of soil phosphorus have also been shown to increase the phytotoxicity of simazine (Adams, 1965), atrazine (Stolp and Penner, 1973) and diuron (Bingham and Upchurch, 1959).

Ries *et al.* (1967) reported that activity of simazine was greater in rye when nitrogen was given as nitrate rather than ammonium and this was confirmed by McReynolds and Tweedy (1970). This was a direct interaction with nitrogen form since McReynolds and Tweedy showed in their experiments that the fertilizers did not alter soil pH.

All of the above experiments were made in the glasshouse where careful control of other variables can be achieved. Aberg and Stecko (1976) commented that there are few quantitative data from field experiments to demonstrate effects of nutrient level on herbicide phytotoxicity, although there are some qualitative indications. Adams (1965) reported that observations from weed control demonstrations in Minnesota suggested that atrazine gave better weed control on soils high in available phosphorus, an effect which he subsequently confirmed in the glasshouse experiments referred to above. LeBaron (1970) reported unpublished data of Buchholtz (1966) who observed that oat injury from atrazine residues was lowest on plots fertilized with high rates of nitrogen and little or no phosphorus and injury was greatest on plots which had received low rates of nitrogen and high rates of phosphorus. In the same experiments, variable rates of potassium showed no consistent effects. Ashley and Rahn (1967) showed that nitrogen at 55 kg/ha reduced injury to oats from atrazine residues but Doll and Meggitt (1969) reported that atrazine injury to oats increased as nitrogen level increased.

In general, the limited evidence from field experiments suggests that effects of soil nutrient level on activity are small compared with those from other soil properties and since fertilizers are themselves important in crop production it is unlikely that variation in their use could be employed specifically to modify the activity of herbicides.

## D. Variability in Application and Soil Surface Conditions

The discussion so far has concentrated on the properties of the soil and the environment which affect herbicide activity and has suggested several reasons why activity in the field will be more variable than in glasshouse experiments. Another factor which might account for some of the increased variability in the field is the unevenness of distribution of herbicide over the soil surface. Riley and Morrod (1976) suggested that in laboratory or glasshouse tests, the rate of application is usually within 10% of the intended dose. Even in carefully prepared small scale field trials, variations much greater than this can occur. Fryer and Kirkland (1970) reported that recovery of linuron from soil immediately after application varied from 31 to 144% of the nominally applied rate in 15 separate field experiments. The mean recovery was 78% and the coefficient of variation 40%. In one of these experiments, a number of filter papers was distributed over the soil surface prior to herbicide application and the mean deposition on these was 73% of the nominal dose and the coefficient of variation was 23%. Similar variations in point to point soil concentrations of herbicide have been reported by Wauchope *et al.* (1977). Given this inherent variability in spray deposition, it is surprising that soil-applied herbicides work as reliably as they do. Wauchope *et al.* (1977) commented that it is fortunate that weeds apparently "sample" much larger volumes of soil than standard soil borers and suggested that vertical redistribution of herbicide in the soil plays an important role in producing a homogeneous weed response. It also seems probable that for many weed species/herbicide combinations, the dose required for control will be small compared with the total available amount so that high susceptibility will mask effects of uneven distribution. A critical assessment of variability in application relative to the level of weed control achieved might determine whether improvements in application could achieve more consistent results.

Another aspect of herbicide application that may be significant to its subsequent activity is the nature of the soil surface to which the compound is applied. Riley and Morrod (1976) suggested that if the surface was very uneven at the time of application, subsequent weathering would lead to an even greater inequality in herbicide distribution. They argued that some weeds, particularly those which germinate within soil clods would be exposed to very low concentrations of herbicide and therefore not controlled. Larson *et al.* (1958) studied the effects of soil surface conditions on the action of various pre-emergence herbicides in soya bean and cotton.

They found that rolling to produce a very smooth surface reduced effectiveness of chlorpropham, dinoseb, and PCP against broad-leaved and grass weeds mainly because of increased germination of weeds in rolled plots. Similar results were obtained by Roberts and Hewson (1971) who examined weed control by dichlormate and propachlor. They produced a rough surface by harrowing, a surface of intermediate roughness by raking the harrowed plots and a very fine surface by rotovating the harrowed plots. They then superimposed a rolled or non-rolled treatment on all three types of surface. There was a consistent pattern of increased seedling numbers with increasing fineness of the surface soil and rolling always increased numbers compared with those on comparable plots which were not rolled. Of the two herbicides examined, dichlormate did perform more efficiently on a percentage basis on finer, firmer seedbeds but there was no difference in the density of surviving weeds. With propachlor, percentage weed kill was similar in all treatments; the numbers of surviving weeds were proportional to those which emerged. From a practical point of view there was no advantage in providing a fine even surface prior to herbicide application, and this proved to be a positive disadvantage with propachlor.

It is difficult to make generalizations about all herbicides on the basis of this limited experimental evidence, but it appears that the popular concept that herbicides work more efficiently when applied to a smooth soil surface is open to question since in practical terms it is the number of weeds remaining that is important.

## III. Selectivity

Various aspects of herbicide selectivity were reviewed by Holly (1976) who suggested three mechanisms by which selectivity of soil-applied herbicides can be achieved.

1. By variations in the position of roots or other organs taking up herbicide in relation to the location and availability of the herbicide in soil.
2. By differential absorption of chemical.
3. Following uptake, by differential translocation, differential rates of herbicide metabolism, or by selective effects on metabolic systems at the site of action.

Aspects of the mode of action of herbicides and of their degradation in plants in relation to selectivity have been the subjects of intensive review in recent years (Ashton and Crafts, 1973; Kearney and Kaufman, 1975) and further comment here is not necessary. This discussion will concentrate on aspects of differential availability in soil and possible ways in which activity or selectivity might be modified.

Most soil-applied herbicides are sprayed on the soil surface and either left there undisturbed or mechanically incorporated into surface soil layers. If weeds germinate near the soil surface and the crop is deep sown, the herbicide may reach the roots of the weed but not of the crop while it is at a susceptible stage of growth. This type of selectivity is often referred to as depth protection, and when selectivity rests on this basis it is dependent on conditions not favouring extensive downward movement. It seems doubtful that certain selectivity can ever be achieved solely by depth protection and in almost all crop-herbicide situations there must be some degree of physiological tolerance involved as well. One example where depth protection may be important is the use of simazine for weed control in field beans. Because of the risk of crop damage, light soils are excluded from the recommendations. Depth protection can also occur with herbicides which enter the shoot, for example the use of di-allate and tri-allate for selective control of wild oat in cereals. Parker (1963) showed that inhibition of shoot growth by di-allate resulted from direct entry into the shoot and that the greatest zone of sensitivity was just above the coleoptile node near the apical meristem. Shallow incorporation of these herbicides gives selective weed control if the cereal is sown in untreated soil about 2 to 3 cm below the herbicide layer. This is because the mesocotyl of wild oat moves the coleoptile node upwards into treated soil at an early stage of growth whereas this zone of sensitivity in wheat and barley remains close to the seed in untreated soil. In all examples where depth protection is important in selectivity, the recommendations for herbicide use call for specified depths of untreated soil above the crop seed.

Other examples of selectivity based on depth protection were reviewed by Holly (1976).

## IV. Modification of Activity and Selectivity

Although there are over 150 herbicides available throughout the world, there are still problems of obtaining selective control of all the weeds which occur in different crops. If an increase in the margin of selectivity could be achieved this might allow use of herbicides in crops where normally the risk of damage is too great (Blair *et al.*, 1976).

### A. Adsorbents

One possible way to enhance selectivity is to inactivate the herbicide with a suitable adsorbent. The action of this adsorbent however must be localized to the immediate area around the crop plant in order not to affect overall weed control. Activated charcoal has been the most widely used adsorbent and its first reported use was to inactivate the soil effects of 2,4-D (Lucas and Hamner, 1947). Subsequently, Arle *et al.* (1948) demonstrated that treatment of sweet potato sprouts with a slurry of activated charcoal prior to transplanting into 2,4-D treated soil gave complete protection from the herbicide. Gibson (1964) dipped roots of 33 different herbaceous species into activated charcoal before transplanting and found that some species were protected from subsequent simazine treatments whereas others were not. He suggested that only those species which showed some inherent tolerance to the herbicide could be protected. Root dipping of strawberry runners in dry charcoal powder has been shown to give protection from simazine (Robinson, 1965) and Kratky *et al.* (1970) demonstrated that root treatment of strawberry plants in this way also conferred some protection to daughter runners produced by the treated plants. The runners were best protected against subsequent simazine treatments but some protection from terbacil and chlorthal-dimethyl was also achieved.

There have been several studies on the use of bands of charcoal, either on the soil surface or as a layer between the seed and the soil surface to protect direct-seeded crops. Ripper (1956) described special equipment which sowed the seed, deposited a subsurface charcoal band above the seed and then sprayed herbicide on the soil surface. He reported excellent protection to sugar beet from phytotoxic effects of 2,4-D, 2,4,5-T, propham and EPTC although the method gave little protection from TCA. Examples of band applications of charcoal to the soil surface over the drill row have been described for protection of direct-seeded alfalfa from a triazine herbicide (Linscott and Hagin, 1967) for protection of Italian ryegrass from diuron (Burr *et al.*, 1972) and for protection of direct-seeded asparagus from chloramben, chlorbromuron, linuron and nitralin (Ogg, 1978). Chandler *et al.* (1978) showed that only carbon on the immediate soil surface was effective in improving selectivity of diuron in cotton. It did not prove effective when applied with the seed via the seed hopper or when incorporated into the surface soil.

Ogg (1978) observed reduced weed control in the crop row as a result of band application of charcoal which agreed with earlier observations of Kratky and Warren (1971). Kratky and Warren also noted that heavy rainfall could erode and displace a surface charcoal layer sufficient to significantly reduce its protective effect. They developed an alternative technique in order to overcome these two disadvantages. They buried seeds in holes 2 cm deep and 2·5 cm diameter which were filled with a charcoal-vermiculite mixture. This gave protection to direct-seeded cucumbers and tomatoes from surface applications of simazine. They subsequently modified the technique by using wafers comprising seed embedded in a mixture of activated charcoal, vermiculite and fertilizer which allowed the possibility of mechanized field sowing.

A third way in which charcoal has been used to increase crop tolerance is by production of pelleted seed. In general, this has not been successful because the seedlings rapidly grow away from the treated zone (Croxford *et al.*, 1975). Success has been reported for the protection of *Rosa* spp. against simazine and neburon (Stryckers and van Himme, 1967) and rice against butachlor, chloramben and oxidiazon (Nangju *et al.*, 1976).

One of the features of the above results is that in several instances, protection was afforded against some compounds and not others. This in part reinforces the suggestion that the crop must have some inherent tolerance of the herbicide (Gibson, 1964). It probably also reflects that any one charcoal will not be equally adsorptive for all herbicides and that all charcoals are not equally efficient adsorbents. For example, Coffey and Warren (1969) showed that the relative adsorption of eight herbicides by two different charcoals was chlorpropham > trifluralin > 2,4-D > diphenamid > chlorthal-dimethyl > dinoseb > chloramben > paraquat. They also showed that steam activated charcoal was more adsorptive than charcoal which had been chemically activated. Jordan and Smith (1971) compared the adsorptive properties of 24 different charcoals and found the best to be 100 times more adsorptive than the poorest, but there was no clear reason for this difference.

## B. Chemical Protectants

The use of specific chemicals as antidotes or protectants to enhance herbicide selectivity was reviewed by Blair *et al.* (1976) and the chemistry and mode of action of these materials was reviewed by Pallos and Casida (1978). The idea of using specific herbicide antidotes was first suggested by Hoffman (1962) who showed that seed dressings of 4-chloro-2-hydroxyiminoacetanilide, 2,4-dichloro-9-xanthenone or *N*-methyl-3,4-dichlorobenzene sulfonamide protected wheat against damage by barban. Hoffman (1969) subsequently reported a similar effect from 1,8-naphthalic anhydride. Burnside *et al.* (1971) showed that treatment of maize seed with 0·5% by weight of this compound conferred protection against EPTC and butylate and subsequently various other uses of naphthalic anhydride for different herbicide-crop combinations were reported (Chang *et al.*, 1973; Jordan and Joliffe, 1971; Blair and Dean, 1976; Parker and Dean, 1976). Naphthalic anhydride requires high doses in the soil to be effective and can therefore only be used economically as a seed treatment. The newer compound, *N,N*-diallyl-2,2-dichloroacetamide (R-25788) has also been used as a seed dressing in maize and has been shown to give somewhat better protection against EPTC damage than naphthalic anhydride (Chang *et al.*, 1973). Because its effects are more specific and it is effective at lower concentrations than naphthalic anhydride, R-25788 can be applied directly to the soil and formulated mixtures with EPTC, butylate and other thiol-carbamates such as vernolate are commercially available.

Although naphthalic anhydride and R-25788 are the only two compounds which have so far been developed commercially, other chemicals have shown potential as protectants. Examples are 2,2-dimethyl-3-dichloro-acetyloxazolidine (Spotanski and Burnside, 1973), 3-hydroxy-5-methyl-isoxazole (Ogawa and Ota, 1975) and an experimental compound, CGA-43089 which is currently being evaluated for its effect as a protectant against metolachlor injury in sorghum (Ellis *et al.*, 1978; Nyffeler *et al.*, 1978).

## C. Formulation

The main effects of formulation on the activity of soil-applied herbicides are indirect through control of persistence. For example, losses of compounds which are volatile or photochemically unstable can be reduced by formulation as granules rather than wettable powders or emulsifiable concentrates. It has also been shown that certain surfactants added to normal commercial formulations can directly affect the activity of some herbicides. In greenhouse trials, control of *Lolium perenne* in cotton by diuron and in corn by atrazine was improved without increased crop damage by addition of non-ionic polyoxyethylene surfactants to the spray solution (Kramer and Manning, 1971), and trifluralin phytotoxicity to *Echinochloa crus-galli* var. *frumentacea* and *Panicum dichotomiflorum* was increased by addition of 1% or 2% Soil Penetrant 80—a mixture of polyoxyethylene ether surfactants (Hatfield and Sweet, 1973). Yapel (1972) and Van Valkenburg and Yapel (1972) studied the effects of polyoxyethylene sorbitan monooleate (Tween 80) at 0·1% (w/v) on the activities of 15 monosubstituted trifluoromethanesulphonanilides and demonstrated that the biological activity of some compounds was enhanced whereas with others it was reduced. The effects varied with the test species and in some instances, when activity against a grass weed (*Setaria* sp.) was increased, activity against a broadleaf weed (*Brassica kaber*) decreased and vice-versa. Addition of Tween 80 increased the hydrophilicity of all 15 compounds and, in general, the changes in biological activity caused by the surfactant could be related to the changes induced in the octanol/water partition coefficients of the various compounds. Mercer and Hill (1977, 1978) have reported increased water solubility and reduced adsorption together with increased mobility and activity in soil of simazine plus 5% Lissapol NDP.

They suggested that addition of surfactant might permit use of simazine on organic soils where activity is normally too low.

All of the above experiments were made under laboratory or glasshouse conditions; the few results available from field trials have shown less promise. Hatfield and Sweet (1973) reported little benefit from additional surfactants on the activity of atrazine or chlorpropham and Parochetti *et al.* (1977) demonstrated that several surfactants did not significantly improve the activities of a number of soil-applied herbicides.

It is difficult to draw conclusions from this limited experimental evidence but it appears that the improvements in activity of soil-applied herbicides by modification of formulation with additional surfactants may not be sufficiently large to be of practical significance.

## V. Conclusions and Suggestions for Future Research

It is often stated that one of the main disadvantages of soil-applied herbicides is their unreliable performance but when the large number of soil, plant and environmental factors which interact to determine activity is considered, it is surprising that they work as reliably as they do. In the introduction to a paper published in 1958—shortly after the first soil-applied herbicides became available—Larson *et al.* commented that "variable results have been attributed in part at least to a large number of factors, a few of which are: soil type, soil moisture, soil surface conditions, rainfall, plant varietal differences, and application variation and error". Most of these have been considered in the present review and there is little doubt that our understanding of the relative importance of each has increased. In many situations, none of these factors dominates over the others, so all must be taken into account when we try to explain or predict activity in a given situation (Riley and Morrod, 1976). These interactions often make interpretation of results from field experiments difficult and there is a need for more quantitative data on individual factors from experiments designed to minimize effects from others. Good examples of this approach were described by Best *et al.* (1975) and Ladlie *et al.* (1976) who studied effects of pH on herbicide activity. The pH of adjacent field plots was adjusted with suitable amendments and the influence of variable pH on activity could then be assessed in one soil under one set of weather conditions. A similar approach to the study of herbicide-soil nutrient interactions should be possible. This would help to define whether the few qualitative indications of effects are of real significance to practical herbicide use.

Because weather conditions, particularly rainfall, play an important role in influencing herbicide performance there is again a need for more quantitative data. For example, data from experiments on a range of soil types to evaluate the amounts of rainfall or irrigation needed and its timing for optimum weed control are required. This might suggest ways in which irrigation could be used to ensure herbicide "activation" under dry conditions. A more basic understanding of the interactions between herbicide distribution/time of exposure/age of plant in relation to biological response is also essential to interpretation of rainfall effects on activity.

There have been many attempts over the years to define precise adjustments of herbicide application rates based on soil properties. In practice, commercial rates are still assessed following extensive field testing and are only varied within large groups of soils. More precision in rate adjustment at present seems impracticable for a number of reasons. Weather conditions can dominate over soil properties within groups of soils (e.g. Hance *et al.*, 1968); there are often marked variations in soil properties within a field and it would not be possible in practice to vary the rate from one small area to another; increased precision in defining rates would serve little purpose, unless application itself could be made more precise (e.g. Fryer and Kirkland, 1970).

Modification of crop tolerance or herbicide activity by use of protective adsorbents or chemical protectants would appear to offer the best opportunities for improving herbicide performance since this might extend the use of herbicides to situations where normally the margin of selectivity is too small.

## References

Aberg, E. and Stecko, V. (1976). Internal factors affecting toxicity. In *Herbicides: Physiology, Biochemistry, Ecology*. Vol. 2. (L. J. Audus, Ed.) pp. 175–200, Academic Press, London and New York.

Adams, R. S. (1965). Phosphorus fertilization and the phytotoxicity of simazine. *Weeds*, **13**, 113–116.

Arle, H. F., Leonard, O. A. and Harris, V. C. (1948). Inactivation of 2,4-D on sweet potato slips with activated carbon. *Science, N.Y.*, **107**, 247–248.

Ashley, R. A. and Rahn, E. M. (1967). Persistence of atrazine at two locations as affected by soil incorporation and certain additives. *Proc. 21st NEast Weed Control Conf.*, pp. 557–562.

Ashton, F. M. and Crafts, A. S. (1973). *Mode of Action of Herbicides*. Wiley, New York.

Best, J. A., Weber, J. B. and Monaco, T. J. (1975). Influence of soil pH on *s*-triazine availability to plants. *Weed Sci.*, **23**, 378–382.

Bingham, S. W. and Upchurch, R. P. (1959). Some interactions between nutrient level (N, P, K, Ca) and diuron in the growth of cotton and Italian ryegrass. *Weeds*, **7**, 167–177.

Blair, A. M. and Dean, M. L. (1976). Improvements in selectivity of perfluidone against *Rottboellia exaltata* in maize with herbicide protectants. *Weed Res.*, **16**, 47–52.

Blair, A. M., Parker, C. and Kasasian, L. (1976). Herbicide protectants and antidotes. *PANS*, **22**, 65–74.

Brockman, F. E. (1975). *Factors affecting activity and selectivity of methazole in seedling alfalfa.* Ph.D. Thesis, Cornell University.

Burnside, O. C., Wicks, G. A. and Fenster, C. R. (1971). Protecting corn from herbicide injury by seed treatment. *Weed Sci.*, **19**, 565–568.

Burr, R. J., Lee, W. O. and Appleby, A. P. (1972). Factors affecting use of activated carbon to improve herbicide selectivity. *Weed Sci.*, **20**, 180–183.

Chandler, J. M., Wooton, O. B. and Fulgham, F. E. (1978). Influence of placement of charcoal on protection of cotton (*Gossypium hirsutum*) from diuron. *Weed Sci.*, **26**, 239–244.

Chang, F. Y., Stephenson, G. R. and Bandeen, J. D. (1973). Comparitive effects of three EPTC antidotes. *Weed Sci.*, **21**, 292–295.

Coffey, D. L. and Warren, G. F. (1969). Inactivation of herbicides by activated carbon and other adsorbents. *Weed Sci.*, **17**, 16–19.

Corbin, F. T., Upchurch, R. P. and Selman, F. L. (1971). Influence of pH on the phytotoxicity of herbicides in soil. *Weed Sci.*, **19**, 233–239.

Croxford, D. E., Elkins, D. M. and Kapusta, G. (1975). Crop protectants and herbicides for orchardgrass—alfalfa establishment. *Weed Sci.*, **23**, 414–418.

Day, B. E., Jordan, L. S. and Jolliffe, V. A. (1968). The influence of soil characteristics on the adsorption and phytotoxicity of simazine. *Weed Sci.*, **16**, 209–213.

Doll, J. D. and Meggitt, W. F. (1969). Uptake and effectiveness of herbicides at different nutrient levels and combinations. *Abs. Mtg. Weed Sci. Soc. Am.*, 212.

Dubey, H. D. and Freeman, J. F. (1964). Influence of soil properties and microbial activity on the phytotoxicity of linuron and diphenamid. *Soil Sci.*, **97**, 334–340.

Eagle, D. J. (1976). Soil texture classification for the adjustment of herbicide dose. *Proc. 1976 Br. Crop Prot. Conf.—Weeds*, pp. 981–988.

Ellis, J. F., Peek, J. W., Boehle, J. and Muller, G. (1978). A new herbicide safener which permits effective grass control in sorghum. *Abs. 1978 Mtg. Weed Sci. Soc. Am.*, 44.

Fryer, J. D. and Kirkland, K. (1970). Field experiments to investigate long-term effects of repeated applications of MCPA, tri-allate, simazine and linuron: Report after six years. *Weed Res.*, **10**, 133–158.

Fryer, J. D. and Makepeace, R. J. (1978). *Weed control Handbook.* Vol. 2. *Recommendations.* Blackwell Scientific Publications, Oxford.

Gibson, W. B. (1964). Pointers to weed control in herbaceous plants. *Comml. Grow.*, 3594, 973.

Graham-Bryce, I. J. (1972). Herbicide movement and availability in soils. *Proc. 11th Br. Weed Control Conf.*, 1193–1202.

Grover, R. (1968). Influence of soil properties on phytotoxicity of 4-amino-3,5,6-trichloropicolinic acid (picloram). *Weed Res.*, **8**, 226–232.

Hance, R. J., Hocombe, S. D. and Holroyd, J. (1968). The phytotoxicity of some herbicides in field and pot experiments in relation to soil properties. *Weed Res.*, **8**, 136–144.

Harris, C. I. and Sheets, T. J. (1965). Influence of soil properties on adsorption and phytotoxicity of CIPC, diuron and simazine. *Weeds*, **13**, 215–219.

Hatfield, H. H. and Sweet, R. D. (1973). Adjuvants for pre- and post-emergence herbicide application. *Proc. 27th NEast. Weed Control Conf.*, pp. 170–176.

Hoffman, O. L. (1962). Chemical seed treatments as herbicidal antidotes. *Weeds*, **10**, 322–323.

Hoffman, O. L. (1969). Chemical antidotes for EPTC on corn. *Abs. Mtg. Weed Sci. Soc. Am.*, 12.

Holly, K. (1976). Selectivity in relation to formulation and application methods. In *Herbicides: Physiology, Biochemistry, Ecology*. Vol. 2. (L. J. Audus, Ed.) pp. 249–277, Academic Press, London and New York.

Jordan, L. S. and Day, B. E. (1962). Effects of soil properties on EPTC phytotoxicity. *Weeds*, **10**, 212–215.

Jordan, L. S. and Jolliffe, V. A. (1971). Protection of plants from herbicides with 1,8-naphthalic anhydride as illustrated with sorghum. *Bull. Environ. Contam. and Toxicol.*, **6**, 417–421.

Jordan, P. D. and Smith, L. W. (1971). Adsorption and deactivation of atrazine and diuron by charcoals. *Weed Sci.*, **19**, 541–544.

Kearney, P. C. and Kaufman, D. D. (1975). *Herbicides: Chemistry, Degradation and Mode of Action*. Marcel Dekker Inc., New York.

Kramer, J. A. and Manning, D. T. (1971). Effect of non-ionic surfactants on the pre-emergence herbicidal activity of atrazine and diuron. *Proc. N. Cent. Weed Control Conf.*, **26**, 107–108.

Kratky, B. A. and Warren, G. F. (1971). Activated carbon-vermiculite mixture for increasing herbicide selectivity. *Weed Sci.*, **19**, 79–81.

Kratky, B. A. and Warren, G. F. (1973). Water-soil-plant interactions with terbacil *Weed Sci.*, **21**, 451–454.

Kratky, B. A., Coffey, D. L. and Warren, G. F. (1970). Activated carbon root dips on transplanted strawberries. *Weed Sci.*, **18**, 577–580.

Ladlie, J. S., Meggitt, W. F. and Penner, D. (1976). Effect of pH on metribuzin activity in the soil. *Weed Sci.*, **24**, 505–507.

Lambert, S. M. (1966). The influence of soil-moisture content on herbicidal response. *Weeds*, **14**, 273–275.

Larson, R. E., Klingman, D. L. and Fletchall, O. H. (1958). The effect of soil smoothing devices on the action of pre-emergence herbicides in soybeans and corn. *Weeds*, **6**, 126–132.

LeBaron, H. M. (1970). Ways and means to influence the activity and the persistence of triazine herbicides in soils. *Residue Rev.*, **32**, 311–353.

Leefe, J. S. (1968). Effect of soil pH on the phytotoxicity of simazine to strawberries. *Can. J. Pl. Sci.*, **48**, 424–425.

Linscott, D. L. and Hagin, R. D. (1967). Protecting alfalfa seedlings from a triazine with activated charcoal. *Weeds*, **15**, 304–306.

Lucas, E. H. and Hamner, C. L. (1947). Inactivation of 2,4-D by adsorption on charcoal. *Science N.Y.*, **105**, 340–342.

McGlamery, M. D. and Slife, F. W. (1966). The adsorption and desorption of atrazine as affected by pH, temperature and concentration. *Weeds*, **14**, 237–239.

McReynolds, W. D. and Tweedy, J. A. (1970). Effect of nitrogen form on simazine accumulation in corn, soybeans and rye. *Weed Sci.*, **18**, 270–272.

Menges, R. M. and Tamez, S. (1973). Effect of soil incorporation on selectivity, movement and persistence of herbicides in onion plantings. *J. Amer. Soc. Hort. Sci.*, **98**, 390–393.

Mercer, E. R. and Hill, D. (1977). Influence of surfactants and dispersants on the herbicidal activity of simazine. *ARC Letcombe Laboratory, Annual Report 1976.* pp. 94–96.

Mercer, E. R. and Hill, D. (1978). Herbicides which are applied to the soil. *ARC Letcombe Laboratory, Annual Report 1977* pp. 76–77.

Mindreboe, K. J. (1970). Weed control with pre-emergence herbicides in relation to rainfall. *Weed Res.*, **10**, 71–74.

Nangju, D., Plucknett, D. L. and Obien, S. R. (1976). Performance of carbon-coated rice seed on flooded soil. *Weed Sci.*, **24**, 73–80.

Nearpass, D. C. (1965). Effects of soil acidity on the adsorption, penetration and persistence of simazine. *Weeds*, **13**, 341–346.

Neururer, H. (1972). Studies on the behaviour of herbicides in soil. *Bodenkultur.*, **23**, 43–73. (*Weed Abstracts*, No. 934, 1973).

Nyffeler, A., Gerber, H. R. and Hensley, J. (1978). Laboratory studies on the behaviour of the safener CGA-43089. *Abs. 1978 Mtg. Weed Sci. Soc. Am.*, 45.

Ogawa, M. and Ota, Y. (1975). Effects of 3-hydroxy-5-methylisoxazole on the reduction of herbicide injury to rice seedlings. *Abstracts 5th Asian Pacific Weed Control Conference*, 91.

Ogg, A. G. (1978). Herbicides and activated carbon for weed control in direct-seeded asparagus (*Asparagus officinalis*). *Weed Sci.*, **26**, 284–286.

Pallos, F. M. and Casida, J. E. (1978). *Chemistry and Action of Herbicide Antidotes.* Academic Press, New York and London.

Parker, C. (1963). Factors affecting the selectivity of 2,3-dichloroallyl diisopropylthiolcarbamate (di-allate) against *Avena* spp in wheat and barley. *Weed Res.*, **3**, 259–276.

Parker, C. and Dean, M. L. (1976). Control of wild rice in rice. *Pestic. Sci.*, **7**, 403–416.

Parochetti, J. V., Wilson, H. P. and Beste, C. E. (1977). Effects of several adjuvants on pre-emergence and post-emergence herbicides in 1976. *Proc. 31st NEast. Weed Sci. Soc.*, pp. 105–112.

Ries, S. K., Chmiel, H., Dilley, D. R. and Filner, P. (1967). The increase in nitrate reductase activity and protein content of plants treated with simazine. *Proc. natn. Acad. Sci. USA*, **58**, 526–532.

Riley, D. and Morrod, R. S. (1976). Relative importance of factors influencing the activity of herbicides in soil. *Proc. 1976 Br. Crop Prot. Conf.—Weeds*, 971–980.

Ripper, W. E. (1956). A new method of selective weed control for related plants, in particular, broadleaf weeds in beet. *Proc. 3rd Br. Weed Control Conf.*, 225–232.

Roberts, H. A. and Hewson, R. T. (1971). Herbicide performance and soil surface conditions. *Weed Res.*, **11**, 69–73.

Robinson, D. W. (1965). The use of adsorbents and simazine on newly planted strawberries. *Weed Res.*, **5**, 43–51.

Schmidt, R. R. (1973). Calculations on the herbicidal activity of metribuzin in relation to various soil properties. *Proc. European Weed Research Council Symposium, Herbicides and the Soil*, 24–30.

Schnappinger, M. G., Trapp, C. P., Boyd, J. M. and Pruss, S. W. (1977). Soil pH and triazine activity in no-tillage maize as affected by nitrogen and lime application. *Proc. 31st NEast Weed. Sci. Soc. Am.*, 116.

Selman, F. L. and Upchurch, R. P. (1961). The herbicide–phosphorus interaction. *Proc. 14th sth. Weed Control Conf.*, 480.

Selman, F. L. and Upchurch, R. P. (1970). Regulation of amitrole and diuron toxicity by phosphorus. *Weed Sci.*, **18**, 619–623.

Sheets, T. J., Crafts, A. S. and Drever, H. R. (1962). Influence of soil properties on the phytotoxicities of the *s*-triazine herbicides. *J. agric. Fd Chem.*, **10**, 458–462.

Splittstoesser, W. E. and Derscheid, L. A. (1962). Effects of environment upon herbicides applied pre-emergence. *Weeds*, **10**, 304–307.

Spotanski, R. F. and Burnside, O. C. (1973). Reducing herbicide injury to sorghum with crop protectants. *Weed Sci.*, **21**, 531–536.

Stickler, R. L., Knake, E. L. and Hinesley, T. D. (1969). Soil moisture and effectiveness of pre-emergence herbicides. *Weed Sci.*, **17**, 257–259.

Stolp, C. F. and Penner, D. (1973). Enhanced phytotoxicity of atrazine-phosphate combinations. *Weed Sci.*, **21**, 37–40.

Stryckers, J. and Van Himme, M. (1967). Protection against soil-acting herbicides in seedbeds of *Rosa* seedlings. *Med. Fac. Landbouww. Rijksuniv. Gent,* **32**, 964–976.

Upchurch, R. P. and Mason, D. D. (1962). The influence of soil organic matter on the phytotoxicity of herbicides. *Weeds*, **10**, 9–14.

Upchurch, R. P., Selman, F. L., Mason, D. D. and Kamprath, E. J. (1966). The correlation of herbicidal activity with soil and climatic factors. *Weeds*, **14**, 42–49.

Valkenburg, W. Van. and Yapel, A. F. (1972). The effect of penetrant aid on pre-emergent herbicidal activity of trifluoromethanesulfonanilides. In *Biological Correlations—the Hansch Approach.* Advances in Chemistry Series No. 114. pp. 252–267. American Chemical Society, Washington D.C.

Walker, A. (1971). Effects of soil moisture content on the availability of soil-applied herbicides to plants. *Pestic. Sci.*, **2**, 56–59.

Walker, A. and Roberts, H. A. (1975). Effects of incorporation and rainfall on the activity of some soil-applied herbicides. *Weed Res.*, **15**, 263–269.

Wauchope, R. D., Chandler, J. M. and Savage, K. E. (1977). Soil sample variation and herbicide incorporation uniformity. *Weed Sci.*, **25**, 193–196.

Weber, J. B. (1970). Mechanisms of adsorption of *s*-triazines by clay colloids and factors affecting plant availability. *Residue Rev.*, **32**, 93–130.

Weber, J. B., Perry, P. W. and Ibaraki, K. (1968). Effect of pH on the phytotoxicity of prometryne applied to synthetic soil media. *Weed Sci.*, **16**, 134–136.

Wiese, A. F. and Smith, D. T. (1970). Herbicidal activity as affected by soil incorporation and rainfall. *Weed Sci.*, **18**, 515–517.

Yapel, A. F. (1972). Structure-activity correlations for meta- and para-substituted trifluoromethanesulfonanilide pre-emergence herbicides. In *Biological Correlations—the Hansch Approach.* Advances in Chemistry Series No. 114. pp. 183–251. American Chemical Society, Washington D.C.

# 9 Effects on Soil Microflora

*M. P. Greaves*
Agricultural Research Council Weed Research Organization,
Begbroke Hill,
Yarnton, Oxford OX5 1PF,
UK

*H. P. Malkomes*
Biologische Bundesanstalt für Land und Forstwirtschaft,
Institut für Unkrautforschung,
3300 Braunschweig,
F.R. Germany



## I. Introduction

Despite many years of study, we are unable unequivocally to identify and quantify those effects of pesticides which may harm soil microbial function and, in particular, soil fertility.

For many reasons we must attempt to improve urgently our understanding of herbicide/microorganism interactions. Not the least is the introduction of the requirement for data on "effects on the soil microflora" in the registration procedures of various countries, notably the USA (The Environmental Protection Agency (EPA) Guidelines for Registering Pesticides in the US, Anon., 1978), UK (Pesticide Safety Precautions Scheme Wildlife Data Guide, Anon., 1971) and the Council of Europe (Agricultural Pesticides, Anon., 1973). There is already concern that the escalating costs of obtaining the environmental data for new products is bound to curtail the development of new chemicals (Braunholtz, 1977; Norman, 1978). Thus, it is imperative that such requirements are based on sound scientific principles and that the data obtained are valid and indicate reliably the true potential of the chemical for causing harm in the environment.

It is the intention of this chapter to illustrate the deficiencies in our knowledge and, more particularly, to indicate where we believe future research might be directed to improve the situation. Detailed information on methods and effects can be found in Greaves *et al.* (1976), Grossbard (1976), Anderson (1978a, b), Johnen (1978), Simon-Sylvestre and Fournier (1979).

## II. Current Concepts in Methodology

A brief outline of current concepts in methodology is presented by Anderson and Greaves (1977) and further discussion is provided by Greaves (1978). A detailed consideration of available methods (Anderson, 1978a) vividly illustrates both the diversity of methods used and the many problems they involve.

Curl and Rodrigues-Kabana (1971, 1977), Johnen and Drew (1977), Greaves *et al.* (1978a) and Atlas *et al.* (1978) have published programmes of methods for identifying side-effects of pesticides on non-target organisms. These programmes cover most of the methods and approaches which are considered to be the best available at present.

The magnitude and duration of side-effects are influenced greatly by the persistence of the pesticide. The reader is referred to specialist chapters in this book for details of herbicide persistence and microbial degradation.

### A. Microbial Processes

The most important consequences of effects of herbicides on soil microorganisms will be those relating to soil fertility. Thus, the most useful information will come from the study of major microbial processes in soil. The use of radioactive isotopes has greatly improved the detailed study of substrate transformation in soils but many problems still exist, especially in relating the statistical significance of observed changes to their biological significance.

#### 1. *Nitrogen transformations*

The most popular of the methods used at present are those which measure the mineralization of soil nitrogen, whether native in the soil or freshly added as plant material. Undoubtedly, the status of mineral nitrogen as a major plant nutrient gives impetus to such studies. However, modern agriculture uses considerable quantities of nitrogenous fertilizers and, in some instances, contributions from soil nitrogen become small in comparison. On the other hand, the benefits from mineralization of soil nitrogen in many soils, the escalating costs of fertilizers, and their limited use in developing countries, mean that continuing attention must be paid to ensure that the soil's capacity to provide at least some of the plant's nitrogen requirement is maintained. It is right, therefore, that effects of herbicides on mineralization of soil nitrogen remain a prime target for investigation.

The chemical analytical methods for following nitrogen transformations, especially for measuring nitrate-, nitrite- and ammonium-nitrogen, are simple, sensitive, accurate and reproducible. In addition, they are readily automated (Marsh *et al.*, 1979) and thus allow optimum replication within experiments. Indeed, these advantages may partly explain the frequency with which the methods are used (Anderson, 1978b).

Anderson (1978b) concluded that "ammonification seems to benefit from the application of all groups of pesticides to soil". Similarly, he showed that the effects of herbicides on nitrification and denitrification were most often stimulatory. Though this information may be regarded as indicating a general lack of harmful effects by herbicides, many problems remain. For example, a herbicide may stimulate ammonification while inhibiting nitrification (Marsh and Greaves, 1979) so that the balance between ammonium and nitrate ions could be altered. It remains to be decided whether this is good or bad for agricultural crops in general or for individual species in particular.

Despite the problems associated with measuring effects of herbicides on nitrogen transformations, which have been detailed by Anderson (1978a, b) it seems that effects of herbicides on nitrogen transformation are most appropriately studied using soil incubation techniques such as that described by Greaves *et al.* (1978a). Additional tests using added substrate such as plant material might also be useful.

### 2. *Nitrogen fixation*

It is well established that many pure cultures of free-living microbial species indigenous to soil can fix atmospheric nitrogen. What is less clear is whether these organisms function significantly in the field. Jensen (1965) suggested that the low levels of available carbon and energy sources in soils effectively limit nitrogen-fixation by free-living organisms. Perhaps the most important organisms in this context are the blue-green algae, particularly where soil disturbance by tillage is minimized. Thus, in a natural system like sand dunes, or agricultural systems using minimum or no-tillage techniques or in paddy soils, these algae may contribute significant amounts of mineral nitrogen. Reproduction of such conditions in laboratory tests is difficult if not impossible and this makes interpretation of experimental data difficult, especially where they are to be used for registration purposes. However, the increasing use of minimum tillage systems, and the importance of paddies in large parts of the world, suggests a need for satisfactory laboratory tests.

Nitrogen-fixation by symbiotic organisms associated with legumes is of unarguable importance throughout the world. The susceptibility of *Rhizobium* spp. to herbicides has been investigated by a variety of methods, ranging from plate counts of *Rhizobium* cells to examination of nodulation of legumes (Anderson, 1978a). The findings are that the susceptibility of the host plant to herbicides is generally greater than that of the bacterium. Reductions in nodulation and nitrogen-fixation are usually ascribed to plant damage rather than to direct effects of the herbicide on the microorganisms.

The methods of measuring nitrogen-fixation by both free-living and symbiotic microorganisms are usually the same. The acetylene-reduction method is used most often despite some serious disadvantages. For example, it is implied from the work of Knowles *et al.* (1973) that the nitrogen-fixing activity of anaerobic organisms may be inhibited by the acetylene substrate. In addition, Witty (1979) has suggested that ethylene production may occur from sources other than nitrogen fixation, thus giving spurious results.

Johnen *et al.* (1978) and Greaves *et al.* (1978b) have highlighted the problems of assaying side-effects of herbicides on legumes. Frequently, large standard errors are associated both with the acetylene-reduction method and with measurements of nodule numbers and weights. Plant weights are a more satisfactory means of assessing effects on the legume.

In spite of these and other problems, examinations of herbicide effects on legume nodulation and growth are important. At present, it seems that the most valuable information comes from measurement of plant growth rather than from more specialized measurements of *Rhizobium* populations or function.

### 3. *Respiration*

Respiration measurements are probably the commonest methods of studying microbial activity in soil. As pointed out by Stotzky (1965) respiration usually correlates well with other soil activities, such as transformations of carbon, phosphorous and nitrogen. Domsch *et al.* (1979) have shown similar correlations. Certainly the use of radioactive isotopes has alleviated many of the problems of non-specificity involved in traditional methods of measuring respiration. At the same time, these have been succeeded by problems of induced changes in microbial population composition and in enzyme synthesis associated with the introduction of specific substrates into the soil. Perhaps the use of uniformly-labelled $^{14}C$-plant material provides the best opportunity for assessing the impact of herbicides on organic matter breakdown.

One of the most serious criticisms of respiration measurements, whether of $CO_2$ production or $O_2$ uptake, is that a measured nil-effect may hide large, counterbalancing, stimulation and inhibition of different major parts of the soil microflora. This criticism certainly leads to the conclusion that respiration measurements should be used in conjunction with other methods of assessing more specific side-effects on microbial function.

Where respiration measurements on soil are required, and especially where they are to last for long periods, the preferred method is to incubate the soil under test in a flow of $CO_2$-free air and to measure $CO_2$ production by titrimetry (Greaves *et al.*, 1978a) or more sophisticated analytical techniques (Anderson, 1978a). Short term measurements of $CO_2$ production or $O_2$ uptake often have little value in the general assessment of side-effects, though they may be useful in more academic investigations.

### 4. *Organic matter decomposition*

McLaren and Peterson (1967) pointed out that organic matter plays an essential part in securing soil structure and high fertility. It also acts as a major source of nutrients and energy for the soil microflora. For these reasons, and because of the increasing use of systemic chemicals which may enter soil in plant residues, many methods of determining the effects of herbicides on organic matter decomposition exist. Many of these involve the addition to soil of specific substrates, such as protein and starch which are intended as models of plant residues, and measurement of product formation, often production of $CO_2$. Amongst the many other methods reviewed by Anderson (1978a) are those based on serial dilution of soil into media containing the substrate or soil incubations followed by measurement of products such as mineral nitrogen or measurements of substrate loss. A method which receives little attention by Anderson (1978a), although it has been a standard method for many years, is the use of the litter bag or of buried substrates such as straw (Latter and Howson, 1977; Malkomes, 1980) or cloth made from pure cellulose fibres (Greaves *et al.*, 1978a). As herbicides kill plants, whose residues must decompose in soil, it seems appropriate to use methods of following decomposition which mimic the natural deposition of plant residues in or on the soil. The introduction of specialized cellulose cloth (e.g. Shirley Test Cloth, The Shirley Institute, Manchester, UK), which is specifically designed for burial tests, increases their convenience. The flexibility and simplicity of these methods also allows both laboratory experiments and field tests. As they rely on determination of weight loss during degradation, they can be used by relatively unskilled personnel, a considerable advantage in many situations.

### 5. *Mineralization of other elements*

While the importance of nitrogen may explain the emphasis accorded to it in tests of herbicide effects on soil microbial function, it is puzzling that similar emphasis has not been given to other important plant nutrients. In most mineral soils, one half to two thirds of the total phosphorus present is in organic forms, so it is an important potential source of plant and microbial nutrients (Cosgrove, 1967). Indeed, microbial mineralization has been considered so important as a source of phosphorus that attempts have been made to exploit microbiological release of phosphate to increase crop yield. In the USSR, strains of *Bacillus megatherium* have been used to inoculate seeds or soil in attempts to achieve increased phosphate release (Cooper, 1959). Generally, these have been unsuccessful. It is thought that immobilization of organic phosphate by iron and aluminium effectively stops microbial mineralization in most soils.

The methods of following mineralization of organic phosphorus have been reviewed by Anderson (1960) and Cosgrove (1967). It seems that the most suitable methods for assessing the effects of herbicides, like those for nitrogen transformations, involve simple soil incubations accompanied by extraction and measurement of inorganic phosphorus at suitable intervals. While it must be accepted that such methods will probably only detect major changes in mineralization, it is only such changes which are likely to be of significance to plant nutrition.

In soil, transformations of sulphur, an element essential to all living cells, seem to be mainly a microbiological process. While there have been many studies of sulphur transformations in soil (Freney, 1967; Simon-Sylvestre, 1972) there are few included in investigations of the effects of herbicides on microbial function in soil (Simon-Sylvestre and Fournier, 1979). This is undoubtedly because of the considerable gaps in our basic understanding of sulphur mineralization in soils. Until more is known of the nature of sulphur-containing organic matter, and the conditions affecting its stability, it seems reasonable to conclude that sulphur mineralization is not a suitable function to study in relation to herbicide effects.

Similar conclusions can be drawn concerning elements such as calcium, magnesium, sodium, silicon and chlorine. Although all these elements contribute to the fertility of soil, it is not yet feasible to include studies of herbicide effects of their interrelationships with soil microorganisms as part of a regulatory package.

## B. Measurement of Enzymes

It has been said frequently that the measurement of enzyme activities in soil allows a rapid, effective and cheap means of assessing whether soils have been affected by chemical treatments. Indeed, the EPA guidelines (Anon., 1978) specifically state the need for measurements of the activities of enzymes such as phosphatase. There are reports of investigations of the effects of herbicides on many different soil enzymes, some of which attempt to draw conclusions regarding the potential effects on soil fertility. Certainly, some enzymes in soil, such as phosphatase and urease, do have an apparent link to nutrient cycling. Others, such as the dehydrogenases, may be correlated with respiratory activity of soil. However, recently there has been increasing criticism of the use of enzyme measurements. Skujins (1967) reviewed the techniques available for enzyme assay in soil and concluded that enzyme activity rarely correlates with soil fertility or, indeed, with microbial activity. Anderson (1978a, b) has reviewed the interrelationships between herbicides and soil enzymes and concluded that most of the methods have serious drawbacks, often compounded by the ill-defined nature of the enzymes involved. Thus, soil phosphatase activity is a combination of several different enzymes with no clearly defined optima for activity. In addition, other enzymes such as urease, may be strongly bound in soil organic matter and protected from herbicides and are, thus, insensitive as indicators of effect. Often it is unclear which component of the soil's enzymic capacity is being measured. It could be that in viable cells, in plant debris or that associated with soil colloids.

In general, therefore, soil enzyme activity is of little direct value in assessing the importance of herbicide effects on soil microorganisms and their contribution to soil fertility.

## C. Measurement of Populations

While it is recognized that enumeration of the species in soil may be a valuable contribution to understanding the ecosystem and its metabolic capacity, accumulation of data on microbial population change does little to predict or assess the impact herbicides may have on the soil. This is recognized by most regulatory authorities, which favour the provision of data on effects on processes in soil rather than on populations. An exception to this may be the assessment of effects on soil-borne plant pathogens and their specific antagonists, when population density may have an important bearing on infectivity and so on crop loss. This will be discussed further in a later section of this chapter.

The measurement of total biomass holds more promise as an indicator of major change. That such measurements only reveal major change must be stressed. Many inhibitions of parts of a population will be cancelled out by the increased growth of surviving members of the community and thus biomass will not necessarily change. In reviewing the methods available Anderson (1978a) concludes that most are as yet insufficiently tested for general application in herbicide investigations. Recent work by Jenkinson and Powlson (1976) and Lynch and Panting (1980), on mineralization of microbial carbon in fumigated soil, has produced a most promising method of biomass measurement but it also needs exhaustive testing (Domsch *et al.*, 1979) especially in work with herbicides, before it can be recommended.

## D. Pure Culture Studies

It is accepted that studies of the effects of chemicals on pure cultures of soil microorganisms offer little help in assessing the potential of the chemicals to cause harm in the soil. Rather, their main function is to help explain modes of action of the chemicals and to indicate possible interactions between herbicides, organism and, for example, nutrient availability or pH. One exception to this may be the use of pure cultures in preliminary screening of the toxicity of chemicals to important or common species in soil (Cooper *et al.*, 1978). Even if this method is used only to identify those compounds with no or very little inherent toxicity, it can serve a valuable service by reducing the numbers of chemicals which are tested by conventional, time-consuming, labour-intensive, and therefore very expensive, means.

## E. Standardization of Methods

In an ideal system, the methods used in different laboratories should be identical in essential detail. This is particularly important when data may be required for registration of the herbicide. Unfortunately, the very nature of soil microbial ecology precludes such standardization and creates innumerable problems in providing data which have value in predicting or interpreting effects of chemicals in the soil. Also, inevitable variations in facilities and technical expertise add further complications. The problems inherent in test methods, and which preclude standardization, have been reviewed by Greaves *et al.* (1976) and Anderson (1978a). The essential conclusion to be drawn is that, with the exception of certain chemical and physical conditions to be applied, particularly in analytical techniques associated with tests, standardization of test methods is generally not possible. Thus, it is essential that there should be flexibility in registration requirements which allows some freedom in the choice of method. Such flexibility is, for example, implied in sub-part D of the EPA Guidelines

(Anon., 1978) which contain little if any guidance on methodology. On the other hand, in a draft of a further guideline (sub-part J), as yet (1979) unpublished, dealing with "Hazard evaluation, non-target plants and micro-organisms", there is a noticeable degree of inflexibility in choice of method. This is especially true for the evaluation of hazard to aquatic algae. Such a change is a retrograde step.

### F. Field Versus Laboratory Investigation

Removal of soil from the field, its transport to the laboratory, storage and preparation for laboratory-based investigations causes large, sometimes irreversible, changes in the soil chemistry and microbiology. This being so, it is highly debatable whether the results obtained in the laboratory can be used in any way to predict what is likely to happen in the field. The counter arguments suggest that laboratory tests only need to indicate some potential of a chemical to cause harmful effects. Where such potential is discovered, the actual risk is assessed by further investigation in the field. The extreme problems associated with microbiological studies in the field (Anderson, 1978a) however lead to the unavoidable conclusion that, at present, laboratory-based tests are to be preferred.

There is one over-riding disadvantage to laboratory tests which cannot be dismissed lightly. That is that they invariably use root-free soils. As herbicides are used in crops, and plant roots are known to dominate the soil's microbiology, this is a serious deficiency. The use of organic amendments such as lucerne meal to soils used for laboratory experiments cannot mimic the rhizosphere effect. It seems logical that considerably more research effort should be directed towards studying herbicide/microorganism interactions in planted systems. This will be discussed further in a later section.

## III. Evaluation and Interpretation of Results

Despite the inadequacies of the methods available for measuring effects of herbicides on the soil microflora, there are considerable data in the literature. The reviews by Anderson (1978a, b) include hundreds of references, mostly published in the past 20 years. Much of this published material shows a conspicuous lack of any attempt to evaluate or interpret the data in practical terms, hardly surprising, considering the paucity of quantitative data relating soil microbial activity to soil fertility. Nonetheless, it remains of paramount importance that we should have some means of deciding, objectively, whether observed effects of herbicides, in laboratory or field experiments, have any implications for the continued practical use of the chemical.

## A. Prediction of Effects in the Field

Interpretation of anti-microbial effects in relation to soil fertility is difficult. There is not enough information available to determine whether high microbial activities directly result in increased crop yield or whether they are indicators of environmental conditions which are favourable for successful crop growth (Grossbard, 1972). There is no correlation of chemical and physical soil conditions with microbial reactions following pesticide application (Parr, 1974) and the significance of many results of pesticide effects on soil microorganisms is not known.

A change in species composition of soil microorganisms may occur after pesticide application but elimination of a single species is very unlikely (Domsch, 1972). After disturbance, nature will try to restore the former equilibrium quickly, as there is rarely total exposure of soil microorganisms to a biologically active concentration of a herbicide. As far as we know in most cases, a single exposure will cause only short-term effects.

In all trials to interpret pesticide effects on soil microorganisms the "sterilization phenomenon" must be considered, as subsequent stimulation of microbial populations and activities will be much greater the stronger the preceding biocidal action (Domsch, 1972). Therefore, stimulatory effects will be difficult to interpret too.

Limited substrate in the soil causes relatively long generation times in microorganisms. Hence, measurable effects on decomposing and synthetic activities can be expected mainly when luxury amounts of substrates are available, as in the autumn or spring, or in microhabitats such as the root region of crops (Domsch, 1972). This is why Greaves *et al.* (1976) recommended that measurement of herbicidal effects should be, preferably, in the root region, where more actively growing microbes occur.

The difficulties of selecting adequate criteria for measuring herbicidal effects on the soil microflora and their activities result mainly from fundamental problems of soil microbiology. Absolute measurement of microbial activities and determination of qualitative changes of the huge populations of thousands of species connected in many metabolic processes, is impossible (Grossbard, 1970). The methods available are of different sensitivity. Thus, respiration is insensitive and non-specific and does not necessarily reflect changes in the soil microflora (Parr, 1974). Short-term inhibition of $CO_2$ production may be more or less important depending on whether the reduction period coincides with critical phases in crop growth or not. Its importance also depends on the correlation between respiration and soil fertility (Grossbard, 1976). In addition, the effects of interactions in the root zone between microbes, herbicides and plant roots on plant growth and health are mainly unknown.

The amount of herbicide coming into physical contact with microbes is of greater importance to side-effects than the dose applied per unit area (Grossbard, 1976). In field trials especially, the situation is obscure because, in most cases, herbicide concentration in the soil or soil solution is unknown when herbicides are applied to the soil surface (Grossbard, 1976). Many microorganisms will not contact herbicides as most compounds do not penetrate more than 1 to 2 cm into the soil. Crude estimates of herbicide concentrations in soil can be made based on factors like herbicide volatility, solubility and adsorption.

Results from field trials may be difficult to interpret, because the microflora or its activities may respond mainly to changes in vegetation rather than to direct herbicide effects (Grossbard and Davies, 1976).

Selective or toxic effects of herbicides towards microorganisms depend on the composition of the nutrient supply as well as on soil characteristics (Grossbard, 1976). In other cases, the initial level of the activities being measured may influence the extent of their reaction to the chemical (Parr, 1972). Johnen (1977) mentioned four main factors which can change the size of herbicide side-effects to microorganisms: physicochemical soil characteristics, the biological situation in the soil, climatic influences and agricultural practices. Thus, herbicides are limited frequently to the surface (top 2 cm) layer of soil. Crop roots, however, are mostly located at lower depths and it is these roots which determine, to a great extent, the composition and activity of the soil microflora. Therefore, it could be argued strongly that testing for side-effects of herbicides should be limited to those compounds which are known to penetrate to the rooting zone in appreciable quantities. Side-effects confined to the surface layers may be safely ignored as the large volume of underlying, rooted soil will maintain normal biological processes and will, subsequently, inoculate the surface layers.

At the moment all attempts to evaluate pesticide side-effects on soil microorganisms can only be seen as leading to the discovery of more objective criteria.

## B. Evaluation of Data

Even if there were criteria which were absolutely appropriate as indicators of potential serious harm to the soil microflora, there still remains the problem of deciding the relative importance of each observed effect a chemical may have. More than this, it has yet to be decided how to interpret or evaluate the importance of the sum of effects, both inhibitory and stimulatory, which could occur when investigating multiple microbial activities.

During a series of international symposia initiated by the Federal Biological Research Centre for Agriculture and Forestry, together with the Federal Research Centre for Agriculture in the Federal Rupublic of Germany, Professor K. H. Domsch presented a basic system which goes a long way to allowing the sort of objective decision making which is necessary. A synthesis of the system, contributed to this chapter by Professor Domsch, is presented in the next section.

## 1. *Types of microbial responses*

Chemicals with biocidal properties can produce a limited number of negative responses in reactive biological systems (Fig. 1). Microbial processes

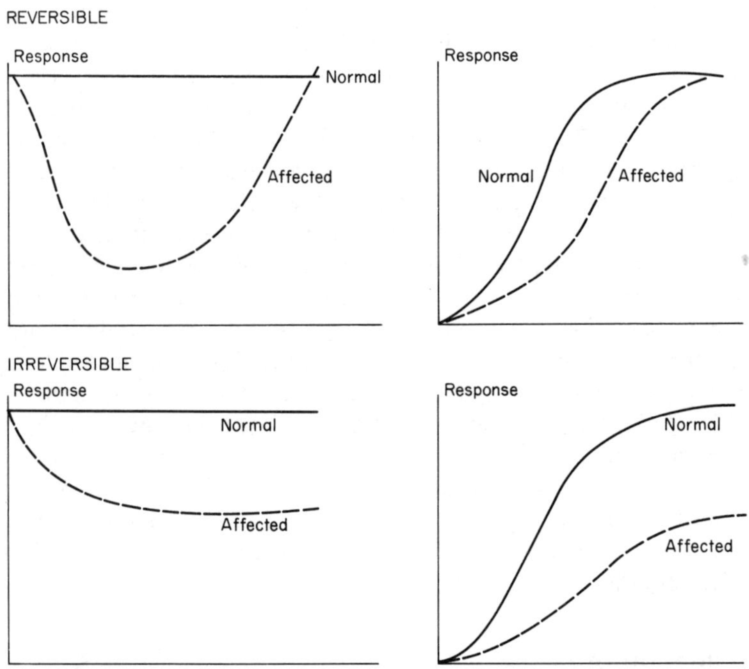


Fig. 1. Principal types of reversible and irreversible negative responses.

or populations may be established at a high level and they may normally continuously maintain this level, or may start at a low level and increase during the monitoring period. The effects themselves may be reversible or irreversible within this period. The four resulting types cover most of the responses thus far observed.

It is evident that any monitoring system which allows such an uncomplicated recognition of reaction types requires high quality data. Since a chemical which is incorporated into the soil reacts in a dynamic system, which has a multitude of ways and means for repair and restitution, the time factor has a key position. Side-effect data, without information on the time-dependent changes of the microbial response, are only of limited use for evaluation purposes.

Reversible effects may be analysed more specifically in order to recognize their quantitative aspects. There are two main aspects of all reversible negative responses: (a) the amplitude of the maximal depression and (b) the phase-shift due to a retarded process expressed as maximal time deficit (Fig. 2). It should be mentioned here that, from an ecological point of view, irrespective of the magnitude of depression, the delay in the re-

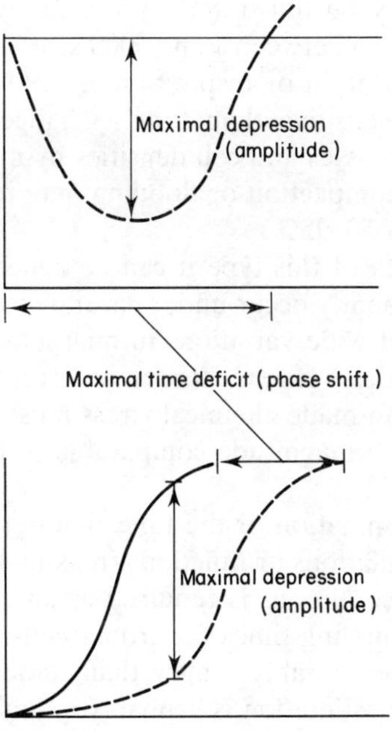


Fig. 2. Criteria used for the evaluation of reversible side-effects.

establishment of a negatively influenced microbial structure or function ranks higher than its magnitude in the hierarchy of evaluation criteria.

## 2. *Interpretation of results*

(*a*) *Reversible effects.* The recognition of response types is a prerequisite for further evaluation, but does not allow a decision to be made on which effects might be ecologically insignificant, tolerable or even critical. In order to approach this problem, it is proposed that, in situations where the application of herbicides is justified from an agricultural point of view, the magnitude of the response to man-made chemical stress should be compared with that of naturally occurring stress situations. From studies in microbial ecology a wealth of information is available on the consequences of naturally occurring catastrophes in soil microhabitats: water availability is frequently reduced to critical levels or soil is flooded and gas exchange is limited; the temperature often deviates from optimum conditions or nutrients are lacking and cells are eliminated by soil animal grazing, microparasites or other adverse conditions.

The magnitude of depressions in soil microbial systems under perfectly normal conditions may be illustrated by the following examples. When water potentials decrease between 1 and 5000 kPa, microbial functions like ammonification, nitrification or respiration are reduced by 10–90%, with bacteria being more sensitive than fungi. Changes in soil temperature frequently reduce biomasses and cell densities by more than 70%; limited gas exchange due to compaction or flooding may reduce cell densities or microbial functions by 50–95%.

From ecological data of this type it can be concluded that depressions of 50% or more, frequently occur under natural conditions. Soil meteorological data prove that wide variations in microenvironmental conditions are normal events. If these statements can be accepted, the importance of depressions due to man-made chemical stress must not be overestimated, as long as they are of a magnitude comparable to those found in natural situations.

Likewise, the determination of the time required for the restitution of normal microbial populations or functions (maximal time deficit) after the end of natural stress conditions is required as an ecological yardstick. A key parameter is the doubling-time of microbial cells. Under field conditions doubling-times are considerably longer than under optimum laboratory conditions. A realistic estimation is probably a doubling-time of about 10 days. In an assumed stress situation, where a reduction from a high level (e.g. 100) to a low level (e.g. 12·5) has taken place, a total of three doublings is required to bring the population back to the original level. In fact, three doublings or $3 \times 10$ days have frequently been observed in soil populations which recover from the exposure to drying (and remoistening) or freezing (and thawing).

If this basic consideration can be accepted, delay periods or phase shifts of 30 days following a *c.* 90% depression are still a normal and natural phenomenon. Applied to the situation of chemical stress, a delay of 30 days should be regarded as being of no ecological significance. A delay of 31–60 days can still be grouped as tolerable while more than 60 days might indicate critical situations.

(*b*) *Irreversible effects.* The side-effects which are apparently irreversible require separate consideration from the two above-mentioned response-types. These side-effects are characterized by deficits of populations or functions when herbicide-treated soils are compared with non-treated soils at the end of the monitoring period. The deficits may be due to (i) the persistence of the toxic agent or (ii) the inability of a population to recover from a depression. The general potential hazard can be recognized by high deficits (>50%) or long-lasting deficits (>60 days) at the end of the monitoring period, or both.

If herbicides should cause side-effects of this irreversible type, the first indication of such cases must be followed in two steps: (a) confirmation by repeated tests under a set of different environmental conditions and (b) field testing.

This scheme of Professor Domsch seems logical. It is interesting to note that Marsh (1978) has made similar, though less detailed, proposals and Laskowski (1979) took a similar approach in a paper presented to the Weed Science Society of America meeting in San Francisco. Certainly, it would seem reasonable to have a system which judges the acceptability of the effects of herbicides on soil microflora, by comparing them to what already occurs naturally. There are, of course, many necessary refinements to the proposed evaluation systems. Fournier (personal communication) has suggested that such a comparative system may be more applicable to specific microbial functions. For more general tests, such as those of soil respiration, the microbial species affected may vary according to the stress applied, and comparison becomes difficult. Moreover, disappearance of an effect does not prove that soil has returned to normal. For example after partial sterilization, soil respiration may return to its former level quite quickly, whereas pesticide degradation may not be re-established for a long period. It has certainly seemed odd to many soil microbiologists that the usually small effects of herbicides on the soil microflora should cause concern to many people when "partial sterilization" of soils in horticulture, using heat or chemicals, is regarded as highly desirable, if not essential.

## IV. Future Research Needs

The foregoing part of this chapter, inescapably, has painted a gloomy picture of our lack of knowledge of soil microbial ecology and our consequent inability to test adequately and predict the potential important side-effects of herbicides in the field. Clearly, there is considerable need for continued urgent research in the area of herbicide–soil–microbe–plant interaction if we are to be able to continue to use pesticides within the tolerances of risk which are politically acceptable at the moment. What is less clear is where the limited resources available should be best deployed.

The limitations of space preclude a comprehensive cover here of all possibilities and the topics presented obviously reflect a degree of personal bias. However, it is hoped that they represent important areas of research deserving of continuing and increased support.

### A. Rhizosphere Studies

Plant roots form intimate associations with soil microorganisms and, through the production of root exudates, stimulate an active microbial population that dominates the soil of cropped fields. In turn, the microflora of plant roots can affect crop growth in many different ways. Microbial production of organic compounds has been reviewed by McCalla and Norstadt (1974) and Lynch (1976). While many of these products are never sufficiently concentrated to affect plant growth, others undoubtedly are. Plant growth regulators (e.g. auxins), phytotoxins (e.g. volatile fatty acids) and antibiotics are all formed near plant roots. In addition, microorganisms may compete with plants for available nutrients or promote their supply to the root.

The evidence confirms that plant root–microorganism interactions are an essential feature of so-called soil fertility. Bearing in mind that herbicides, and other pesticides, are normally applied to a crop, or to soil about to receive a crop, it seems essential that our future research should include efforts directed to investigating the effects of herbicides on root–microbe interaction. It is less evident where the research should be channelled specifically. There would seem to be two principal areas of work: (i) *Basic research*, that is the study of the influence of herbicides on the general biology of the root region and (ii) *Applied research*, directed towards understanding the effects of herbicides on organisms of known economic significance. Thus, *Rhizobium*, mycorrhizae, root pathogens and organisms involved in the biological control of pathogens are obvious targets. These will be discussed in a later section.

With the enormous range of herbicides available it is difficult to know where to start in basic research. Nonetheless, it is essential to define, even imprecisely, some goals. Otherwise, future research will be piecemeal, a criticism which can be applied justifiably to much of the work in the past.

Some limitation of the numbers of herbicide–plant interactions to be investigated can be achieved by considering, initially at least, only those chemicals used on large areas of major crops. Further, attention should be focussed on those which are translocated within the plant, especially those exuded by the roots or which may change root morphology and physiology.

Once the range of herbicides and crops has been narrowed, similar narrowing of the microbial subjects studied must be attempted. As root exudates are the driving force behind root region microorganisms it is most important that we know more about the effects herbicides may have in changing their quantity and composition. The methods used by Barber and Martin (1976) would seem to be particularly appropriate. Other methods are described in the review of root exudates from crops by Rovira (1969).

There is still a great need for further information on microbial numbers, species composition, distribution and growth rates in the root region. The methods available are numerous and it is not appropriate to review them here. Suffice it to say that all of them, from the simple to the most sophisticated, can provide important basic information.

Of more immediate importance than basic research, especially in the area of developing registration requirements, is the development of applied research. This is discussed most conveniently in separate sub-headings.

### 1. *Rhizobium*

Although the requirement for data on effects of herbicides on *Rhizobium* is already included in some registration guidelines, there is a need for development and establishment of better methods allowing unequivocal identification of side-effects. As pointed out earlier in this chapter, present methods have serious drawbacks.

### 2. *Mycorrhizae*

Interactions between pesticides and mycorrhizal fungi have been reviewed by Papavizas and Lewis (1979). It is generally accepted that vesicular–arbuscular (VA) mycorrhizae have importance to the nutrition of crop plants, especially those grown in soils of low fertility. In these soils, response of plants to mycorrhizae is crudely related to percentage infection, and this in turn is related to endophyte population in the soil (Mosse, 1977; Black and Tinker, 1977). Obviously, the crop being examined and the level of soil fertility will be modifying factors (Bowen and Bevegè, 1976; Mosse, 1977). Infection intensity would also need to be related to the stage of growth of the host crop (Smith, 1978).

Effects of herbicides on both infection of crop plant and on populations of mycorrhizal fungi in the soil must be studied. Soil populations are usually studied using a sieving method to obtain and count spores from an aliquot of soil (Mosse and Bowen, 1968). However, this assumes that all propagules are spores and that these are all above the size collected. This is a particularly dangerous assumption as one type of VA endomycorrhiza, the 'fine endophyte', *Rhizophagus* (*Glomus*) *tenuis* which is very common, may not form spores readily, and then not large ones. Thus correlations between spore numbers in sievings and infection of crop may be poor. More recently, dilution to extinction methods have received more favour. These involve preparing dilutions of soil with the same soil which has been sterilized, and then planting with a highly susceptible crop such as clover or medic. Results are assessed by counting infection points on roots, or by simply scoring roots for infection, and using the most probable number technique. As yet, these methods have not been published (Bowen and Smith, personal communication).

Root infection has been studied by various methods and these are summarized by Becker and Gerdemann (1977). Becker and Gerdemann (1977) and Hepper (1977) used colorimetric or biochemical methods. Though these may be useful experimentally, their validity for survey or monitoring purposes over a range of locations is suspect. The methods assume, for example, that the same intensity of root colouring is given by all endophytes under all conditions.

Reduction in plant growth following herbicide application to a mycorrhizal-responsive plant (e.g. onion in a low fertility soil) might be a further method. Reduction in plant growth due to direct effects on the test plant could be accounted for, as it would also occur in high fertility soil.

Other areas worthy of study include the possibility that total plant destruction, as when swards are killed prior to direct drilling, may result in a decline in mycorrhizal population in the soil. This, and overkill soil treatments such as methyl bromide and other soil sterilants, may cause stunting of subsequent plantings, especially in plants with low rooting density such as citrus (Kleinschmidt and Gerdemann, 1972; Hattingh and Gerdemann, 1975). One advantage is that such a treatment may allow subsequent introduction of a more beneficial strain of mycorrhiza than that originally present.

### 3. *Soil-borne pathogens*

There have been several comprehensive reviews of herbicide interactions with soil-borne plant pathogens in recent years (Katan and Eshel, 1973; Kavanagh, 1974; Altman and Campbell, 1977, 1979; Papavizas and Lewis, 1979). As with literature on herbicide effects on other non-target organisms, that dealing with pathogens reveals many discrepancies, even with the same herbicide. These probably arise from variations in technique, pathogens, soils and other experimental factors (Papavizas and Lewis, 1979). This frequency of discrepancies is particularly evident in the results of greenhouse studies and Papavizas and Lewis (1979) suggest, quite rightly, that "few meaningful conclusions can be drawn from greenhouse studies alone". Discrepancies also exist between data from field experiments, though less frequently than in greenhouse work. The majority of field studies indicate that many herbicides increase soil-borne diseases. Whether these effects are the result of direct interaction between chemical and pathogen, or whether they are indirect results from weed control is less clear. There is a wide variety of physiological and biochemical effects which herbicides can have on crops at doses far below those which cause effects serious enough to affect yield or health. These apparently minor effects may influence the crop's susceptibility to disease.

There is a considerable wealth of information regarding herbicide effects on pathogen physiology and survival. Most of the work, however, has used laboratory cultures and extrapolation to the field is impossible. The situation is made more complex by the inevitable variation in results reported. The literature suggests that herbicides may affect pathogen growth, sporulation, propagule germination, inoculum density, survival in soil and competitive saprophytic ability. These effects could be brought about by several different mechanisms, such as direct effects on the pathogen and its virulence, effects on host susceptibility and effects on other competitive or inhibitory microorganisms (Altman and Campbell, 1977, 1979; Papavizas and Lewis, 1979). However, analysis of the published work fails to reveal unequivocal evidence of increases in virulence or indirect effects resulting from changes in non-target microorganisms, particularly in field studies. Evidence of predisposition of crop plants to disease following herbicide use is more convincing (Papavizas and Lewis, 1979).

The general conclusions from the bulk of the literature available are not encouraging. The concern that herbicides may cause increases in disease due to soil-borne pathogens is justifiable in view of the amount of data apparently showing such increases. However, as yet there is a conspicuous lack of data which really contribute further to our present understanding of effects on pathogens. In this context, the conclusions drawn by Papavizas and Lewis (1979) seem appropriate.

> First, positive preliminary results should be verified in the field, if possible. Second, other parameters should be investigated before dissemination of results.

> Such parameters should include use of different soils, different isolates of the test pathogen, and different environmental conditions. We can learn whether a pesticide effect is consistent, or whether it varies with soil type, climate or time of pesticide application. Presentation of data with a clear indication of soil types used, inoculum density and rate, method and time of pesticide application would be of considerable importance and value.

Papavizas and Lewis (1979) further conclude that

> It would be a disservice to our attempts to elucidate pesticide–plant pathogen interactions in nature if we derive our conclusions from such [*in vitro*] observations only. Studies of this kind should be discontinued or reduced.

In these days of increasing pressures to regulate further the use of herbicides by increasing the data for registration purposes, we must maintain a sense of perspective. This can only be done if we understand potential side-effects of herbicides and their consequences more thoroughly than at present. This is particularly true in the area of soil-borne pathogens and it is obvious that we need considerably more information before effective means of predicting real side-effects can be developed.

## B. Microcosms

In the foregoing section, and elsewhere in this chapter, it has been said that laboratory studies do not give results which can be extrapolated to the field. On the other hand, field experiments in microbiology are difficult to design so that they produce unequivocal results. One means of obtaining a compromise in this conflict is to use microcosms, that is small-scale analogues of an ecosystem, of one form or another. Indeed, microcosms have been used increasingly to reveal how pollutants can affect microbial communities and their activities (Bull and Brown, 1979).

Draggan (1976) and Draggan and Van Voris (1979) have discussed the use of terrestrial microcosms as a tool for estimation of environmental transport of toxic materials and their role in ecological research. Rightly, both Draggan (1976) and Bull and Brown (1979) point out that microcosms should be functionally similar to the ecosystem being mimicked. Thus, the processes deemed to be essential by the experimenter should be similar in value to those in the intact ecosystem and should be expected to react to external and internal factors as does the natural system. This places great emphasis on careful design of the microcosm and choice of the function for study.

The proceedings of the symposium on "Terrestrial Microcosms and Environmental Chemistry" (Witt and Gillett, 1977) amply illustrate the complexities involved in microcosm experimentation. The obvious conclusion to be drawn from these proceedings, and the round-table discussions reported in them, as well as from the reviews by Draggan (1976) and Draggan and Van Voris (1979) is that, like other research on herbicides in the soil, the present state of knowledge is fragmentary and inconclusive. Microcosms, so far, are able only to indicate large effects. As ecosystems, particularly the terrestrial ecosystem, are so complex this is probably all we will ever be able to indicate and both scientists and the regulatory authorities must accept this. On the other hand, terrestrial microcosm research may provide much data to improve our understanding of ecosystems and, therefore, it should be encouraged and supported. It must be stressed that, as yet, microcosms are only research tools and are not suitable for providing data for registration purposes.

One of the more exciting developments in investigations of microcosms has arisen from the increasing use of fermentation technology. The use of the chemostat has many practical advantages in extensive screening of interactions between microbial communities and chemicals (Bull and Brown, 1979). As with terrestrial microcosms, however, the main use of the chemostat is to provide basic information to aid our understanding of the ecosystem being investigated. The results, like all results using pure or mixed cultures of microorganisms, cannot be extrapolated to the field and, thus, can act only as indicators of potential effects which must be researched further in the field itself, if possible. Even so, our lack of basic information means that such fundamental research should be pursued actively.

## C. Soil Structure

There are comparatively few reports concerning effects of herbicides on soil structure. Some consider the effects of changed soil cultivation following herbicide usage (Clay and Davison, 1976) and others, direct effects of herbicides (Hance, 1976; Yates *et al.*, 1976). Sometimes soil structure was affected, in certain soils, especially by high herbicide concentrations.

The distribution of microorganisms in soil is related strongly to structure, bacteria preferring to colonize soil aggregates, whereas fungi are associated largely with extraneous organic matter (Hattori, 1973). Microorganisms affect structure through processes such as weathering of inorganic soil components, production of adhesive compounds like polysaccharides or binding of particles by fungal hyphae. In the plant root region, such microbial contributions to structure may be very important, especially in crops with intensive root systems (Russell, 1971).

Although there is no information concerning the relationships between microbial populations and activity, and soil structural changes following herbicide usage, it is clear that alteration of the microbial contribution to structural stability is possible, especially in terms of the stabilizing effects of algae on surface soils. Consequently, it is highly desirable that research be initiated to improve our knowledge of the situation. Such research should be coordinated with that on effects of herbicides on organic matter decompositions, a major contributor to soil structure. One of the first steps should be to compare the effects of herbicides with those of other, established, cultural practices.

## D. Microbial Antagonists

As yet, little is known of those microorganisms in soil which are antagonistic towards other microorganisms, especially to soil-borne plant pathogens, and which colonize the root region of crops. The subject is covered briefly by Anderson (1978b) who also suggests some methods which may be used (Anderson, 1978a). It is an extremely difficult area of research, but also one which has great potential benefits. As such it is desirable that research is encouraged. Perhaps the area of greatest priority is to elucidate whether the antagonisms recorded in laboratory experiments do occur in the field to any significant extent and what mechanisms are involved.

## E. Long-term Use, Sequences and Combinations of Herbicides

Long-term effects of repeated applications of herbicides have been reviewed briefly by Greaves (1979). At present, little is known of such effects and those recorded are restricted to few herbicides and to reports of field trials. As a result, it is difficult to draw firm conclusions. The indications are, however, that effects which have been recorded are generally indirect, resulting for example, from decrease in organic matter input to the soil (Grossbard, 1971; Greaves, 1979). Many observed effects involve changes in single microbial activities or populations, which are regarded as having no importance in terms of overall activity (Voets *et al.*, 1974). Nevertheless, long-term experiments, particularly field experiments such as those established in 1963 at the Weed Research Organization, Oxford, UK (Fryer and Kirkland, 1972; Fryer *et al.*, 1980), can make an important contribution to our knowledge of effects on plant growth, vigour, health, quality and yield. All these may reflect, to some extent, effects on the soil microbial activity. It is important, therefore, to maintain some work in this area, particularly with the more persistent herbicides.

The effects of successive applications of a herbicide within a short time are unknown. Such applications need not necessarily have a greater effect than a single application, especially if the compound has short persistence. For example, using the fungicide benomyl, Foster and McQueen (1977) found no greater depression of the fungal population after successive applications than after a single one. However, the regeneration of the fungi took longer. This would be significant in evaluating the overall importance of the change (see earlier section on evaluation of results). Although inhibitory effects of pesticide sequences have been reported (Jones *et al.*, 1974; Verstraete and Voets, 1974) it is difficult to determine the causative factor(s). This difficulty serves to emphasize the need for more intensive investigation.

Unequivocal results from investigations of the effects of combinations of herbicides, or of combinations of herbicides with other pesticides, are very rare. Unfortunately, most investigations have omitted the essential comparisons with the effects of each single component of the combinations. Again the need for increased research is obvious.

## F. Environmental Factors

Investigations of the effects of long-term use, sequences and combinations of herbicides would be one step nearer to understanding what might happen in practical agricultural situations. There are, of course, many other factors which will interact and thus need to be considered. Pesticide use over a period of years almost certainly involves crop rotation and changing pesticide sequences. These may also cause change in the soil biocoenosis (Dobrovodsky and Longauer, 1977). It is also evident that, in some instances, crop changes may cause greater effects on soil organisms than long-term cultivation of, and pesticide use in, the same crop (Jones *et al.*, 1974). This is a vital consideration to be accounted for when evaluating the importance of observed herbicide effects on the microflora.

Other environmental and agricultural factors are also important. Herbicide effects on microorganisms may be smaller after addition of fertilizers (Bakalivanov and Hlebarova, 1977) or larger in the presence of organic material (Vladutu and Soreanu, 1976). Humidity, temperature, pH, and other soil factors can also modify the response of organisms to herbicides (Wingfield *et al.*, 1977; Gaur and Misra, 1978; Marsh and Greaves, 1979). Any attempt to predict or evaluate potential side-effects of herbicides on the soil microflora and its activities must take account of environmental factors and of variability within these factors. This will only be possible if future research pays attention to these aspects. Again, it must be stressed, the need is for accumulation of basic information, not for introduction of such factors into tests for registration purposes.

## G. Indirect Effects

It is beyond the scope of this chapter to deal with this topic in detail. Nonetheless, the importance of recognizing indirect effects when interpreting the results of laboratory or field experiments is great and it is appropriate to mention the subject, if only briefly.

Greaves *et al.* (1976) described the increase in *Fusarium* foot rot of narcissus crops following weed control with herbicides. This was due to increased soil temperature following removal of the weed canopy and not due to direct effects on the pathogen. Increased susceptibility of the host to pathogens, inhibition of antagonists and reduction of organic matter input to soil have been mentioned in other sections of this chapter. There are many other possible indirect effects which may outweigh the direct effects of herbicides. It would be impractical to suggest that considerable research efforts should be made to elucidate these indirect effects, if only because many are already known. What is vital is that they should be recognized and accounted for in experiments. One of the best ways of ensuring this is through coordinated research involving a range of specialists.

## H. Coordinated Research

Just as indirect effects may be overlooked if microbiological research is done in isolation, many other potentially important interactions may be missed. It is axiomatic that microorganisms are only a part, albeit an important one, of the soil ecosystem. They interact with every component of soil and react, in sympathy or antipathy, to changes in all these components.

To suggest that microbiological research on herbicide effects should only be done in concert with research on chemical and physical changes in soil, investigation of soil animals, protozoa, with weed biology, crop nutrition and so on would be totally unrealistic. Often, it is difficult enough to finance the relatively sparse amount of herbicide research being done at present. However, there are many things which can be done with little additional cost. The role of the chemist is vital. Herbicide degradation during experiments will obviously affect the effects of the chemical on organisms through declining concentrations of the parent compound or the appearance of metabolites. The distribution of the chemical in the soil should be known and whether it is adsorbed on organic matter or soil minerals. Where rhizosphere studies are made, it is often more important to know how much chemical is retained on the plant than how much was applied. Similarly, other specialists can make essential contributions to our interpretation of cause and importance of effects apparently due to herbicides. It seems unnecessary to stress the importance of employing such specialists in microbiological investigations, even in an advisory capacity only, whenever it is possible. Unfortunately, all too often, reports of microbiological investigations reveal clearly that this has not been done and the reports are frequently, therefore, of little value.

## V. Conclusions

In common with all recent reviews of this subject, this chapter must conclude that, as yet, we know all too little of soil microbial ecology and that considerable research effort is essential. This research must be principally basic in nature but the applied aspects must not be neglected.

Applied research should be directed at ways of improving our ability to identify, quantify and evaluate potential harmful effects of herbicides to soil microbial activities which are essential to the maintenance of soil fertility. While much has been done in this area, there remains an obvious need for continued research. Agriculture is dependent on the continuing supply of effective, safe chemicals for weed control. Any requirement which adds significantly to development costs, without making a valuable contribution to environmental safety, must be evaluated most critically. Until research has provided unequivocal means of assessing effects on the soil microflora, registration authorities might be wiser to pay more attention to those responses of the crop which reflect soil fertility and the microbial contribution to it.

Basic research is the only way in which the information essential to the applied scientists can be obtained. As such it must take a high priority in competition for funding. It is essential that we understand microbial behaviour in soils more thoroughly. In particular, the actual microbial contribution to nutrient cycling must be clarified. Equally, the interactions between herbicide, soil, crop plant and microorganisms require continued exhaustive research.

The choice of important specific areas of research depends, to a great extent, on individual bias. However, the following topics seem to warrant special attention with regard to the effect of herbicides.

(i) Microbial aspects of nutrient cycling including decomposition of crop residues.

(ii) Rhizosphere studies—in particular nodulation of economically important legumes and mycorrhizal associations.

(iii) Soil-borne plant pathogens including microbial antagonism.

(iv) Microbial aspects of soil structure.

These subjects can be studied effectively only if appropriate valid methods are developed and applied. In this context, the use of microcosms may be advantageous and this should receive attention. Both terrestrial microcosms and chemostat models should be considered.

In all research, particular attention should be paid to coordination with other specialists, wherever possible.

There is a pressing need for the development of means of interpreting experimental data. Such interpretation should pay heed to the normal behaviour of microbial communities under what may be termed "acceptable stress".

It is vital that those responsible for the introduction and imposition of regulatory requirements take full note of the advice of microbiologists and other specialists. Hopefully, future requirements then will be based more on sound fundamental scientific principle rather than on political or emotive considerations.

## VI. Acknowledgements

Although this chapter has dealt with many subjects only superficially, even this would have been impossible without the help of many colleagues throughout the world. With their help, it is hoped that at least some personal bias has been avoided. In particular, we would like to thank Professor K. H. Domsch (Federal Research Centre for Agriculture, Braunschweig, Federal Republic of Germany) for providing the section on evaluation of results, Dr G. D. Bowen (CSIRO, Australia) for his contribution on mycorrhizae, Dr A. D. Rovira (CSIRO, Australia) for comments on rhizosphere research. Dr G. Papavizas (US Dept. of Agriculture) allowed the use of an unpublished review manuscript on soil-borne pathogens and Dr S. Draggan (National Science Foundation, Washington, USA) contributed much useful help in the microcosm section. Many other colleagues contributed greatly in discussions and they are gratefully acknowledged. Last, but not least, members of staff at our two institutes, particularly Mrs H. A. Davies and Mrs S. L. Cooper, have been patient and tireless in checking manuscripts and references.

## References

Altman, J. and Campbell, C. L. (1977). Effect of herbicides on plant diseases. *A. Rev. of Phytopathol.*, **15**, 361–385.

Altman, J. and Campbell, C. L. (1979). Herbizide und Umwelt: eine Übersicht über fördernde und hemmende Interaktionen mit Pflanzenkrankheiten. *Z. Pflkrankh. Pflschutz.* **86**, 290–302.

Anderson, G. (1960). Factors affecting the estimation of phosphate esters in soil. *J. Sci. Fd. Agric.*, **11**, 497–503.

Anderson, J. R. (1978a). Some methods for assessing pesticide effects on non-target soil micro-organisms and their activities. In *Pesticide Microbiology*. (I. R. Hill and S. J. L. Wright, Eds.), pp. 247–312, Academic Press, London and New York.

Anderson, J. R. (1978b). Pesticide effects on non-target soil microorganisms. In *Pesticide Microbiology*. (I. R. Hill and S. J. L. Wright, Eds.) pp. 313–533, Academic Press, London and New York.

Anderson, J. R. and Greaves, M. P. (1977). The effects of herbicides on soil microorganisms. In *Weed Control Handbook Vol. 1 Principles*. (J. D. Fryer and R. J. Makepeace, Eds.) pp. 78–82, Blackwell Scientific Publications, Oxford.

Anon. (1971). *Pesticide Safety Precaution Scheme*, Appendix D. Wildlife Data Guide. UK Ministry of Agriculture, Fisheries and Food.

Anon. (1973). Agricultural Pesticides 3rd Edition. Council of Europe, Strasbourg.

Anon. (1978). Environmental Protection Agency. Registration of pesticides in the United States, *Federal Register*, **43**, (132), Part II, 29696–29741.

Atlas, R. M., Pramer, D. and Bartha, R. (1978). Assessment of pesticide effects on non-target soil microorganisms. *Soil Biol. Biochem.*, **10**, 231–239.

Bakalivanov, D. and Hlebarova, S. (1977). Relationship between soil microflora and Afalon and Patoran herbicides under fertilizer application. *Acta Phytopathol. Acad. Sci. Hung.*, **12**, 101–108.

Barber, D. A. and Martin, J. K. (1976). The release of organic substances by cereal roots in soil. *New Phytol.*, **76**, 69–80.

Becker, W. N. and Gerdemann, J. W. (1977). Colorimetric quantification of vesicular-arbuscular mycorrhizal infection in onion. *New Phytol.*, **78**, 289–295.

Black, R. L. B. and Tinker, P. B. (1977). Interaction between effects of vesicular-arbuscular mycorrhiza and fertiliser phosphorus on yields of potatoes in the field. *Nature*, **267**, 510–511.

Bowen, G. D. and Bevege, D. I. (1976). Microorganisms, the third dimension in phosphate nutrition and ecology. In *Prospects for Improving Efficiency of Phosphorus Utilization*. Proc. Symp. at Univ. New England, Armidale, N. S. W., Australia, 1976. (G. J. Blair, Ed.), pp. 103–112, Reviews in Rural Science, III.

Braunholtz, J. T. (1977). The crop protection industry; products in prospect. *Proc. 1977 Br. Crop Prot. Conf.—Pests and Diseases*, 659–670.

Bull, A. T. and Brown, C. M. (1979). Continuous culture applications to microbial biochemistry. In *International Review of Biochemistry. Microbial Biochemistry*, **21**, (J. R. Quayle, Ed.), pp. 177–226, University Park Press, Baltimore.

Clay, D. V. and Davison, J. G. (1976). The effect of different soil management systems and herbicides on the growth and cropping of raspberries. *Proc. 1976 Br. Crop Prot. Conf.—Weeds*, 241–248.

Cooper, R. (1959). Bacterial fertilisers in the Soviet Union. *Soils Fertil.*, **22**, 327–333.

Cooper, S. L., Wingfield, G. I., Lawley, R. and Greaves, M. P. (1978). Miniaturized methods for testing the toxicity of pesticides to microorganisms. *Weed Res.*, **18**, 105–107.

Cosgrove, D. J. (1967). Metabolism of organic phosphates in soil. In *Soil Biochemistry*. (A. D. McLaren and G. H. Peterson, Eds.), pp. 216–228, Edward Arnold (Publishers) Ltd., London.

Curl, E. A. and Rodriguez-Kabana, R. (1971). Microbial interactions. In *Research Methods in Weed Science.* (R. E. Wilkinson, Ed.) pp. 161–194, Southern Weed Science Society, Pop Enterprises, Inc. Atlanta, Georgia.

Curl, E. A. and Rodriguez-Kabana, R. (1977). Herbicide-plant disease relationships. In *Research Methods in Weed Science.* (B. Truelove, Ed.), pp. 173–191, Southern Weed Science Society, Auburn, Alabama.

Dobrovodsky, J. and Longauer, S. (1977). (Effects of single crop corn cultivation with the use of herbicides on agrocoenoses and soil microflora). *Agrochémia.*, **17**, 249–252.

Domsch, K. H. (1972). Einfluß von Pestiziden auf mikrobielle Prozesse und ökologische Beziehungen im Boden. *Ber. Landwirtsch.* **50**, 392–403.

Domsch, K. H., Beck, Th., Anderson, J. P. E., Söderström, B., Parkinson, D. and Trolldenier, G. (1979). A comparison of methods for soil microbial population and biomass studies. *Z. Pflanzenernaehr. Bodenkd.*, **142**, 520–533.

Draggan, S. (1976). The microcosm as a tool for estimation of environmental transport of toxic materials. *Int. J. Environ. Stud.*, **10**, 65–70.

Draggan, S. and Van Voris, P. (1979). The role of microcosms in ecological research. *Int. J. Environ. Stud.*, **13**, 83–85.

Foster, M. G. and McQueen, D. J. (1977). Multiple applications of Benomyl and effects on non-target soil fungi. *Bull. environ. Contam. Toxicol.*, **17**, 468–476.

Freney, J. R. (1967). Sulfur-containing organics. In *Soil Biochemistry* (A. D. McLaren and G. H. Peterson, Eds.) pp. 229–259, Edward Arnold (Publishers) Ltd., London.

Fryer, J. D. and Kirkland, K. (1972). Field experiments to investigate long-term effects of repeated applications of MCPA, tri-allate, simazine and linuron: report after six years. *Weed Res.*, **10**, 133–158.

Fryer, J. D., Smith, P. D. and Hance, R. J. (1980). Field experiments to investigate long-term effects of repeated applications of MCPA, tri-allate, simazine and linuron: II Crop performance and residues 1969–78. *Weed Res.*, **20**, 103–110.

Gaur, A. C. and Misra, K. C. (1978). Dynamics of microbial populations in soil as influenced by Simazine and ecological factors. *Zbl. Bakteriol. 2. Abt.*, **133**, 357–361.

Greaves, M. P. (1978). Problems and progress in the evaluation of herbicide to the soil microflora. *Report A.R.C. Weed Research Organization, 1976–1977*, **7**, 95–103.

Greaves, M. P. (1979). Long-term effects of herbicides on soil microorganisms. *Ann. appl. Biol.*, **91**, 129–132.

Greaves, M. P., Davies, H. A., Marsh, J. A. P. and Wingfield, G. I. (1976). Herbicides and soil microorganisms. *C.R.C. Crit. Rev. Microbiol.*, **5**, 1–38.

Greaves, M. P., Cooper, S. L., Davies, H. A., Marsh, J. A. P. and Wingfield, G. I. (1978a). Methods of analysis for determining the effects of herbicides on soil microorganisms and their activities. *Technical Report Agricultural Research Council Weed Research Organization*, **45**, pp. 55.

Greaves, M. P., Lockhart, L. A. and Richardson, W. G. (1978b). Measurement of herbicide effects on nitrogen fixation by legumes. *Proc. 1978 Br. Crop Prot. Conf.—Weeds*, 581–585.

Grossbard, E. (1970). An appraisal of the criteria by which to measure the effect of herbicide on the soil microflora. *Meded. Fac. Landbouw, Rijksuniv. Gent.*, **35**, 515–530.

Grossbard, E. (1971). The effects of repeated field applications of three herbicides on the evolution of carbon dioxide and mineralization of nitrogen in soil. *Weed Res.*, **11**, 263–275.

Grossbard, E. (1972). Do herbicides affect the microorganisms of the soil? *Report A.R.C. Weed Research Organization, 1969–71*, **4**, 72–83.

Grossbard, E. (1976). Effects on the soil microflora. In *Herbicides; Physiology, Biochemistry, Ecology*, Vol. 2, (Ed. L. J. Audus), pp. 99–147, Academic Press, London.

Grossbard, E. and Davies, H. A. (1976). Specific microbial responses to herbicides. *Weed Res.*, **16**, 163–169.

Hance, R. J. (1976). Herbicide usage and soil properties. *Pl. Soil.*, **45**, 291–293.

Hattingh, M. H. and Gerdemann, J. W. (1975). Inoculation of Brazilian sour orange seed with an endomycorrhizal fungus. *Phytopathology*, **65**, 1013–1016.

Hattori, T. (1973). *Microbial life in the soil.* Marcel Dekker Inc., New York.

Hepper, C. M. (1977). A colorimetric method for estimating vesicular-arbuscular mycorrhizal infection in roots. *Soil Biol. Biochem.*, **9**, 15–18.

Jenkinson, D. S. and Powlson, D. S. (1976). The effects of biocidal treatments on metabolism in soil—V. A method for measuring soil biomass. *Soil Biol. Biochem.*, **8**, 209–213.

Jensen, H. L. (1965). Non-symbiotic nitrogen fixation. In *Soil Nitrogen* (W. V. Bartholomew and F. E. Clark, Eds.), p. 440, American Society of Agronomy, Madison.

Johnen, B. G. (1977). Nebenwirkungen von Pflanzenschutzmitteln auf die Populationsdynamik von Bodenorganismen. *Mitt. Biol. Bundesanst., Berlin-Dahlem*, **178**, 53–65.

Johnen, B. G. (1978). Recent advances in the study of effects of pesticides on the population dynamics of non-target microorganisms. *Proc. 1978 Br. Crop Prot. Conf.—Weeds*, 1037–1046.

Johnen, B. G. and Drew, E. A. (1977). Ecological effects of pesticides on soil microorganisms. *Soil Sci.*, **123**, 319–324.

Johnen, B. G., Drew, E. A. and Castle, D. L. (1978). Studies on the effect of pesticides on symbiotic nitrogen fixation. In *Soil-Borne Plant Pathogens* (B. Schippers and W. Gams, Eds.), pp. 513–523. Academic Press, London and New York.

Jones, L. S., Anderson. O. E. and Dowler, C. (1974). Effects of herbicides in a crop-herbicide rotation on sulfur oxidation in Tifton soil. *Agron. J.*, **66**, 744–747.

Katan, J. and Eshel, Y. (1973). Interactions between herbicides and plant pathogens. *Residue Rev.* **45**, 145–177.

Kavanagh, T. (1974). The influence of herbicides on plant disease. 2. Vegetables, root crops and potatoes. *Scient. Proc. R. Dubl. Soc. Ser. B.*, **3** (**18**), 251–265.

Kleinschmidt, G. D. and Gerdemann, J. W. (1972). Stunting of citrus seedlings in fumigated nursery soils related to the absence of endomycorrhizae. *Phytopathology*, **62**, 1447–1453.

Knowles, R., Brouzes, R. and O'Toole, P. (1973). Kinetics of nitrogen fixation and acetylene reduction, and effects of oxygen and of acetylene on these processes in a soil system. In *Modern Methods in the Study of Microbial Ecology*. (R. Rosswall, Ed.), pp. 255–262, Bull. Ecol. Res. Comm. (Stockholm) 17.

Laskowski, D. A. (1979). Effects of herbicides on microorganisms and microbial processes. *Abs. 1979 Meeting of the Weed Science Society of America*, Abstract No. 248.

Latter, P. M. and Howson, G. (1977). The use of cotton strips to indicate cellulose decomposition in the field. *Pedobiologia*, **17**, 145–155.

Lynch, J. M. (1976). Products of soil microorganisms in relation to plant growth. *C.R.C. Crit. Rev. Microbiol.*, **5**, 67–107.

Lynch, J. M. and Panting, L. M. (1980). Cultivation and the soil biomass. *Soil Biol. Biochem.*, **12**, 29–34.

Malkomes, H. P. (1980). Strohrotteversuche zur Erfassung von Herbizid-Nebenwirkungen auf den Stoffumsatz im Boden. *Pedobiologia*, **20**, 417–427.

McCalla, T. M. and Norstadt, F. A. (1974). Toxicity problems in mulch tillage. *Agric. Environ.*, **1**, 153–174.

McLaren, A. D. and Peterson, G. H. (1967). Introduction to the biochemistry of terrestrial soils. In *Soil Biochemistry*. (A. D. McLaren and G. H. Peterson, Eds.), pp. 1–15. Edward Arnold (Publishers) Ltd., London.

Marsh, J. A. P. (1978). Natural variation in some soil nutrients as a means of assessing the importance of herbicide effects on microbial activity. *Proc. 1978 Br. Crop Prot. Conf.—Weeds*, 617–623.

Marsh, J. A. P. and Greaves, M. P. (1979). The influence of temperature and moisture on the effects of the herbicide dalapon on nitrogen transformation in soil. *Soil Biol. Biochem.*, **11**, 279–285.

Marsh, J. A. P., Kibble-White, R. and Stent, C. J. (1979). Apparatus for the automatic preparation of soil extracts for mineral-nitrogen determination. *Analyst*, **104**, 136–142.

Mosse, B. (1977). Plant growth responses to vesicular-arbuscular mycorrhiza. X. Responses of *Stylosanthes* and maize to inoculation in unsterile soils. *New Phytol.*, **78**, 277–288.

Mosse, B. and Bowen, G. D. (1968). The distribution of *Endogone* spores in some Australian and New Zealand soils and in an experimental field soil at Rothamsted. *Trans. Br. mycol. Soc.*, **51**, 485–492.

Norman, R. F. (1978). *Reported in* Hindering the agrochemical innovators. *Chem. Br.*, **14**, 478.

Papavizas, G. C. and Lewis, J. A. (1979). Side effects of pesticides on soil-borne plant pathogens. In *Soil-Borne Plant Pathogens* (B. Schippers and W. Gams, Eds.), pp. 483–505. Academic Press, London and New York.

Parr, J. F. (1974). Effects of pesticides on microorganisms in soil and water. In *Pesticides in soil and water*, (W. D. Guenzi, Ed.), pp. 315–340, Soil Sci. Soc. Am., Madison.

Rovira, A. D. (1969). Plant root exudates. *Bot. Rev.*, **35**, 35–57.

Russell, E. W. (1971). Soil structure: Its maintenance and improvement. *J. Soil Sci.*, **22**, 137–151.

Simon-Sylvestre, G. (1972). Le soufre du sol—son évolution. *Ann. Agron.*, **23**, 181–199.

Simon-Sylvestre, G. and Fournier, J-C. (1979). Effects of pesticides on the soil microflora. *Adv. Agron.*, **31**, 1–92.

Skujins, J. J. (1967). Enzymes in soil. In *Soil Biochemistry*. (A. D. McLaren and G. H. Peterson, Eds.), pp. 371–414. Edward Arnold, London.

Smith, T. F. (1978). Some effects of crop protection chemicals on the distribution and abundance of vesicular-arbuscular endomycorrhizas. *J. Aust. Inst. agric. Sci.*, **44**, 82–88.

Stotzky, G. (1965). Microbial respiration. In *Methods of Soil Analysis Pt. 2. Chemical and Microbiological Properties*, (C. A. Black *et al.*, Eds.), pp. 1550–1572. American Soc. Agron., Madison.

Verstraete, W. and Voets, J. P. (1974). Impact in sugarbeet crops of some important pesticide treatment systems on the microbial and enzymatic constitution of the soil. *Meded. Fac. Landbouw, Rijksuniv. Gent*, **39**, 1263–1277.

Vladutu, I. and Soreanu, I. (1976). Efectul unor ingrasaminte organice asupra recoltei de porumb si activatii enzymatice a solului in functie de aplicarea erbicidului Gesaprium 50. *Analele Inst. Cercetari Cereale Plante Tehnice Fundulea*, **41**, 193–202.

Voets, J. P., Meerschman, P. and Verstraete, W. (1974). Soil microbiological and biochemical effects of long-term atrazine applications. *Soil Biol. Biochem.*, **6**, 149–152.

Wingfield, G. I., Davies, H. A. and Greaves, M. P. (1977). The effect of soil treatment on the response of the soil microflora to the herbicide dalapon. *J. appl. Bact.*, **43**, 39–46.

Witt, J. M. and Gillett, J. W. Eds. (1977). *Terrestrial Microcosms and Environmental Chemistry*. National Science Foundation/RA 79-0026, Washington D.C.

Witty, J. F. (1979). Acetylene reduction assay can overestimate nitrogen-fixation in soil. *Soil Biol. Biochem.*, **11**, 209–210.

Yeates, G. W., Stout, J. D., Ross, D. J., Dutch, M. E. and Thomas, R. F. (1976). Long-term effects of paraquat-diquat and additional weed control treatments on some physical, biological, and respiratory properties of a soil previously under grass. *N.Z. J. agric. Res.*, **19**, 51–61.

# 10 Effects on Soil Fauna

*H. Eijsackers*
Research Institute for Nature Management
Kemperbergerweg 67
6816 RM Arnhem
Holland

*C. F. van de Bund*
Plant Protection Service
Geertjesweg 15
6706 EA Wageningen
Holland



## I. Introduction

### A. Soil as an Ecosystem

The evaluation of the influence of herbicides on the soil fauna requires information about the soil as a biological entity. Soil is not only the substrate for a large part of the biosphere, it is also an ecosystem in itself. The soil ecosystem comprises an abiotic and a biotic part, the biotic part being composed of soil organic matter, plant roots, soil microflora and soil fauna. On the basis of dry weight, microflora and fauna are only minor parts of the soil (Table I). For organic matter turnover and energy transfer they are of great importance however. Organic matter is broken down stepwise by microflora and fauna in cooperation. Soil structure (and hence aeration and drainage) is greatly influenced by burrowing activity of e.g. earthworms, beetles and millipedes. These species also mix organic material

Table I. Composition of a soil of a grassland on dry weight base

| | | | | | |
|---|---|---|---|---|---|
| Soil | mineral soil 94% | | | | |
| | organic matter 6% | dead org. m. 85% | | | |
| | | living org. m. 15% | living roots 8·5% | | |
| | | | edaphon 6·5% | bacteria + actinomycetes | 50% |
| | | | | fungi | 25% |
| | | | | worms | 14% |
| | | | | macrofauna | 5% |
| | | | | mesofauna | 2·5% |
| | | | | microfauna | 3·5% |

and mineral substances thus preparing a substrate for roots, fungi, bacteria and actinomycetes.

In addition the soil fauna transports microorganisms, useful as well as pathogenic species (see p. 297). Also microorganisms may be grazed by soil fauna species which themselves may be predated. Mineral soil and organic matter do not form an amorphous mass but have a distinct spatial distribution: the soil profile. Soil profiles develop under the influence of climate, parent material, vegetation cover and the soil biota. Under natural conditions this results in distinct forest soil profiles such as Mull and Mor. Under agricultural conditions the top layer is frequently a fairly homogeneous plough layer or a layer of organic matter and grass roots. Forest soils, meadow and agricultural soils each have their own specific soil fauna composition, with forest soil having the most diverse and agricultural land, the most restricted composition (Fig. 1). These differences between the

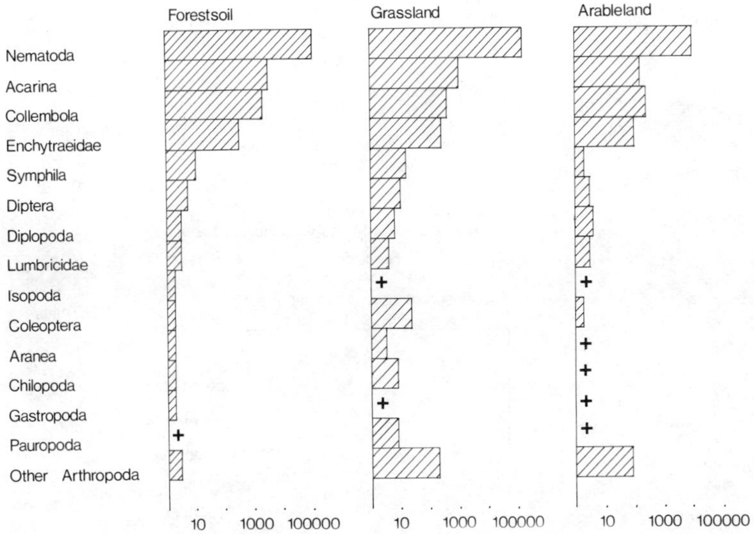


Fig. 1. Mean composition of the soil fauna in a forest, meadow and arable soil, mean numbers per $dm^2$ (0–30 cm depth), + means present.
(compiled from various literature sources and our own field sampling).

different soil profiles are not only of interest from a pedological and soil ecological point of view. Pesticide distribution and availability are greatly influenced by vertical distribution of organic matter and clay minerals. Moreover, soil structure and particle size also influence both the distribution of pesticides and soil fauna.

## B. Composition of the Soil Fauna

Soil fauna composition varies greatly between different habitats and in any case forms a highly heterogeneous group. The size of organisms varies from a few hundredths of a millimetre to more than 20 cm. In the series of groups of Lumbricidae, Enchytraeidae, Nematoda, Ciliata and Bacteria there is a ten fold difference in linear size between each. Accordingly the volume of their habitats differ by a factor of 1000. Thus simple measurement of environmental conditions for each of these groups is not possible.

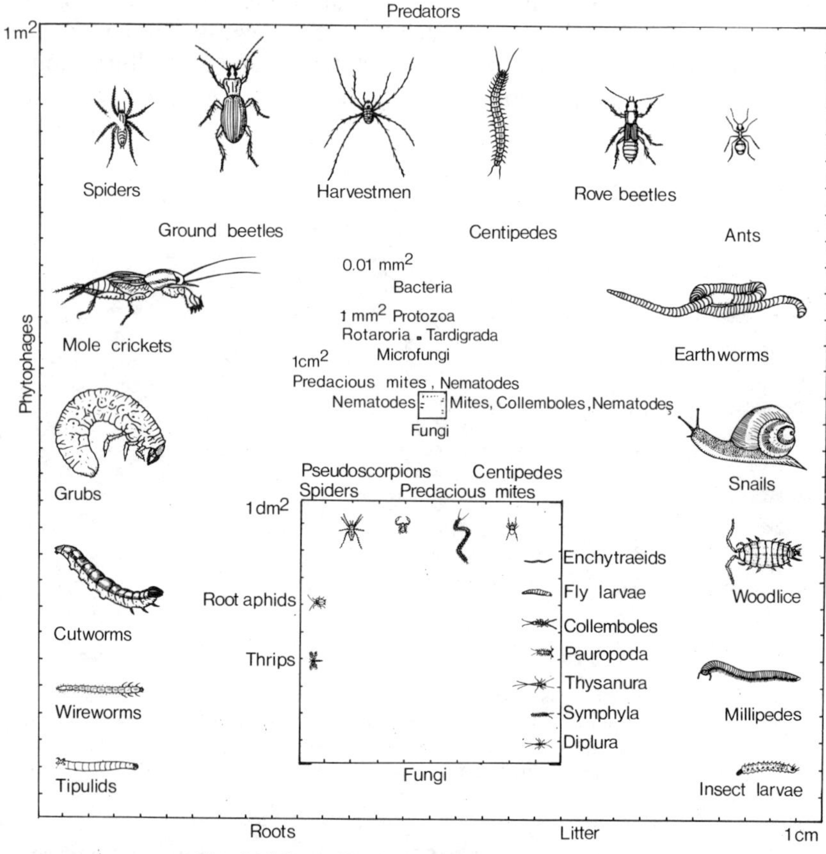


Fig. 2. Schematic survey of soil fauna community.
Species larger than 1 cm are sampled per $m^2$, species from 1–10 mm per $dm^2$, from 0·1–1 mm per $cm^2$ and from 10–100 $\mu$ per $mm^2$. On the base of the figure the food material is mentioned with which phytophages (left) and saprophages (right) feed themselves. Some important species have been drawn according to their size along the margins. (adapted from J. v.d. Drift 1963).

Their types of habitat also vary considerably. Protozoa, Rotatoria, Nematoda and Tardigrada, live in the water-filled spaces between soil particles and organic matter. If the soil dries out and the habitat disappears as a result, the animal may come into a resting state from which it emerges again only in the presence of moisture. Acarina and Collembola live in the air-filled spaces between soil particles and organic matter. Of the bigger specimens of the soil fauna, some live in these spaces too, although soil texture limits the penetration of these animals, and some live on the soil surface. A few groups (e.g. earthworms, diplopods and certain carabids) burrow into the soil. These burrows are important for root growth, drainage and aeration of the soil. Further details are described by Wallwork (1970 and 1976).

Soil fauna can be classified systematically or according to its food type or size. Figure 2 gives a brief survey of a number of important groups, their dimension, food type and the approximate size of their habitat.

## C. Feeding Habits of Soil Animals

With the diversity of soil fauna in mind it is surprising that there is at first sight rather a low degree of food differentiation. Major groups of soil animals as well as different species within these groups share the same food resources (Anderson, 1975). In this context it is important to realize that food ingestion as studied by gut analysis is not necessarily correlated with what is actually assimilated. Moreover, there may be a marked temporal variability. This is of practical importance for polyphagous soil animals occurring in agricultural soils. From Table II one can see that there are groups of species with a clear food specialization, and groups which are polyphagous and can change food utilization. Under certain circumstances these last groups can switch to a phytophagous behaviour. Also, even without such a clearcut switch, feeding behaviour may change from agriculturally useful to deleterious. Fungal hyphae, an important food resource in soil, are grazed by many species of soil fauna. This feeding habit creates new substrates for microorganisms and may remove pathogenic fungal species on roots. On the other hand new substrates for fungal attack may be created by this grazing which can result in consumption of root-tissue and, for instance with sugar beet, in seedling mortality (Heijbroek, 1972). Food utilization can change with management practice and weather conditions. Unfortunately most of these phenomena are assumptions, based on scattered field information and limited laboratory research. Detailed experiments are needed if they are to be confirmed. Special attention must be paid to predator–prey relationships which may play an important role in the regulation of micro- and mesofauna such as nematodes and

Table II. Common groups of arthropods inhabiting arable soil which change or extend their feeding habits and may cause occasional damage to crop plants

| Group of Arthropods | Predacious | Necro-phagous | Fungi-phagous | Sapro-phagous | Copro-phagous | Phyto-phagous |
|---|---|---|---|---|---|---|
| **Acarina** | | | | | | |
| *Acaridae* | * occ. | * | * | * | | * occ. |
| Rhizoglyphus robini Claparède | | | | | | |
| Tyrophagus similis Volgin | | | | | | |
| Tyrophagus longior (Gervais) | | | | | | |
| *Galumnidae* | * occ. | | * | * | | * occ. |
| Galumna spp. | | | | | | |
| *Oribatulidae* | | | * | * | | * occ. |
| Schelerobates spp. | | | | | | |
| *Uropodidae* | * | | * | * | | * occ. |
| Fuscuropoda marginata (Koch) | | | | | | |
| **Collembola** | * occ. | * | * | * | * | * occ. |
| *Onychiuridae* | | | | | | |
| Onychiurus armatus (Tullb.) | | | | | | |
| Onychiurus campatus Gisin | | | | | | |
| Onychiurus fimatus Gisin | | | | | | |
| *Isotomidae* | | | | | | |
| Folsomia fimetaria (L.) | | | | | | |
| *Entomobryidae* | | | | | | |
| Heteromurus nitidus (Templeton) | | | | | | |
| Orchesella spp. | | | | | | |

| | | | | | |
|---|---|---|---|---|---|
| **Isopoda** | | | | | |
| *Armadillidiidae* | | | | | |
| Armadillidium nasutum Budde-Land | | | * | * | * occ. |
| Armadillidium vulgare (Latreille) | | | | | |
| *Oniscidae* | | | | | |
| Oniscus asellus L. | | | * | * | * occ. |
| Porcellio laevis Latreille | | | | | |
| **Diplopoda** | | | | | |
| *Blaniulidae* | | | | | |
| Blaniulus guttulatus (Bosc.) | | * | | * | * occ. |
| Chonoiulus palmatus (Nemec) | | | | | |
| *Julidae* | | | | | |
| Brachyiulus littoralis Verhoeff | | | | * | * occ. |
| Cilindroiulus frisius (Verhoeff) | | | | | |
| *Polydesmidae* | | | | | |
| Brachydesmus superus Latz. | | | | * | * occ. |
| *Strongylosomidae* | | | | * | * occ. |
| Oxidus gracilis (Koch) | | | | | |
| **Coleoptera** | | | | | |
| *Carabidae* | * | * | | | * occ. |
| Amara spp. | | | | | |
| Pseudophonus rufipes (Degeer) | | | | | |
| Harpalus spp. | | | | | |
| Trechus quadristriatus (Schrank) | | | | | |
| Bembidion spp. | | | | | |
| Clivina fossor (L.) | | | | | |
| *Staphylinidae* | * | * | * | | * occ. |
| Trogophloeus spp. | | | | | |

Table II—*continued.*

| Group of Arthropods | Predacious | Necro-phagous | Fungi-phagous | Sapro-phagous | Copro-phagous | Phyto-phagous |
|---|---|---|---|---|---|---|
| *Elateridae* | * | | | * | | * |
| Agriotes spp. | | | | | | |
| Corymbites spp. | | | | | | |
| Athous spp. | | | | | | |
| Hypnoidus spp. | | | | | | |
| *Cantharidae* | * | | | | | * occ. |
| Cantharis spp. | | | | | | |
| **Diptera** | | | | | | |
| *Tipulidae* | | | | | | |
| Tipula spp. | | | | * | | * |
| Nephrotoma spp. | | | | | | |
| *Sciaridae* | | | * | * | | * |
| Lycoriella spp. | | | | | | |
| Bradysia spp. | | | | | | |

| | | |
|---|---|---|
| *Chironomidae* | | |
| Camptocladius stercorarius Degeer | * | * occ. |
| Gymnometriocnemus terrestris Goetg. | | |
| *Bibionidae* | | |
| Dilophus febrilis L. | * | * |
| Bibio spp. | | |
| *Petauristidae* | | |
| Trichocera spp. | * | * occ. |
| *Syrphidae* | | |
| Eumerus strigatus (Fall.) | * | * |
| *Anthomyiidae* | | |
| Delia platura (Mg.) | * | * |

occ. means occasional damage

springtails. Differences in sensitivity of predator and prey species to pesticides may affect the balance between these groups. For example the springtail population, which is insensitive to certain pesticides, has been shown to increase when predatory mites, which are highly susceptible, disappear (Sheals, 1955; Karg, 1961; Van de Bund, 1965). This in turn may influence the amount of damage the springtails can do to crop-seedlings, but here also further experiments are needed before definite conclusions can be drawn.

## II. Types of Effects

### A. Relative Distribution of Herbicide and Soil Fauna

In the field, distribution of the herbicide is uneven especially when measured on the habitat scale of soil fauna. According to the vegetation or crop type the absolute amount and the distribution of pesticide reaching the soil surface will differ. With spraying of shrubs the distribution will be much more uneven than with spraying a cereal crop, it may differ under shrubs by a factor of 30 between the lowest and the highest amount (Eijsackers, 1978e).

Soil fauna distribution is neither even nor random. Most species show a more or less distinctly aggregated distribution. This has been ascribed to such factors as egg-clumping, clustering of animals during bad conditions or local concentrations of food or moisture. Light, moisture, temperature and soil texture are the main factors which influence vertical distribution of soil fauna. Neither horizontal nor vertical distributions are fixed but, depending on diurnal or seasonal fluctuations, the distribution patterns change continuously.

### B. Influence of Weather Conditions

Temperature and precipitation are of great importance for the distribution of herbicides and soil fauna. With a high temperature and moist conditions, soil fauna will be active (also probably on the surface) and therefore there is an increased chance that it will come into contact with the spray liquid. On the other hand microbial and chemical decomposition processes will be accelerated and leaching may be increased under these conditions.

High temperatures combined with low moisture slow down soil microbiological processes, soil animals retreat to lower layers under these conditions, while leaching will be low leaving the herbicide on the soil surface. With cold, moist weather soil fauna activity will be moderate and contact with the herbicide will be low. The microbial decomposition rate will be low and there will be some leaching, hence local concentrations of herbicides will be reduced. During cold, dry weather all process-rates will be minimal. Soil fauna will not be affected immediately but because the herbicide persists the effect may be delayed.

## C. Effect of Mobility and Repellency on Exposure

Because of the irregular distribution of a pesticide in the soil, mobile species, especially spiders, beetles and ants, which are active on the soil surface, are the most likely groups to contact it. Predators can take in the pesticide together with their prey. However, if the herbicide has a repellent action the animals will avoid treated places or discard treated prey. It is not clear to what extent the higher rate of metabolism of an active animal will increase the intake of the herbicide via the integument or the trachea compared with an immobile species.

## D. Level of Biological Effect

Effects can be measured at different levels of biological organization including physiological and biochemical processes, individual reactions, population fluctuations and changes in ecosystem processes.

Biochemical and physiological processes have been used mainly in toxicological experiments. With soil fauna species there has not been much attention paid to this subject, although the possible importance of soil animals in decomposing pesticides is clearly recognized. The physiological condition of the experimental animals may greatly influence the impact of a herbicide. In our own experiments (H.E.) it was observed that during moulting soil animals are highly vulnerable to adverse influences. The same holds for breeding specimens. Nutritional level also influences herbicide impact and in field samples most specimens show empty guts.

Effects on individual animals and populations have been given the most attention. Individual reactions to herbicides are mentioned in Section III.B3. The usual population measurements are the numbers of animals, the fluctuation of numbers in time, the distribution of the animals in the soil and their age distribution. For this last characteristic either the population can be split into juveniles and adults, or a complete cohort-analysis can be made.

Effects on ecosystems have been examined only recently. Theoretical developments with regard to stability and diversity have greatly stimulated research in this direction. However, we are far from being able to evaluate fully the consequences of changes in species diversity on the functioning of soil ecosystems. Nevertheless it is one of our tools for the measurement of side effects. A more direct assessment of species diversity in connection with ecosystems functions can be made by concentrating on certain processes like litter decomposition or predator–prey relations.

## E. Primary and Secondary Effects

Direct effects can be measured by changes in behaviour, growth and moulting, food consumption, mortality or mean longevity and reproduction.

Changes in behaviour can be measured in mobility (a decrease as well as an increase), a changed walking pattern as a result of a repellent action of a pesticide, or as changes in specific behavioural activities. Preening for instance may be a useful activity to observe and it is important for uptake of pesticides.

Growth is usually measured as length or weight increment, or as increase in head width. The rate of moulting can be used as a measurement of growth. Moulting is important for several reasons. During moulting the animal is very vulnerable because the integument is soft and easy to penetrate. In springtails moulting has an excretive function too. The midgut epithelium is "moulted" and Humbert (1974) has suggested that refuse is excreted with the epithelium.

Food consumption can be used as a measure of litter decomposition. As such it may be interpreted as a measure of a direct effect on ecosystem functions. Moreover, it is one of the most important ways soil animals contact herbicides especially for saprophagous animals which, because of low digestive efficiency, take in large quantities of food. So these animals may assimilate relatively large amounts of herbicides through their digestive tract.

Mortality, longevity and reproduction give the most direct measurement of the effect of the herbicide on the population of a species.

Indirect, secondary effects are more difficult to monitor than direct, primary effects, although in field experiments they almost always interfere with the measurement of direct effects. These indirect effects may be changes in vegetation composition, litter composition or soil physical conditions.

Herbicide treatment inevitably changes the composition of vegetation and hence the soil fauna composition too. Curry and Tuohy (1978) clearly showed a close relationship between certain broad-leaved grassland species and soil fauna. Edwards and Lofty (1974) and Edwards *et al.* (1975) in their extensive survey of the soil and surface fauna of the Park Grass Plots at Rothamsted found gross differences between plots with different fertilizer treatments and between those with a markedly differing vegetation.

When the vegetation changes, part of the soil fauna will disappear and part will concentrate on the remaining plant species. This concentration effect may have consequences for the amount of crop damage, as Heijbroek (1972) found after herbicide treatment in sugarbeet.

Litter composition also changes after herbicide treatment. A sudden flush of dead plant material will stimulate certain groups of the soil fauna. However, it is not clear to what extent effects persist over a long period. This depends on the rate of recovery of the vegetation which in turn depends on the methods of vegetation management including whether a single or periodic herbicide treatment is used.

Changes in vegetation and litter composition also change soil physical conditions. The absence of vegetation, by reducing insulation, leads to greater fluctuations in soil temperature and changes rates of water loss. In addition, the lack of plant roots increases erosion even in regions of low rainfall intensity.

Zero-tillage systems (direct drilling, slitseeding) provide a different situation. Here, herbicide treatment replaces soil cultivation and consequently soil structure is influenced. The undisturbed soil structure combined with a more natural distribution of organic matter has a distinct effect on numbers and distribution of soil fauna (Van de Bund, 1970; Edwards, 1975).

It is clear that it is hardly possible to separate the different variables that may account for these indirect effects. Therefore in field experiments only gross effects can be measured. For the detection of more subtle effects one has to design laboratory experiments which in turn can be the basis for further field experimentation.

## III. Laboratory and Field Research

### A. Advantages and Disadvantages of Each

Laboratory and field research both have their advantages and disadvantages. Under laboratory conditions it is possible to set up experiments with a limited number of controlled variables. However, in applied research conclusions must be valid for the situation in the field where there are a great number of free variables. To reduce the gap between laboratory and field conditions one can increase the number of controllable variables in the experiment or approach the abiotic natural conditions as far as possible, for instance by selecting a fluctuating temperature regime. Field observations must, however, give the basic data for manipulating the laboratory conditions in the right way.

We have to define what is the real purpose of evaluating side effects. Is it to measure changes with mean, "normal" environmental conditions, or is it to assess the risk of catastrophic effects in extreme conditions? In our opinion the latter approach must have priority. Laboratory experiments are particularly useful for creating extreme conditions, but field experiments are always necessary to test the results of laboratory experiments and to confirm the prediction based on them. Laboratory experiments have the advantage over field experiments of speed, precision and low cost. They allow a quick screening of the sensitivity of the different soil fauna groups so that field experiments can then be focussed onto the sensitive groups.

## B. Laboratory Research

### 1. *Cultured or specimens collected in the field*

Whether it is better to use cultured or field-collected experimental animals depends on the purpose of the experiment. For screening programmes for direct effects there must be an almost continuous supply of large numbers of specimens, which are physiologically and genetically more or less the same. Testing for side effects requires specimens which are representative of field populations. There are seasonal changes in the composition of field populations in physiological condition, age distribution, etc. so freshly collected samples are preferable. For culturing, a species must be adapted to laboratory conditions, easy to breed, have simple food requirements, be adapted to high population densities, require no complicated conditions for mating and oviposition, have a high innate reproductive rate and be easy to handle (Felton *et al.*, 1977).

For eco-toxicological experiments Kaminski (1975) lists the following criteria for the species to be used: high survival rate in small experimental containers, facility of mass rearing in the laboratory, facility of transportation also in hermetic containers, high longevity, relative inertness to environmental conditions, widespread occurrence in nature, sensitivity to toxicants, pronounced symptomatology of test response. Lofs-Holmin (1979) classifies the suitability of different earthworm species for eco-toxicological experiments according to their abundance in cultivated fields, ease of taxonomic identification of juvenile specimens, ease of culturing, reproduction rate, growth rate, uniformness of growth curve and ease of hand sorting. Almost all these criteria are based on practical reasons, the ecological significance of the different species gets less attention although this should be of prime importance. A disadvantage of cultured specimens is that their adaption to laboratory conditions may have adverse effects on processes such as moulting which can become highly irregular as Van der Drift (1975) showed for the diplopod *Glomeris marginata.* Food quality has a distinct influence on chemical composition and biological characteristics of the springtail *Onychiurus justi* (Snider, 1971). High population densities may result in reduced numbers, especially with species which show cannibalism. However, there may be some advantage in this reduction as Hunter-Jones (1961) found that the natural cannibalistic tendency of the desert locust provided uniform size batches and a homogeneous age-grouping under laboratory conditions. Nevertheless, individuals of certain species have to be reared separately to exclude cannibalism.

Soil animals do not require complicated culture conditions for mating and oviposition. Further details on culture conditions for a number of soil animal groups are given by Tomlin (1977).

Animals collected from the field must have a common and preferably widespread distribution, must be easy to sample and handle, and representative of certain soil faunal groups. Selecting species for field collection on these criteria may conflict with the importance of the role of the selected species in soil functions. Further, field collection is largely dependent on season and weather conditions. In summer during dry conditions and in winter with frost a sufficient supply may be uncertain.

### 2. *Experimental conditions*

Environmental variables and pesticide distribution are of prime importance. Temperature, air humidity, light, structure and moisture content of soil substrate and/or litter layer are the most important environmental variables which of course differ according to season and climate. To simulate temperature conditions a temperature of 10–15°C, 80–90% RH with a moisture content of 25% and only minimal light during daytime will suffice. Addition of litter provides food and shelter, but also changes environmental conditions. However, it is not certain that uniform average conditions are satisfactory. For example in an experiment with the springtail *Onychiurus quadriocellatus* the effect of 2,4,5-T was tested at constant temperatures of 15, 20 and 25°C and also with a temperature regime fluctuating between 15 and 25°C. Under fluctuating conditions the effect was the same as at the highest temperature of 25°C (Eijsackers, 1978e). Joosse *et al.* (1973), Petersen (1971) and Gregoire-Wibo (1976) found, however, that production values and growth rates of springtails at fluctuating temperatures were about equal to values obtained at the mean constant temperature. Substrate moisture content is a critical factor too. Springtails, for instance, require different moisture conditions during different phases of their lifecycle (Verhoef, 1978). The effect of herbicides may be influenced by the substrate moisture content as shown by experiments with the carabid *Notiophilus biguttatus* (Eijsackers, 1978d).

Pesticides can be applied by dipping specimens in herbicide solutions, by a dropwise application from a syringe, by spraying onto the animals or their substrate, or by mixing through the soil. With dipping, dripping or

Table III. Survey of laboratory methods to assess herbicide effects on different soil fauna groups

| Soil fauna group | Test substrate | Measurement[a] | Herbicide | Reference |
|---|---|---|---|---|
| Protozoa | water | r | Diuron | Gel'cer (1967) |
| | water/agar | r | Diuron | Gel'cer and Geptner (1976) |
| Rotatoria | water | m | Aminotriazole, Bromacil, Chlorpropham, Dalapon, Diuron, Paraquat, Sesone | Caseley and Eno (1966) |
| Nematoda | water | m | Aminotriazole, Chlorpropham, 2,4-D, 3,4-D, Dalapon, 2,4-DB, Dinoseb, Diuron, MCPA, Sesone, 2,4,5-T, 2,4,5-TP | Chapell and Miller (1956) |
| | plantroot | gallination | Aminotriazole, Dalapon, MH | Apt *et al*. (1960) |
| | water | a, m | DNOC, MCPA | Witkowski (1973) |
| | water | m | Dinoseb-acetate, DNOC | Witkowski (1974) |
| | water | r | MCPA | Witkowski (1975) |
| | water | a | Aminotriazole, Atrazine, Buturon, 2,4-D, Dichlobenil, Diquat, Diuron, Linuron, MCPA, Monolinuron, Neburon, Simazine, Paraquat, 2,4,5-T | Frey (1976) |
| | cornmeal agar + bacteria | r | Aminotriazole, Atrazine, Buturon, 2,4-D, Dichlobenil, Diquat, Diuron, Linuron, MCPA, Monolinuron, Neburon, Simazine, Paraquat, 2,4,5-T | Frey (1976) |
| | agar | m | Linuron, Monolinuron | Fedorko *et al*. (1977) |
| | soil | | Simazine, TCA | Atlavinyté *et al*. (1977) |
| Acari | filter paper | 1 | Simazine, Paraquat | Baumler *et al*. (1978) |
| | soil | m | DNOC, MCPA | Dziuba (1971) |
| | soil-agar | m | Prometryne | Sanocka and Sanocka-Woloszyn (1970) |

| | | | | |
|---|---|---|---|---|
| Collembola | soil | m | Simazine, TCA | Atlavinyté *et al*. (1976) |
| | soil-agar, soil | a, m, e, r | 2,4,5-T | Eijsackers (1978a) |
| | filter paper + litter, algae, fungal, hyphae | c, m, e, f, g | 2,4,5-T | Eijsackers (1978b) |
| | soil | m, r | AMS | Eijsackers and Chardon (1979) |
| | soil-agar | a, m, r | Aminotriazole, Simazine | Eijsackers and Van Doorn (unpubl. data) |
| | soil-agar | m, r | Chlorpropham, Linuron, Prometryne | Sanocka and Sanocka-Woloszyn (1970) |
| | soil | m | Atrazine, Paraquat | Subagja (1978) |
| | plaster of Paris + yeast | m, r, g | Altrazine, Paraquat | Subagja (1978) |
| Isopoda | soil | a, l | 2,4,5-T | Eijsackers (1978c) |
| | glass + litter | c, g, l | 2,4,5-T | Eijsackers (1978c) |
| | soil | l | AMS | Eijsackers and Chardon (1979) |
| | glass + bark, litter | c | AMS | Eijsackers and Chardon (1979) |
| Diplopoda | glass + litter | c, m | 2,4,5-T | Eijsackers (1978e) |
| | glass + litter | c | AMS | Eijsackers and Chardon (1979) |
| | soil | b | AMS | Eijsackers and Chardon (1979) |
| | filter paper | c, m | 2,4-D | Hoy (pers. comm.) |
| Orthoptera | art. food | r | Aminotriazole, Atrazine, Asulam, Bentazone, Carbetamide, Cycluron, 2,4-D, Dicamba | Rivière (1976) |

Table III—*continued*.

| Soil fauna group | Test substrate | Measurement[a] | Herbicide | Reference |
|---|---|---|---|---|
| Orth. larvae | art. food | m, g | Diphenamid, Diuron, Fenoprop, Ioxynil, Lenacil, Linuron, Oxadiazon, Paraquat, Phenmedipham, Picloram, Chloridazon, Terbutryne, Tri-allate | Rivière (1976) |
| Homoptera | glass, dipping topical | m | Dinoseb, Diuron, MSMA | Stam *et al*. (1978) |
| Coleoptera | soil | m | | Edwards (1973) |
| | soil + agar | m | Aminotriazole | |
| | soil, soil + food | a, l, f | 2,4,5-T | Eijsackers (1974) |
| | soil | a, l | AMS | Eijsackers and Chardon (1979) |
| | soil, topical | m | Linuron, 2,4-D, DNBP, MCPA, Nitrofen + Neburon, Prometryne, Chloridazon, Simazine | Kowalczyk-Rozek and Slizynski (1977) |
| | glass, dipping topical | m | Dinoseb, Diuron, MSMA | Stam *et al*. (1978) |
| | soil, soil + food | a, m, e | Aminotriazole + Simazine, Chlorpropham, 2,4-D, Dalapon, Diquat, DNOC, Na-chlorate, Pyrazon, Simazine | Muller (1971) |
| Coleoptera larvae | art. food | m, g, e | 2,4-D, MH, 2,4,5-T | Davis (1968) |
| Hymenoptera | glass | m | Dinoseb, Diuron, MSMA | Stam *et al*. (1978) |

| | | | | |
|---|---|---|---|---|
| Gastropoda | agar | a, m | 2,4-D | Hoy (pers. comm.) |
| Lumbricidae | soil | f | TCA | Atlavinyté (1975) |
| | soil | m, r, f regeneration | Simazine, TCA | Atlavinyté *et al.* (1977) |
| | soil | m, r | Aminotriazole, Atrazine, Bromacil, Chlorpropham, Dalapon, Diphenamid, Diuron, Endothal, Paraquat, Sesone, 2,3,6-TBA | Caseley and Eno (1966) |
| | soil (injection into haemocoel) | m | Atrazine, Chloramben, Dicamba | Chio and Sanborn (1978) |
| | water | m | Aminotriazole, Atrazine, 2,4-D, Monuron, Simazine | Ghabbour and Imam (1967) |
| | soil | m, r | TCA | Lofs-Holmin (1980) |
| | soil | m | Chlorpropham, Chlorpropham + DNOC, DNOC, Propham, Propham + DNOC | Van Rhee (1972) |
| | soil | a | Aminotriazole, Paraquat | Van Rhee (1972) |

[a] a = activity, c = food consumption, e = ecdysis, f = feeding rate, g = growth rate, l = longevity, m = mortality, r = reproduction.

spraying methods we must distinguish between dosage and dose. The dosage of herbicide applied in the test solution is not necessarily the same as the dose which the test animal receives. When spraying the soil or with mixing, the herbicide can be distributed homogeneously or heterogeneously. This may be important for instance when the herbicide is present only in the upper soil layer as earthworms could be unaffected, but breakdown and burying of litter from the soil surface may be hampered.

In the next two sections the different methods described in the literature are discussed in more detail. A survey of these methods and the applied herbicides is given in Table III. The size of effects has already been reviewed by Eijsackers and Van der Drift (1976). The effects of pesticides on the soil fauna are discussed by Edwards and Thompson (1973).

### 3. *Experiments with one species*

(*a*) *Effects by direct contact.* The simplest method for testing the effects of direct contact is by topical application. Edwards (1973) found this method very satisfactory with carabids. For the smaller animals this method becomes rather laborious. Experiments on effects by direct contact can be conducted in water, with an artificial substrate or with a soil substrate. Solutions of herbicides in water can be used for faunal groups which live in the water-filled spaces between soil particles like protozoa, rotifers, tardigrades and nematodes. For other groups such as arthropods or annelides this would be a rather artificial habitat. These groups can be tested in glass dishes, with a substrate of filter paper, soil, or a mixture of soil and agar.

Mortality, activity, repellency or attraction, longevity, reproduction, growth and moulting can be used to measure responses of organisms. Atlavinyté *et al.* (1977) also studied the effects on amputated earthworms, so combining the possible adverse effects of herbicide and ploughing on the capacities of the affected earthworms to regenerate. Mortality, activity and repellency can be used in short-term experiments, whereas reproduction, growth and moulting require a prolonged experimental period. The classical $LD_{50}$-bioassay has been reviewed thoroughly by Hoskins and Craig (1962) so will not be discussed here. Greenway *et al.* (1977) discuss some of the problems associated with experiments on behaviour controlling chemicals. Measurements of activity and repellency are made by direct observations or by time lapse photography. They can be evaluated by means of a distinct key response or a continuous activity scale. Greenway *et al.* (1977) give an elegant example of the application of both types of approach. With nematodes Frey (1976) used the absence of body movement as a measure of herbicide effect. Eijsackers (1978a, 1978b and 1978d) observed the behaviour of springtails, isopods and carabids on a substrate partly treated with water (control) or with different amounts of 2,4,5-T solutions. Therefore it was possible to score both behaviour on the two substrates and also the preference or aversion for the substrates. Hoy (pers. comm.) did the same with millipedes and also scored activity in tests with slugs. For measuring the burrowing activity of earthworms Van Rhee (1972) used cages constructed of two sheets of perspex about 1 cm apart, with the space between the plates filled with soil. Activity was studied by measuring the length of the burrowing holes.

Longevity measurements partly overcome one of the disadvantages of $LD_{50}$ which is the difficulty in choosing the optimum experimental period. Daily assessment of treated and untreated animals of either a mixed or a homogeneous age distribution (depending on whether they were field-collected or cultured) are time consuming however. Pilot experiments are useful to deduce the extent to which the mean longevities with different treatments coincide with $LD_{50}$ or $LD_{10}$ values.

Measurements of reproduction have been made by counting numbers of eggs, cysts and cocoons using elutriation methods (Murphy, 1962). In a few studies age group distribution has been assessed, or the sensitivity of juvenile specimens has been assessed separately. For earthworms age group distribution is difficult to determine. Earthworms can be separated into juveniles, sub-adults and adults using the presence of a clitellum to identify the adults. However, sub-adults cannot be defined as the group older than juveniles but lacking a clitellum as adult earthworms in bad conditions can temporarily loose their clitellum whereas degenerating earthworms loose it permanently.

Growth measurements have been used in only a few studies. Davis (1968), Eijsackers (1978c) and Lofs-Holmin (1980) measured weight gain of carabid larvae, isopods and lumbricids respectively. With a great number of soil fauna species, weighing is complicated by the water loss which occurs during weighing. An alternative method for measurement of growth is to estimate length or head-width. However, to do this the more active species have to be anaesthetized with diethyl ether or $CO_2$ and this may adversely affect them.

Moulting by counting exuviae has been studied only in experiments with beetle larvae (Davis, 1968; Müller, 1971) and with springtails (Eijsackers, 1978b). Because several springtail species consume their exuviae, these animals have to be cultured separately, moreover it becomes possible then to calculate not only the mean moulting frequency but also the exact period between moultings.

*(b) Effects following consumption of treated food.* Food consumption interacts with a herbicide in several ways: the herbicide may be ingested with food, its effects vary with the state of nutrition of the animal and direct

contact with the herbicide may cause a decreased rate of food consumption and thus hamper litter breakdown.

The effects of herbicides ingested with food can be assayed using treated plant material (algae included), treated litter (soaked or sprayed), treated prey, or with an artificial diet. In experiments with earthworms which ingest large amounts of soil, effects by direct contact complicate the effects through food consumption. This combined influence can also occur in experiments with a non-sterilized soil substrate in which the microflora may adsorb significant amounts of herbicide. In this case microarthropods grazing the microflora may ingest the herbicide or become affected by direct contact with it.

Measurements used in feeding experiments include: food weight loss or reduction of the surface of the food, pellet production, analysis of gut content, and the effects of the consequences of the uptake of herbicide via the food such as mortality and activity. However, none of the measurements of food uptake is easy to quantify satisfactorily. Weighing food gives uncertain results because of weight loss by evaporation. Measurement of the area of the unconsumed food (for instance litter discs) overcomes this problem but is only of use with animals which cut out complete litter fragments, so the method is not suitable for superficial grazers like isopods. Consumption of fungal hyphae growing on a substrate of organic matter can be classified only by subjective grading of the fungal growth.

Pellet production is easy to measure, but counting the pellets is time consuming and correlation with the amount of food eaten is not always very good. Moreover, numbers and sizes of the pellets vary with the age of specimens. Therefore, it is necessary to use a randomized block design (Campbell, 1974). Gut content gives a quick impression of the eagerness with which the food is eaten, but cannot be quantified to estimate total consumption during a period of time. An extra problem in feeding experiments is the irregular feeding pattern, which with some species is correlated with moulting. With certain springtail species moulting can by synchronized or with other species experiments have to be preceded by a few days starvation.

Only a few authors have described experiments to study the consumption of plant or microbial material or have stated explicitly how they fed their experimental animals. Frey (1976) used corn meal agar with a bacterial suspension for his experiments with nematodes. Sanocka-Woloszyn and Woloszyn (1970) and Eijsackers (1978a) used a soil–agar substrate for experiments with springtails, but did not distinguish effects by direct contact from those by injection of contaminated organic matter or microflora. In further experiments with springtails Eijsackers (1978b) used algae, larch needles, birch litter discs and pellets of isopods, fed sprayed leaves of

*Prunus serotina* as food. Effects proved to be rather variable. One of the reasons may be the preference soil animals show for the litter of certain plant species (Curry and Tuohy, 1978).

#### 4. *Experiments with mixed populations*

Experiments with more than one species have not been very numerous, though they are promising for hazard evaluation. They include predator–prey and food chain experiments, studies on competitive ability, and model ecosystem (microcosm) experiments. Predator–prey and food chain experiments deal with effects through food consumption as described in Section III B3b, but pay attention to the way the herbicide influences the prey as well as the predator. For instance a change in the mobility of the prey caused by the herbicide may effect its chance of capture. Accumulation or adsorption of the herbicide by the prey may have a repellent action and may change the transport of the herbicide in the ecosystem. Some experiments of this kind are reported by Eijsackers (1978d).

Competition, although one of the basic adaptive forces in evolution, has not yet been included in hazard evaluation experiments. The experiments of Christiansen (1967) on competition among springtails are promising for application in hazard evaluation experiments.

Model ecosystem or microcosm experiments have been set up to assess the transport, fate and effects of a chemical upon ecosystem components, specifically populations and trophic web members. Terrestrial systems have included synthetic simplified systems, synthetic complex systems and natural core systems. In the first vermiculite is often used for substrate, with the second type the substrate varies between a homogenized sandy soil and an artificial profile built of standardized natural components, whereas in the third type intact profiles are used.

With these different types of systems, transport and fate in the abiotic environment can be measured, as Jackson and Levin (1979) and Jackson *et al.* (1979) showed with arsenite. The main problems arise with the introduction of soil animals which can be gnotobiotic, standardized or natural. Working with gnotobiotic animals is extremely difficult and only possible with microfauna (amoebae, ciliata). Standardization by selection of certain species and adding them in a prescribed order is always arbitrary. In natural profiles the composition of the soil fauna will vary although Van Voris *et al.* (1978) found little variations in arthropod and nematode density between replicate microcosms. A review of the application of undisturbed soil core systems is given by Ausmus *et al.* (1979). They conclude that artificial substrates or homogenized soils reinoculated with biota are not to be recommended because of the instability of substrates with respect to the characteristics monitored, the limited replicability and accuracy of units, and the proportionately higher cost of the system. However, they base this conclusion mainly on the measurement of integral characteristics like $CO_2$-efflux. Confirmatory research with soil fauna species has still to be done.

A good review of the state-of-the-art of terrestrial microcosm research is given by Gillett and Witt (1979) and Witt and Gillett (1979). As advantages or potential capabilities they mention:

(i) the ability to demonstrate interactions, and to test these interactions in a closed, "safe" system;
(ii) the provision of indices of temporal and spacial distribution;
(iii) the cost-effectiveness of testing chemicals compared to field experiments;
(iv) the usefulness as an intermediate step between simple laboratory tests and the complex processes in the field;
(v) the possibility to get insight in fundamental processes;
(vi) the provision of more information for safe use than from laboratory tests alone.

On the other hand, microcosms have definite limitations:

(i) they are short-lived and not self sustaining (although it is questionable whether this is an absolute prerequisite as natural soil systems are triggered by the yearly influx of organic matter during leaf-fall. H.E.);
(ii) simple properties may be measured more readily in less complex systems;
(iii) microcosms may include only part of all the significant processes of a given environment;
(iv) biological effects and transport processes can only be measured in a relative sense.

In order to overcome these problems they suggest that "future research should pay attention to standardization of soil animals, to a series of systems with a gradual built up complexity, and to verification of laboratory tests against field studies."

As a useful research tool litter fragmentation may be chosen as this comprises limited scaling, simple and complex interactions and is suitable for field verification for instance with litter bags. Research on effects of herbicides in microcosms has been very sparse. Metcalf *et al.* (1979) have tested a number of herbicides in their system (Metcalf *et al.*, 1971) which is of the synthetic complex type. From their data it can be concluded that the herbicides tested (simazine, trifluralin, PCP and 2,4,5-T) negatively affected the soil fauna species under study (an earthworm, isopod and slug). Moreover, the herbicides were detected in the bodies of the soil fauna species but gut content and total body content were not separated.

### 5. *Concluding remarks*

Laboratory experiments are needed to assess: (i) the metabolism of herbicides by soil fauna species, (ii) the toxicity of herbicides for soil fauna species and the resulting effects upon soil biological processes, (iii) the toxicity of herbicides for ecosystem processes. Because of their metabolic rate, representatives of the microfauna (Protozoa and Rotatoria) are suitable for metabolic studies. Moreover, their short generation time makes them suitable for studies on reproduction. For this purpose also Acari and Collembola may be useful. Functional processes like litter fragmentation can best be studied with representatives of the macrofauna like Isopoda, Diplopoda and Lumbricidae. Burrowing and soil mixing activities are easily assessed with Lumbricidae which are also convenient experimental animals to assess direct toxic effects. However, their size and the size of the necessary experimental cages are a disadvantage, so Collembola, Isopoda and Diplopoda are more suitable. With these groups and with Acari there is the added advantage of a clear age distribution enabling the analysis of effects on the composition of a population (cohort analysis). Research about ecosystem processes is still in a preliminary phase. Much further work is urgently needed if we are to assess effects of herbicides upon natural ecosystems.

In the experiments, the irregular distribution of the herbicide on the soil surface and the possibility that mobile soil species can fly has to be taken into account. The experiments may be run under mean environmental conditions or under extreme conditions. The selection of conditions will depend on the practical use of a particular herbicide.

## C. Field Research

### 1. *Design of field trials*

The effect of pesticides on soil organisms in the field varies largely in response to factors such as soil type, soil structure, tillage, use of organic manure and fertilizers, previous rotational crops, weather conditions and soil moisture. Each of these can have a large effect on the soil fauna and on the activity of the applied pesticides. Without comparison with valid controls it is nearly impossible to distinguish the effect of pesticides on the soil fauna from the effects of other factors. For this reason if the effects of pesticides are to be assessed reliably, great care is necessary to ensure that the pesticide application is the only difference between treated and control areas. For micro-arthropods we suggest a plot of $3 \times 3$ m; with a strip of at least 1 m between plots. Treatments should be replicated at least three times and preferably laid out in a Latin square. The herbicide under study should be incorporated into the soil after the usual tillage and a standard crop such as perennial ryegrass, white clover, siletta or turnips sown. Incorporation of the herbicide ensures the maximum effect on the soil fauna. For larger fauna species a plot size of 10 × 10 m is suggested. In order to prevent exchange between plots they can be fenced or strips of 10 m can be laid out between them. In practice the herbicide should be applied onto the crop if it is to be used post-emergence.

## 2. *Sampling methods*

Because of the irregular distribution pattern of the soil fauna it is often very difficult to decide how adequate sampling can be achieved. Therefore an overall sampling of the field has to be carried out before the trial field is laid out. Usually with micro-arthropods (<2 mm) like springtails and mites, 36–48 soil samples have to be taken. The content of each sample ought to be at least 50 ml but does not need to be larger than 150 ml. For the macrofauna the content varies between 125 and 500 ml. After analysis of the micro-arthropods the number of soil samples needed from each plot can be decided. Generally in any collection of micro-arthropods extracted from soil samples there are a few dominant species, in a relative abundance of at least 5% of all sampled individuals, a number of subdominant species in a relative abundance of 2–5% and a larger number of recedent species in a relative abundancy of 1–2%. Most of these occur in at least 10% of the soil samples and can be considered as common species. Some species will aggregate in a few or a single soil sample in very large numbers and will have a great influence on the total number of arthropods. In an objective sample the frequently occurring species and the very abundant aggregated species must be represented in a set of soil samples. A useful criterion is to take the number of soil samples in which all those species are present which occur in a frequency of at least 10% in the whole set of preliminary samples. Another criterion is the number of samples in which all species are present which have a relative abundance of at least 5% of the total catch of the presampling. For the statistical background the reader is referred to Murphy (1962). In many cases six samples per trial plot are sufficient. For soil sampling, augers with a diameter of 5–10 cm are preferred. To prevent compression of the soil cores during sampling the auger has to be slightly conical 0·5–0·75 cm wider at 5 cm from the mouth than at the mouth. The size of the auger, the sampling technique and the method of extracting of the soil samples depend on the dimension and nature of the group of animals that will be investigated, a generalized scheme is given in Table IV. In Table V a survey is given of the different field studies on herbicide effects on the soil fauna, including type of herbicide, application method and dose, habitat substrate, sampling and extraction meth-

Table IV. Arrangement of soil groups according to their size and corresponding sampling methods and sample sizes

| Microfauna | Mesofauna including micro-arthropods | Macrofauna meso-arthropods | Macrofauna macro-arthropods |
|---|---|---|---|
| <0·2 mm | 0·2–2 mm | 2–20 mm | >20 mm |
| soil sample + culturing 5 ml | soil sample (auger) 50 ml | soil sample (auger) 500 ml | soil sample (digging) 5000 ml |
| Protozoa<br>Rotatoria | Nematoda<br>Tardigrada<br>Acari<br>Collembola<br>Thysanura<br>Protura | Symphyla<br>Thysanoptera<br>Pauropoda | Isopoda<br>Araneida<br>Chilopoda<br>Orthoptera<br>Dermaptera<br>Isoptera<br>Heteroptera<br>Homoptera<br>Lepidoptera<br>Coleoptera<br>Hymenoptera<br>Diptera |

ods, characteristics to measure the effect and the amount and duration of the effect.

Although most soil animals can be sampled by extraction in Tullgren funnels, it must be stressed that there is no universal method of sampling and extraction which is equally efficient for all groups of soil animals. The Tullgren funnel for instance is efficient for extracting small arthropods, but is inefficient for extracting Enchytraeidae and large fly larvae like leatherjackets. When all major groups of soil animals are to be sampled, appropriate methods of sampling and extracting must be used for each group of the soil fauna.

Euedaphic macro-arthropods, meso- and micro-arthropods, Enchytraeidae, Lumbricidae and Nematoda can be best sampled by soil sampling. Nematodes and enchytraeids can be separated by elutriation techniques using subsamples of 20 and 100 $cm^3$, respectively. For arthropods, desiccation methods are suitable, appropriate sample sizes for micro-, meso, and macro-arthropods (respectively <2 mm, 2–20 mm and >20 mm) being 50, 500 and 5000 $cm^3$. Large arthropods like spiders and beetles can best be caught with pitfall traps buried in the soil. For earthworms, and large euadaphic arthropods the best results are obtained by digging down to an

Table V. Survey of field methods to assess herbicide effects on different soil fauna groups.
In this table coding is arranged for each herbicide as follows:

Effect of the herbicide: + distinct increase; 0 no effect; – distinct decrease; (–) effect insignificant. When variable effects were found a combination of the symbols is used.

Measurements: 1. number of individuals of the large taxa (orders, families); 2. relative abundance of the large taxa; 3. number of individuals of the small taxa (genus, species); 4. relative abundance of the small taxa; 5. frequency of occurrence of the small taxa in the soil samples; 6. total number of species; 7. average number of species per soil sample; 8. index of similarity of the constitutional species of the soil fauna; 9. index of diversity; 10. response by changes in functional groups; A. analysis of variance.

| Herbicide | Application Rate | Sampling method | Extraction method | Habitat substrate | Measurement | Effect on group of soil fauna | Other remarks | Author |
|---|---|---|---|---|---|---|---|---|
| Diethylhydrazide | 6–8 kg/ha | soil sampling | | arable soil | 1,2,3,4,5,6,10 | Collembola +<br>Nematoda + | | Witkowski and Losinski (1978) |
| Maleic hydrazide | 9–18 kg/ha | sampling of plants | | grassland | 1,10 | Nematoda – | sprayed on herbage | Apt *et al.* (1960) |
| Benzoylprop-ethyl | 1·5 kg/ha | soil sampling | Modified Tullgren | arable soil | 1 | Acarina 0<br>Collembola 0<br>Coleoptera 0<br>Diptera 0<br>Enchytraeidae 0<br>Lumbricidae 0 | | Stafford (1973) |
| Chlorpropham | 2·5 kg/ha | pitfalls | | surface arable land | 1,2,3,4,5,8<br>sex index<br>biomass | Carabidae 0 | | Müller (1972) |

| | | | | | | | | |
|---|---|---|---|---|---|---|---|---|
| TCA | | pitfalls soil sampling | | arable soil | 1 | Carabidae (0)<br>Stanhylinidae (0)<br>other Coleoptera (0)<br>Collembola (0) | | Honczarenko (1974) |
| TCA | 89·6 kg/ha | soil sampling | Tullgren | grassland soil | 1 | Coleoptera −<br>Diplopoda +<br>Collembola ±<br>Acarina ±<br>Lumbricidae − | sprayed on herbage | Fox (1964) |
| Dalapon | 49·7 kg/ha | soil sampling | | grassland soil | A1 | Collembola 0<br>Acarina 0 | Comparence control, mowed, mowed + herbicide herbicide | Malone (1970) |
| Dalapon | 5·5 kg/ha | sampling of plants | | grassland | 1,10 | Nematoda − | Sprayed on herbage | Apt *et al.* (1960) |
| Dalapon | 17 kg/ha | soil sampling | Modified Tullgren high gradient cylinder extraction | grassland soil | A1,2,3,4,6,8 | Collembola −<br>Acarina −<br>Coleoptera −<br>Diptera −<br>Pauropoda −<br>Protura 0<br>Symphyla − | | Curry (1970) |
| Dalapon | 22·4–45 kg/ha | soil sampling | Tullgren | grassland soil | 1 | Elateridae (+)<br>Coleoptera (+)<br>Diplopoda (+)<br>Collembola (+)<br>Acarina (+)<br>Lumbricidae (+) | Sprayed on herbage | Fox (1964) |

Table V—*continued.*

| Herbicide | Application Rate | Sampling method | Extraction method | Habitat substrate | Measurement | Effect on group of soil fauna | Other remarks | Author |
|---|---|---|---|---|---|---|---|---|
| Simazine + TCA | | pitfalls | | arable soil | 1 | Carabidae –<br>Staphylinidae –<br>Collembola –<br>other Coleoptera – | | Honczarenko (1974) |
| Simazine + TCA | 2·5 kg/ha | pitfalls | | Surface arable land | 1,2,3,4,5,8 sex index biomass | Carabidae ± | | Müller (1972) |
| Simazine | 1 kg/ha | pitfalls | | arable soil | 1 | Oligochaeta –<br>Carabidae – 0<br>Curculionidae (–) 0 | | Honczarenko (1969) |
| Simazine | | pitfalls soil sampling | | arable soil | 1 | Collembola –<br>Carabidae (–)<br>Staphylinidae (–)<br>other Coleoptera (–) | | Honczarenko (1974) |
| Simazine | 3·36 kg/ha | soil sampling | Mod. Tullgren | arable soil | 1 | Collembola –<br>Onychiuridae –<br>Isotomidae –<br>Acarina –<br>Gamasina –<br>Enchytraeidae – | Cultivated in the soil, spraying on herbage | Edwards and Arnold (1964) Edwards (1970) |

| | | | | | | | | |
|---|---|---|---|---|---|---|---|---|
| Simazine | 4–8 kg/ha | soil sampling | Tullgren | arable soil | A 1,2,3,4,5,6,9 | Acarina –<br>Collembola – | Clear change of dominance structure | Helmecke *et al*. (1977) |
| Simazine | 4–8 kg/ha | soil sampling | Tullgren | arable soil | 1,2,3,4,5,6,9 | Acarina –<br>Collembola – | Sprayed on herbage | Prasse (1978) |
| Simazine + Aminotriazole | 10–20 kg/ha | soil sampling | | orchard soil | 1 | Nematoda – | | Frey and Wilcke (1978) |
| Simazine + Aminotriazole | 6–8 kg/ha | soil sampling | Method of Cichorius | orchard soil | 1 | Nematoda –<br>Rotatoria – | | Wilcke (1968) |
| Atrazine | 9 kg/ha | soil sampling | Tullgren | grassland, soil | 1 | Coleoptera –<br>Diplopoda –<br>Collembola –<br>Acarina ±<br>Lumbricidae ± | Sprayed on herbage | Fox (1964) |
| Atrazine | | pitfalls<br>soil sampling | | arable soil | | Carabidae (–)<br>Staphylinidae (0)<br>other Coleoptera (–)<br>Collembola – | | Honczarenko (1974) |

Table V—*continued.*

| Herbicide | Application Rate | Sampling method | Extraction method | Habitat substrate | Measurement | Effect on group of soil fauna | Other remarks | Author |
|---|---|---|---|---|---|---|---|---|
| Atrazine | 8 kg/ha | | | | 1,2 | Carabidae –<br>Elateridae –<br>Collembola –<br>Acarina –<br>Enchytraeidae –<br>Protozoa – | | Popovici *et al.* (1977) |
| Prometryne | 1 kg/ha | pitfalls | | arable soil | 1 | Oligochaeta –<br>Carabidae – 0<br>Curculionidae – | | Honczarenko (1969) |
| Cyanazine | 2 kg/ha | soil sampling | Modified Tullgren | arable soil | 1 | Acarina 0<br>Collembola 0<br>Coleoptera 0<br>Diptera 0<br>Enchytraeidae 0<br>Lumbricadae 0 | | Stafford (1973) pers. communication |
| Cyanazine + 2,4-D | | | | arable soil | 1 | Collembola –<br>Acarina – | | Bhattacharye and Joy (1977) |
| Linuron | 1 kg/ha | pitfalls | | arable soil | 1 | Oligochaeta –<br>Carabidae –<br>Curculionidae – | | Honczarenko (1969) |

| | | | | | | | | |
|---|---|---|---|---|---|---|---|---|
| Monuron | 11·2 kg/ha | soil sampling | Tullgren | grassland soil | 1 | Coleoptera –<br>Diplopoda 0<br>Collembola –<br>Acarina –<br>Lumbricidae – + | Sprayed on herbage | Fox (1964) |
| Monuron | 9–18 kg/ha | sampling of plants | | grassland | 1,10 | Nematoda – | Sprayed on herbage | Apt *et al.* (1960) |
| Diuron | 1·5 kg/ha | | | arable soil | 2 | Protozoa, Ciliata – | | Gel'cer and Geptner (1976) |
| Monuron | 1·5 kg/ha | | | arable soil | 2 | Protozoa, Ciliata – | | Gel'cer and Geptner (1976) |
| Chloroxuron | 3 kg/ha | pitfalls | | arable soil | 1 | Olichogaeta –<br>Carabidae –<br>Curculionidae – | | Honczarenko (1969) |
| Aminotriazole | 5·6 kg/ha<br>11·2 kg/ha | | | grassland | 1 | Nematoda – | | Apt *et al.* (1960) |
| Paraquat | 11 l/ha | soil sampling | Modified Tullgren high gradient cylinder extraction | grassland soil | A 1,2,3,4,6,8 | Acarina –<br>Collembola –<br>Coleoptera –<br>Diptera –<br>Pauropoda –<br>Protura 0<br>Symphila – | sprayed on herbage afterwards surface cultivation | Curry (1970). |
| DNOC | 1200 kg/ha<br>6 kg/ha | soil sampling | Tullgren | arable soil | 1,2,3,4 | Collembola ±<br>Acarina ± 0 | | Karg (1964) |
| DNOC | 3 kg/ha | | | arable soil | 1 | Collembola (–) | | Bieringer (1968, 1969) |

Table V—*continued*.

| Herbicide | Application Rate | Sampling method | Extraction method | Habitat substrate | Measurement | Effect on group of soil fauna | Other remarks | Author |
|---|---|---|---|---|---|---|---|---|
| MCPA | 2·25 kg/ha | soil sampling | Flotation, Salt and Hollick | arable soil | 1,3 | Collembola −<br>Acarina +<br>Thysanoptera +<br>Araneidae −<br>Diplopoda + | | Davis (1965) |
| MCPA | 1·5–3 kg/ha | soil sampling | Tullgren | arable soil | A<br>1,2,3,4,5,6,9 | Acarina ±<br>Collembola − | | Helmecke *et al.* (1977) |
| MCPA | | soil sampling | Tullgren | arable soil | 1,2,3,4,5,6,9 | Acarina ±<br>Collembola − | sprayed on herb | Prasse (1978) |
| MCPA | | soil sampling | | arable soil | 1 | Collembola 0 | | Bieringer (1968, 1969) |
| 2,4-D/MCPA | 1·5 kg/ha | pitfalls | | arable soil | | Olichogaeta −<br>Carabidae − 0<br>Curculionidae − | | Honczarenko (1969) |
| 2,4-D | 1·5–3 kg/ha | soil sampling | Tullgren | arable soil | A<br>1,2,3,4,5,6,9 | Acarina ±<br>Collembola 0 | | Helmecke *et al.* (1977) |
| 2,4-D | 1·5–3 kg/ha | soil sampling | Tullgren | arable soil | 1,2,3,4,5,6,9 | Acarina 0<br>Collembola 0 | sprayed on herb | Prasse (1978) |
| 2,4-D | | | | arable soil | 1 | Collembola 0 | | Bieringer (1968, 1969) |
| 2,4-D | | | | arable soil | 1 | Collembola +<br>Acarina | | Bhattacharye and Joy (1977) |

| | | | | | | |
|---|---|---|---|---|---|---|
| 2,4,5-T | 0·225 g/m2<br>2·25 g/m2 | litterbag<br>soil sampling | Tullgren | Forest litter, mineral soil | A,1 | Collembola +<br>Acarina + | Gottschalk and Shure (1979) |
| Di-allate<br>Chloridazon | 3·5 l/ha | pitfalls | | arable soil | A 1,9 | Acarina 0<br>Collembola 0<br>Coleoptera 0<br>Diptera 0 | Gregoire-Wibo *et al.* (1975) |
| Phenmedipham<br>Ethofumesate | 8 l/ha<br>20 l/ha | pitfalls | | arable soil | A 1,9 | Acarina 0<br>Collembola –<br>Coleoptera 0<br>Staphylinidae –<br>Diptera 0 | Gregoire-Wibo *et al.* (1975) |
| Phenmedipham<br>Ethofumesate | 8 l/ha<br>10 l/ha | pitfalls | | arable soil | A 1,9 | Acarina 0<br>Collembola –<br>Coleoptera 0<br>Staphylinidae –<br>Diptera 0 | Gregoire-Wibo *et al.* (1975) |
| Ethofumesate<br>Chloridazon | 5 l/ha<br>4 kg/ha | pitfalls | | arable soil | A 1,9 | Acarina 0<br>Collembola 0<br>Coleoptera 0<br>Diptera 0 | Gregoire-Wibo *et al.* (1975) |

appropriate depth over an area of $0{\cdot}25\ m^2$ and hand sorting the soil. For further details of sampling methods the reader is referred to Murphy (1962).

Arthropoda may be extracted from soil samples by a variety of dry and wet extraction methods. Dry methods are specially suitable for extraction of litter. The method is simple, can be carried out easily and quickly and leaves the animals undamaged but it is less suitable for mineral soil. This is especially the case when the soil has a high content of silt and clay, as a considerable number of soil animals will be trapped in aggregates during drying. Another difficulty is that during the extraction procedure small particles of dried soil and organic matter often fall into the fluid trap and may cause difficulties in observing and counting the animals that are caught.

Wet extraction methods give good results with mineral soils, particularly with heavy clay or loess soils they are much better than dry extraction. A disadvantage is that in many cases considerable amounts of foam and particles of organic matter float and cause difficulties in separating and counting the extracted micro-arthropods which are often damaged and in a bad condition. For this reason wet extraction is less appropriate for extraction of litter and soils with a high content of organic matter.

When the soil fauna is scarce, as in many arable soils, normal soil sampling can only give reasonable results when very large numbers of cores are taken. In such cases a trapping method may give better results. This can be carried out by using litterbags (Edwards and Heath, 1963; Gottschalk and Shure, 1979), packets of filter paper (Van de Bund, 1980) or baited drinking straws (Mills, 1971). Standard amounts of organic matter held in bags of nylon gauze are buried 5 cm below the soil surface or are laid down on the surface among the litter. By using a small mesh size, $\pm 1{\cdot}5$ mm, it is possible to exclude the large arthropods and the earthworms if only micro-arthropods are of interests. The buried packets can be recovered for examination after about six weeks during the growing season and after four to five months during the winter, depending on the biological activity in the soil. All sorts of organic material like dead leaves of trees or straw can be used although natural products may have some drawbacks. One is that samples of dead plant material, even from the same species, can vary considerably in structure and chemical constitution. Another is the possibility that it may be contaminated with unknown pesticides or by pollutants from the air. These difficulties can be avoided if filter paper is used. This is a homogeneous industrial product which consists mainly of pure cellulose. The breakdown of cellulose is one of the major biological activities in the soil. However, this method does not measure population densities, but mainly the activity of soil animals which move through the soil into the filter-paper packets and become established there in association with the microorganisms which attack and break down the cellulose. A variation of this method (Van de Bund, 1970) consists of small cylinders perforated with holes in the walls and bottom. They are half filled with coarse sand, pieces of organic matter are laid on top of the sand and then the cylinders are filled completely with sand. These filled cylinders are buried and extracted in the same way as litter bags. In an analogous way Lofs-Holmin (1979) sampled earthworms with the use of perforated plastic pots containing mixtures of organic matter and mineral soil.

Another method is the use of pitfalls (Dunger, 1964; Lewis and Taylor, 1967). These measure the locomotory activity pattern of surface-dwelling animals. The chances of being trapped vary considerably according to the species. The structure of the soil súrface and the density and structure of the plant cover also influence the catch. Therefore the data have to be interpreted with this in mind. The pitfall method can only be used successfully for testing side-effects of pesticides when the falls are placed in fenced plots to prevent emigration or immigration of animals. Without such barriers the treated plot will very soon be repopulated by migrating specimens like carabids and many spider species (Müller, 1972) so that the local influence of the pesticide will be masked by specimens migrating from the neighbourhood.

Surface dwelling species of the macrofauna can also be sampled by hand-sorting within a defined area (preferably 50 × 50 cm) during a standard time. This method measures population density. Like the pitfall it has the disadvantage that migrating animals may obscure the effect of the pesticide. This difficulty can be minimized by using very large plots with the sampling area in the centre. Soil animals may be brought to the surface by applying an irritant fluid within a standard area. A solution of *o*-dichlorobenzene is useful for Tipulid larvae (Milne *et al.*, 1958) and potassium permanganate or formalin is frequently used for earthworms (Evans and Guild, 1947; Raw, 1959). The efficiency of these methods can vary considerably and they are especially sensitive to variations in soil moisture.

### 3. *Preparation of specimens*

High gradient funnels are most suitable for extraction of micro-arthropods. They can use normal tap water as the collecting fluid. If the temperature of the cooling air or water is suitable (~10°C) most of the floating micro-arthropods remain alive and the drowned specimens remain suitable for preparation and identification. Because of the short distance between the sample and the collecting fluid alcohol is less suitable for this purpose, as it may discourage arthropods from leaving the soil sample. Such preservatives can be only used with large apparatus like Tullgren funnels (McFadyen, 1962; Murphy, 1962; Van de Bund, 1965; Karg, 1971; Krantz,

1975). The suitability of preservation fluids also depends on the type of soil animal. Hard bodied specimens like beetles can be preserved easily by placing them directly into 70% alcohol. "Van Emden's fluid" is more suitable for soft bodied insects like Coleoptera larvae, Diptera larvae and Lepidoptera (Lewis and Taylor, 1967). Enchytraeids and lumbricids are usually preserved in 5% formalin.

When identification is limited to the level of the large taxa, like orders and families, specimens can easily be separated and counted under a dissecting microscope. With identification at the lowest taxonomic (species) level the micro-arthropods have to be transferred into cavity slides filled with lactic acid (Van de Bund, 1965; Karg, 1971). For the preparation of nematodes the reader is referred to Southley (1970), and the preparation of the other soil fauna groups is described by Dunger (1964) and Phillipson (1971).

### 4. *Identification of specimen*

This is the main difficulty in most soil ecological research especially with the Acarina and the Collembola. There are very many keys for the different taxonomic groups, see for instance Herrich *et al.* (1978) and more general keys to major groups of soil fauna which are given by Dunger (1964) and Lewis and Taylor (1967).

### 5. *Analysis*

For the statistical aspects of analysis the reader is referred to Murphy (1962). A general analysis on the level of the large taxa, like orders and families involving the number of individuals per taxum and the relative abundance of the taxa, can be carried out without difficulty. However, this is a very crude approach and in the experience of one of us (C. v.d. B.) there may be considerable changes in the relative abundance of the constituents of the small taxa (species, genus) and also in the number of the small taxa when there is no clear difference in the number of individuals of the large taxa (orders, families). The specific composition of associations of animal species can be considered as an expression of the condition of the habitat. The richness in species may decrease and the relative abundance of the constituent species may change when the diversity of the environment decreases or when the environment is polluted. The number of species may increase with the diversity of the environment and in many cases when the environment will not be disturbed during a prolonged period of time. So the response of the separate species in most cases gives a clearer assessment than that of the large unspecified assemblies of individuals combined in orders and families. For this reason it is desirable to carry out an analysis on the level of the small taxa like the genera or if possible the species. This requires specialized knowledge and this type of analysis is time consuming. Therefore analysis at the genus and species level can be carried out easily only with samples from habitats which are relatively poor in species and individuals such as those of arable land. For the selection of genera and species the following criteria may be useful: easy to identify, possible to distinguish age classes, possible to sample quantitatively with different environmental conditions, regular or random distribution. At the moment these criteria cannot be fulfilled, all the material sampled has to be analysed but in the future it may become necessary to concentrate on a few groups selected in this way.

Effects of pesticides on the soil fauna may be indicated by the following characteristics:

(i) the number of individuals of the large taxa (Orders, Families);
(ii) the relative abundance of the large taxa;
(iii) the number of individuals of the small taxa (Genera, Species);
(iv) the relative abundance of the small taxa;
(v) the frequency of occurrence of the small taxa;
(vi) the number of species per sampling date;
(vii) the average number of species per soil sample;
(viii) the index of similarity of the constituent species of the extracted soil fauna of the untreated and treated plots (Sörensen *et al.*, cited in Wallwork (1976) and Mountford (1962));
(ix) the index of diversity (Shannon Weaver *et al.*, cited in Wallwork (1976));
(x) changes in functional groups.

Each of these properties is closely connected to the others so changes in one of them may also occur in the others. The more characteristics which show a significant effect, the larger is the probability that the structure of the soil fauna has been changed. The relation between the total number of individuals observed, the number of species and the proportion of individuals in each species can be expressed by an index of diversity. A high index of diversity means that the number of individuals of a mixed population is large, the number of species present is large, and that the number of individuals is evenly distributed over the different species. A low index of diversity means that a relatively large number of individuals occur in a small number of species and that the individuals are distributed very unevenly through the species.

There are several formulae to calculate such an index, see Wallwork (1976). The Shannon Weaver index may be useful and is used by several soil zoologists. However, the interpretation of diversity indices requires great caution. In general it can be assumed that human interference in the soil such as tillage, cultivation, application of fertilizers, herbicides and other pesticides changes the index of diversity. Usually the number of species decreases and one or a few of the remaining species increase in number so that the index decreases, but this does not always happen. For instance, if the numbers of the dominant species decrease while the numbers of other species are unchanged and so become relatively more abundant, the index of diversity will increase. Hence an increase or decrease of an index of diversity gives useful information only when used together with other characteristics like the frequency of occurrence of the observed species, the average number of species per soil sample, the index of similarity of species and the response of changes in functional groups. The functional groups can be classified according to feeding habits (cf. Table II). DDT is a good example of a chemical which can change these groups largely because it causes the predatory Gamasid mites to disappear or decrease to a considerable extent. In response to this a huge increase has been observed in some species of Collembola (sapro- and phytophagous) and some species of saprophagous mites (Sheals, 1955; Karg, 1961; Van de Bund, 1965).

## 6. *Concluding remarks*

For field research, Collembola and Acarina are the most suitable groups of animals to study the side-effects of pesticides, herbicides, industrial fall-out and other forms of pollution in the soil. Because of their small size, small radius of activity, abundance, ubiquity and their great number of species they can indicate very well certain forms of pollution and the course of events following disturbance in and near the soil core where they occur. Because of their richness in species, Collembola and Acarina are specially suitable for the investigation of possible relations between changes in diversity of species and effects of herbicides.

Carabidae and Aranea are also suitable in abundance and number of species for the investigation of diversity, but are less suitable because of their mobility. Spiders and large ground beetles can migrate tens and hundreds of metres in one day, so the daily catch in pitfalls may largely represent migrating specimens coming far from the experimental plot.

Lumbricidae are especially suitable for investigation of the functional response of groups of soil animals, for instance litter fragmentation, transport of organic material, burrowing activity and accumulation of chemicals in the body. Because of the small number of species Lumbricidae are less suitable to study species diversity. Enchytraeidae and Nematoda are suitable for the study of functional responses but are unsuited to the study of diversity of species as both groups are taxonomically very difficult.

It is important to investigate effects of herbicides some days after its application, after two or three months, half a year and after one year. An observed change in faunal structure is more important the later it occurs after application. A decrease of the number in species of the soil fauna and a change in the representation of the predaceous species can be important indicators of changes in the faunal structure.

## IV. Herbicide Side-Effects in Agricultural Practice

It is very difficult to identify herbicide effects in practical agriculture because the herbicide application is only one component of a system of management. One example of such observations is that made by Van Rhee (1972) in orchards which differed from each other only in the application of DNOC. A single application killed more than 50% of the earthworm population. Generally under practical conditions the effect of herbicides on the total numbers of the soil fauna species is small and transient. Some herbicides like DNOC, TCA, monuron, simazine, atrazine and dalapon may have a significant effect on the soil fauna, although several authors suggest that the effects are largely indirect (Curry, 1970; Edwards, 1970; Helmécke, 1977; Prasse, 1978). A decrease of the number of individuals of the soil fauna could mainly be caused by the destruction of the plant cover so that the soil fauna would be exposed to greater physical stress from extreme temperature and drought. The destruction of vegetation by cultivation without application of herbicides may cause equal or even larger decreases of the soil fauna species (Curry, 1970). Herbicides in general do not cause a serious change in the species structure of the soil although the number of species may decrease slightly (Curry, 1970; Helmecke *et al.*, 1977; Prasse, 1978). However, some herbicides like DNOC and the triazines can have a direct and a considerable effect on the soil fauna (Karg, 1961, 1964, 1966; Edwards, 1970).

The interrelation of cultivation and herbicides is illustrated by studies of the influence of tillage on the soil fauna of arable fields. The soil fauna of zero-tilled fields, particularly earthworms, is more abundant and has a larger biomass than the fauna of ploughed fields (Bakermans, 1970; Edwards, 1975). In an experiment in The Netherlands there was not much difference in the number of individuals of micro-arthropods but there was a tendency for the numbers of mites and the number of species of springtails and mites to decrease in zero-tillage plots. A striking difference between zero-tillage and ploughed land was in the vertical distribution of the micro-arthropods (Van de Bund, 1970). Because of a layer of litter on the soil surface the number of individuals in the top 0–3 cm of the soil, was much larger than in the layer from 3–9 cm. The vertical distribution of the micro-arthropods was similar to that of grassland and heaths. In the ploughed arable soil, with an even distribution of dead organic matter, the arthropods were more evenly distributed in the top soil. It is difficult to measure the effects of herbicides in zero-tillage soil because it has to be compared with soil where ploughing has been applied. The effect of herbicides under zero-tillage conditions could perhaps be studied by comparing zero-tillage plots treated with different herbicides including reference compounds known to have no direct effects on soil fauna.

When the activity of the herbicides ceases, the vegetation recovers and protects the soil and its fauna against extreme temperatures and drought. In many cases this leads to a rapid increase of individuals of the soil fauna, especially Collembola and saprophagous groups of Acari. If there are large amounts of freshly killed plant material, the microflora will be stimulated and will increase to a great extent. This will be associated with a huge increase of Nematoda, Acari and Collembola (Karg, 1964). The relative as well as the absolute abundance of saprophagous soil fauna under these conditions is considerably larger than in soil where no herbicides are present. The recovery of the soil fauna will be accompanied by considerable changes in the faunal structure. These are expressed in many cases in a decrease of an index of diversity mainly caused by the very uneven distribution of individuals of the species present. After decomposition of the extra amount of dead organic matter the large number of saprophagous soil animals decrease and the faunal structure may become more or less similar to that of untreated soil. These conclusions are based on work in arable soil. Arable soil is an unstable environment with a relatively low diversity of species. This instability will be accentuated by the application of herbicides and pesticides and the diversity of species decreases in most cases. This smaller diversity is often associated with a large increase in individuals of one or a few species. This is in many cases a saprophagous species which under certain conditions may change feeding habits and become phytophagous. An example of such a behaviour is the increase of incidence of damage to sugarbeet seedlings by the springtail *Onychiurus armatus* resulting from the impoverishment of the diversity of the arable environment (Heijbroek, 1972).

## V. Significance of Effects on Soil Fauna and Suggestions for Future Research

At the moment very few countries have registration requirements for information on the effects of pesticides on soil fauna. The United Kingdom only mentions field observations on earthworms and predatory carabids as relevant work and the analysis of residue levels in earthworms as being particularly valuable. The Biologisches Bundesanstalt in the Federal German Republic is developing a laboratory $LD_{50}$ test with the earthworm *Eisenia foetida*, but this is unfortunately a typical manure species. The EEC (Edwards, personal communication) is developing field and laboratory tests with three earthworm species which assess mortality and reproduction as well as uptake of chemicals by the earthworm. These requirements and proposals for tests are still not completely satisfactory but, given the present state of knowledge and considering the problems of standardization (Busvine, 1977), it is not really possible to suggest firm alternatives. Hence the basic need is for a better background of knowledge both from laboratory and field studies.

It is difficult to determine the agricultural significance of effects of herbicides on the soil fauna. Despite many investigations it is still neither completely clear what is the importance of the fauna in biological processes in arable soil nor is it quantified to what extent changes in the number of individuals may influence biological processes in the soil and soil fertility. Recently micro-arthropods have been shown to have an indirect influence on the incidence of certain plant disease on crops. Wiggins and Curl (1979) found that micro-arthropods, particularly Collembola, fed to a great extent on the microflora of the rhizosphere. Hence they may help to determine the quantitative and qualitative nature of the rhizosphere microflora by transporting propagules and by feeding on mycelium. These activities might affect the competitive advantage of a pathogen at the root surface and so influence disease incidence. Also mites, especially those belonging to the genera *Caloglyphus* and *Rhizoglyphus* are able to transport fragments and cells of microorganisms adhering on their body. It is also known that *Verticillium conidia* and at least certain cells in microsclerotial fragments can pass through the digestive tract of bulb mites (*Rhizoglyphus* spp.) in a viable state (Price, 1975). Shew and Beute (1979) showed that oospores of *Pythium myriotylum* remained viable after passing the alimentary canal of the mite *Caloglyphus michaeli* and that the presence of soil borne mites of this species in field soil infested with *Pythium myriotylum* significantly increased the incidence of peanut pod rot. In the Netherlands Muller and Hollinger (1979) have shown that the presence of the soil borne mites, *Rhizoglyphus robini* and *Rhizoglypus engeli*, could be associated with a poor development of and damage on *Freesia*, *Gladiolus*, *Lilium*, *Hippeastrum* and *Gloriosa*. The agricultural significance of earthworm activity has been demonstrated by introduction of earthworms into the virgin soil of a recently reclaimed polder area in The Netherlands (Van Rhee, 1969, 1977; Rogaar and Boerwinkel, 1978) and in orchard soils in which earthworms were almost eradicated by use of copper fungicides resulting in a significant deterioration of the soil (Van de Westeringh, 1972). Earthworm activity resulted in a changed soil structure, changed root development and increased productivity (Van Rhee, 1977). In normal agricultural practice the burrowing activity of the earthworms is overshadowed by ploughing, which even causes severe mechanical damage to the earthworms. In zero-tillage earthworms are indispensable for the maintenance of a good soil structure. Under these conditions earthworms increase channeling and infiltrability (Ehlers, 1975) reduce compaction and crust formation (Lal, 1976) and maintain germination, plant and root growth on the same level as in ploughed soil (Edwards and Lofty, 1978).

Besides the agricultural implications, studies of soil fauna may also be necessary to assess hazards from environmental protection and nature conservation points of view. For environmental protection a possible relationship between leaching and volatilization and soil fauna activity may be important, for instance. Although a relationship between soil fauna activity and soil structure has been established (cf. I.B.), we do not know of any study reporting a relationship between leaching or volatilization and soil fauna activity however.

In nature conservation, soil fauna species have seldom been considered as endangered species nor has soil fauna species diversity been used in conservation practice (except effects of trampling). Nevertheless for effective operation of soil biological processes the maintenance of species diversity is of prime importance. In addition the soil fauna is an important source of food for birds and mammals so that decreases in numbers or poisoning of soil faunal elements may harm protected wildlife. The study of the causes of the decrease of the partridge in England is an excellent example (Potts, 1970). Suggestions about similar effects with other species like black grouse and meadow birds (Niewold and Beintema, personal communication) emphasize the need for more research in this direction.

## VI. Acknowledgements

We would like to thank Drs Van der Drift and Van Rossum for reviewing the text. Mr H. Jacobs prepared the figures and Ms A. Polman and Ms B. Soplanit typed the different versions of the manuscript.

## References

Anderson, J. M. (1975). The enigma of soil animal species diversity. In *Progress in Soil Zoology* (J. Vaněk, Ed.), 51–58. Proc. 5th Coll. Soil Zoology, Prague, 1973.

Apt, W. J., Austenson, H. M. and Courtney, W. D. (1960). Use of herbicides to break the life cycle of the bentgrass nematode, *Anguina agrostis* (Steinbuck, 1949), Filipjev 1936, *Pl. Dis. Reptr.*, **44**, 524–526.

Atlavinytė, O. (1975). The effect of chemical substances on the activity of Lumbricidae in the process of straw disintegration. In *Progress in Soil Zoology* (J. Vaněk, Ed.), 515–519, Proc. 5th Coll. Soil Zoology, Prague, 1973.

Atlavinytė O., Dečiulytė, J. and Lugauskas A. (1977). The effect of Lumbricidae on plant humification and soil organism biocenoses under application of pesticides. In *Soil Organisms as Components of Ecosystems*. (U. Lohm and T. Person, Eds.), Proc. 6th Int. Coll. Soil Zool., Ecol. Bull. (Stockholm) **25**, 222–228.

Ausmus, B. S., Jackson, D. R. and van Voris P. (1979). The accuracy of screening techniques. In *Terrestrial microcosms and environmental chemistry* (J. M. Witt and J. W. Gillett, Eds.). pp. 123–130, National Science Foundation, NSF/RA report 79–0026.

Bakermans, W. A. P. (1970). Jversl. Inst. biol. scheik. Onderz. Landb. Gewass. 1970, 17.

Bäumler, W., Doppelreiter, H., Elbrady, E. Hassanen, F. and Kiziroglou, I. (1978). Über den Einfluss der Herbizide Paraquat und Simazine auf die Fauna forstlich genützter Böden. *Anz. Schädlingskde, Pflanzenschutz, Umweltschutz*, **51**, 1–5.

Bhattacharya, T. and Joy, V. C. (1977). Effect of Banvel-D, a herbicide, on the microarthropod population of soil. *Indian Biologist*, **9**, 47–51.

Bieringer, H. (1968). Untersuchungen in Herbizid-Dauerversuchen, 4. quantitative Untersuchungen an den Bodencollembolen des Hohenheimer Dauerherbizid-versuches. *Z. Pfl. Krankh. Pfl.Path. Pfl.Schutz*, **4**, 157–162.

Bieringer (1969). Unkrautbekämpfungsmittel und Bodenfauna, *Umschau*, **69**, 244.

Bund, C. F. van de (1965). Changes in the soil fauna caused by the application of insecticides. *Boll. Zool. agr. Bachic.*, s II, v. **7**, 186–189.

Bund, C. F. van de (1970). Influence of crop and tillage on mites and springtails in arable soil. *Neth. J. Agric. Sci.*, **18**, 308–314.

Bund, C. F. van de (1980). Mesofauna in arable soil. In *Integrated control of insect pests in the Netherlands* (A. K. Minks and P. Gruys, Eds.), pp. 87–92. Pudoc Wageningen.

Busvine, J. R. (1977). International standardization of tests for detecting and measuring resistance of pests to pesticides. In *Crop Protection Agents—their Biological Evaluation* (N. R. McFarlane, Ed.), 557–567, Academic Press, London and New York.

Campbell, R. C. (1974). Statistics for Biologists. Cambridge University Press, London.

Caseley, J. C. and Eno, C. F. (1966). Survival and reproduction of two species of earthworm and a rotifer following herbicide treatments. *Soil Sci. Soc. Am. Proc.*, **30**, 346–350.

Chappell, W. E. and Miller, L. I. (1956). The effects of certain herbicides on plant pathogens. *Pl. Dis. Reptr.*, **40**, 52–56.

Chio, H. and Sanborn, J. R. (1978). The metabolism of Atrazine, Chloramben, and Dicamba in Earthworms (*Lumbricus terrestris*) from treated and untreated plots. *Weeds Sci.*, **4**, 331–335.

Christiansen, K. (1967). Competition between collembolan species in culture jars. *Rev. Ecol. Biol. Sol.*, **4**, 439–462.

Curry, O. P. (1970). The effects of different methods of new sward establishment and the effects of the herbicides paraquat and dalapon on the soil fauna. *Pedobiologia*, **10**, 329–361.

Curry, J. P. and Tuohy, C. F. (1978). Studies on the epigeal microarthropod fauna of grassland swards managed for silage production. *J. appl. Ecol.*, **15**, 727–741.

Davis, B. N. K. (1965). The immediate and long-term effects of the herbicide MCPA on soil arthropods. *Bull. ent. Res.*, **56**, 357–366.

Davis, G. R. F. (1968). Effects of 2,4-D, maleic acid hydrazide, and 2,4,5-T on survival and growth of the prairie grain wireworm, *Ctenicera destructor* Brown (*Coleoptera Elateridae*). *Can. J. Zool.*, **46**, 747–750.

Drift, J. van der (1963). The influence of biocides on the soil fauna. *Neth. J. Plant. Path.*, **69**, 188–199.

Drift, J. van der (1975). The significance of the millipede Glomeris marginata (Villers) for oak-litter decomposition and an approach of its part in energy-flow. In *Progress in Soil Zoology* (J. Vaněk, Ed.), 293–298, Proc. 5th Coll. on Soil Zoology, Prague, 1973.

Dunger, W. (1964). Tiere im Boden. Die Neue Brehm-Bucherei 327. A Ziemsen Verlag Wittenberg, Lutherstadt.

Dziuba, S. (1971). Initial observations upon the effect of Krezamon and Dikotex on the mites of the Gamasides group (*Acarina, Mesostigmata*). Zeszyty naukowe universytetu Mikotaja Kopernika w. Toruniu, Nauki Matematyczno-Przyrodnicze-zes-jt 29-Biologia XIV 19–35. Polish (Eng. summ.).

Edwards, C. A. (1970). Effects of herbicides on the soilfauna. *Proc. 10th Br. Weed Control Conf. 1970*, pp. 1052–1062.

Edwards, C. A. (1973). Rep. Rothamsted exp. Stn. for 1972, 211–214.

Edwards, C. A. (1975). Effects of direct drilling on the soilfauna. *Outlook on Agriculture*, **8**, 243.

Edwards, C. A. and Arnold, M. K. (1964). Rep. Rothamsted exp. Stn. for 1963, 147–148.

Edwards, C. A. and Heath, G. W. (1963). The role of soil animals in breakdown of leafmaterial. In *Soil Organisms* (J. Doeksen and J. van der Drift, Eds), 76–84. North-Holland Publishing Company, Amsterdam.

Edwards, C. A. and Thompson, A. R. (1973). Pesticides and the soilfauna. *Residue Rev.*, **45**, 1–79.

Edwards, C. A. and Lofty, J. R. (1974). The invertebrate fauna of the Park Grass Plots. I. Soil fauna. Rep. Rothamsted exp. Stn. for 1974, part 2, pp. 133–154.

Edwards, C. A., Butler, C. G. and Lofty, J. R. (1975). The invertebrate fauna of the Park Grass Plots. II. Surface fauna. Rep. Rothamsted exp. Stn. for 1975, part 2, pp. 63–90.

Edwards, C. A. and Lofty, J. R. (1978). The influence of arthropods and earthworms upon root growth of direct drilled cereals. *J. appl. Ecol.*, **15**, 789–795.

Ehlers, W. (1975). Observations on earthworm channels and infiltration on tilled and untilled loess soil. *Soil Science*, **119**, 242–249.

Eijsackers, H. (1974). Possible side effects of pesticides for the control of *Prunus serotina*. *Ned. Bosb. T*, **46**, 55–62.

Eijsackers, H. (1978a). Side effects of the herbicide 2,4,5-T affecting mobility and mortality of the springtail *Onychiurus quadriocellatus Gisin* (Collembola) *Z. ang. Ent.*, **86**, 349–372.

Eijsackers, H. (1978b). Side effects of the herbicide 2,4,5-T on reproduction, food consumption and moulting of the springtail *Onychiurus quadriocellatus Gisin* (Collembola). *Z. ang. Ent.*, **85**, 341–360.

Eijsackers, H. (1978c). Side effects of the herbicide 2,4,5-T on the isopod *Philoscia muscorum* Scopoli. *Z. ang. Ent.*, **87**, 28–52.

Eijsackers, H. (1978d). Side effects of the herbicide 2,4,5-T on the carabid *Nothiophilus biguttatus* Fabr., a predator of springtails. *Z. ang. Ent.*, **86**, 113–128.

Eijsackers, H. (1978e). Mogelijke nadelige invloeden van het onkruidbestrijdings middel 2,4,5-T op de bodemfauna, Verhandeling 14, Research Institute for Nature Management, Arnhem.

Eijsackers, H. and van der Drift J. (1976). Herbicides and the soil fauna. In *Herbicides, Physiology, Biochemistry, Ecology* (L. J. Audus, Ed.), Vol. 2, Academic Press, London and New York, pp. 149–174.

Eijsackers, H. and Chardon, W. J. (1979). Side effects of ammoniumsulphamate on soil fauna. *Ned. Bosb. T.*, **51**, 1–6.

Evans, E. C. and Guild, W. J. McL. (1947). Studies on the relationships between earthworms and soil fertility. I. Biological studies in the field. *Ann. appl. Biol.*, **34**, 307–330.

Fedorko, A., Kamionek, M., Kozlowski, J and Mianowska, E. (1977). The effects of some carbamide herbicides on nematodes from different ecological groups. *Pol. Ecol. Stud.*, **3**, 23–28.

Felton, J. C., Watkinson, I. A. and Whitebread, S. E. (1977). The provision of test insects for continuous screening programmes. In *Crop protection agents—the biological evaluation* (N. R. McFarlane, Ed.), 531–540, Academic Press, London and New York.

Fox, C. J. S. (1964). The effects of five herbicides on the numbers of certain invertebrate animals in grassland soil. *Can. J. plant Sci.*, **44**, 405–409.

Frey, F. (1976). Effects of herbicides applied in orchards on *Acrobeloides buetschlie* (de Man, 1884) Steiner and Buhrer, 1933. *Z. Pfl. Krankh. und Pfl. Schutz*, **83**, 434–441.

Frey, F. and Wilcke, D. E. (1978). Untersuchungen über Nebenwirkungen zweier Herbizid-präparate auf bodenbewohnende Nematoden in einer Obstanlage. *Z. für Pfl.krankh. und Pfl.schutz*, **85** (1), 1–9.

Gel'cer, Ju. (1967). Pochvennye prosteihei kak test dlya izucheniva biologiecheski aktivnykh veshchestv. (Soil protozoa as a test for the study of biologically active substances). Vestn. Mosk. Univ. Scr. 6, *Biol. Pochpoved*, **22**, 31–39.

Gel'cer, Ju. and Geptner, V. A. (1976). The protisticide effect of herbicides. *Pedobiologia*, **16**, 171–184.

Ghabbour, S. I. and Imam, M. (1967). The effect of five herbicides on three Oligochaete species. *Rev. Ecol. Biol. Sol.*, **4**, 119–122.

Gillett, J. W. and Witt, J. M. (1979). Terrestrial microcosms. In *Terrestrial Microcosms and Environmental Chemistry* (J. M. Witt and J. W. Gillett, Eds.), National Science Foundation NSF/RA 790034, Washington DC.

Gottschalk, M. R. and Shure, D. J. (1979). Herbicide effects on leaf litter decomposition processes in an oak-hickory forest. *Ecology*, **60**, 143–151.

Greenway, A. R., Lewis, T., Mudd, A., Scott, G. C. and Wall, C. (1977). Some chemical and entomological problems in the investigation and use of behaviour-controlling chemicals. In *Crop protection Agents—the Biological Evaluation* (N. R. McFarlane, Ed.), pp. 167–185. Academic Press, London and New York.

Gregoire-Wibo, C. (1976). Durées de développement et croissance de *Folsomia quadrioculata* (Tullberg) (Insecte Collembole) en élevage à température constante et en rythme nycthéméral. *Rev. Ecol. Biol. Sol.*, **13**, 491–503.

Gregoire-Wibo, C., Lebrun, Ph. and Bourlaco, F. (1975). Effects of pesticides on the soil fauna. Contribution to OILB/WPRS Soil Pest Group Meeting, Zeist March 17–21, 1975.

Helmecke, K., Hickisch, B., Mahn, E. G., Prasse, J. and Sternkopf, G. (1977). Beitrage zur Wirkung des Herbizideinsatzes auf Struktur und Stoffhaushalt von Agro-Ökosystemen. *Hercynia N.F., Leipzig*, **14**, 375–398.

Herrich, G. J., Hawksworth, D. L. and Sims, R. W. (Eds.) (1978). *Key works to the Fauna and Flora of the British Isles and North Western Europe*, Academic Press, London and New York.

Heijbroek, W. (1972). De mogelijkheden voor de bestrijding van de belangrijkste voorjaarsplagen III. De springstaart (Onychiurus armatus Tullb.). *Meded. Inst. rat. Suik. Prod.*, **38**, 48.

Honczarenko, J. (1969). Wstepne badania dzialaniem nicktóryck herbicydów na owady glebowe i dzdzownice. *Poskie Pismo Entomogiczne*, **39**, 567–578.

Honczarenko, J. (1974). The impact of certain herbicides upon the soil entomofauna of foddercrops. Metody Badán w Zoologie Gleby Polski Towarzystwo Gleboznawcze Homisja Biologii Gleby Zespol Fauny Symp. Zool. Gleby Rozow 25–26, March 1974.

Hoskins, W. M. and Craig, R. (1962).Uses of bioassay in Entomology. *Ann. Rev. Entomol.*, **7**, 437–464.

Humbert, W. (1974). Localisation, structure et genèse des concretions minérales dans le mésentéron des collemboles Tomoceridae (Insecta, Collembola). *Z. Morph. Tiere*, **78**, 93–109.

Hunter-Jones, P. (1961). *Rearing and Breeding Locusts in the Laboratory*. Anti-Locust Research Centre, London.

Jackson, D. R. and Levin, M. (1979). Transport of Arsenic in Grassland microcosms and field plots. *Water, Air and Soil Pollution*, **11**, 3–12.

Jackson, D. R., Ausmus, B. S. and Levin, M. (1979). Effects of Arsenic on nutrient dynamics of grassland microcosms and field plots. *Water, Air and Soil Pollution*, **11**, 13–21.

Joosse, E. N. G., Brugman, F. A. and Veld, C. J. (1973). The effects of constant and fluctuating temperatures on the production of spermatophores and eggs in populations of *Orchesella cincta* (Linné), (*Collembola, Entomobryidae*). *Neth. J. of Zool.*, **23**, 488–502.

Kaminski, A. (1975). The use of biological tests for the examination of pesticides. *Pure and applied chemistry*, **42**, 255–284.

Karg, W. (1961). Bisherige Erkenntnisse über die Wirkung von Pflanzenschutzmitteln im Boden. *Nachrichtenblatt, Deutsche Pflanzenschutzd. (Berlin) N.F.*, **19**, 97–105.

Karg, W. (1964). Untersuchungen über die Wirkung von Dinitroorthokresol (D.N.O.C.) auf die Mikroarthropoden des Bodens unter Berücksichtigung der Beziehungen zwischen Mikroflora und Mesofauna. *Pedobiologia*, **4**, 138–157.

Karg, W. (1966). Veränderungen in den Bodenlebensgemeinschaften durch die Einwirkung von Pflanzenschutzmitteln. In *Progress in Soil Biology* (O. Graff and J. E. Satchell, Eds.), 310–319. Friedr. Vieweg und Sohn GmbH Verlag, Braunschweig.

Karg, W. (1971). Die Tierwelt Deutschlands, 59 Teil, Acari (Acarina), Milben, Unterordnung Anactinochaeta (Parasitiformes). Die freilebenden Gamasina (Gamasides), Raubmilben. Gustav Fischer Verlag Jena, 1971, 45–47.

Kowalczyk-Rozek, M. and Slizynski, K. (1977). The effect of some herbicides on the imagines of *Pterostichus cupreus L.* and *Pterostichus vulgaris L.* (*Coleoptera: Carabidae*). *Folia Biol.* (*Krakow*), **25**, 237–243.

Krantz, G. W. (1975). *A manual of Acarology*, 2nd. Ed. Oregon State University Book Stores. Corvallis, 85–93.

Lal, R. (1976). No-tillage effects on soil properties under different crops in western Nigeria. *Soil Sci. Soc. Amer. J.*, **40**, 762–768.
Lewis, T. and Taylor, L. R. (1967). *Introduction to Experimental Ecology*, Academic Press, London and New York.
Lofs-Holmin, A. (1979). A pot method for field experiments with earthworms. Swedish University of Agricultural Sciences, Department of Ecology and Environmental Research, Uppsala, report 6.
Lofs-Holmin, A. (1980). Measuring growth of earthworms as a method of testing sublethal toxicity of pesticides. *Swedish J. Agric. Res.*, **10**, in press.
Macfadyen, A. (1962). Control of humidity in three funnel-type extractors for soil arthropods. In *Progress in Soil Zoology* (P. W. Murphy, Ed.) 158–168, Butterworths, London.
Malone, C. R. (1970). Short-term effects of chemical and mechanical cover management on decomposition processes in a grassland soil. *J. appl. Ecol.*, **7**, 591–601.
Metcalf, R. L., Sangha, G. K. and Kapoor, I. P. (1971). Model ecosystem for the evaluation of pesticide biodegradability and ecological magnification. *Environ. Sci. Technol.*, **5**, 709–713.
Metcalf, R. L., Cole, L. K., Wood, S. G., Mandel, D. G. and Milbrath, M. L. (1979). Design and evaluation of a terrestrial model ecosystem for evaluation of substitute pesticide chemicals. EPA-report 600/3-79-004.
Milne, A., Coggins, R. E. and Laughlin, R. (1958). The determination of numbers of leatherjackets in sample turves. *J. Anim. Ecol.*, **27**, 125–145.
Mills, J. F. (1971). Baited plastic drinking straws for studying soil biota. *Proc. Ent. Soc. Ontario*, **102**, 36–41.
Mountford, M. D. (1962). An idea of similarity and its application to classificatory problems. In *Progress in Soil Zoology* (P. W. Murphy, Ed.) 43–50, Butterworths, London.
Müller, G. (1971). Laboruntersuchungen zur Wirkung von Herbiziden auf Carabiden. *Arch. Pflanzenschutz*, **7**, 351–364.
Müller, G. (1972). Faunistisch-ökologische Untersuchungen der Coleopterenfauna der Küstennahen Kulturlandschaft bei Greifswald. *Pedobiologia*, **12**, 169–211.
Muller, P. J. and Hollinger, Th. C. (1979). Mijten op bol- en knolgewassen in de kas. *Vakblad voor de Bloemisterij*, **27**, 37–39.
Murphy, P. W. (ed.) (1962). *Progress in Soil Zoology*, Part I, Methods of sampling and analysis, statistical aspects, pp. 3–71. Part II, Extraction processes and accessory technics, pp. 75–365. Butterworths, London.
Petersen, H. (1971). Methods for estimation of growth of Collembola in cultures and in the field, exemplified by preliminary results for *Onychiurus furcifer* (Börner). Proc. IV. Coll. pedobiologia Dijon 1970, *Ann. Zool. Ecol. anim.* 235–252.
Phillipson, J. (1971). *Methods of Study in quantitative Soil Ecology: population, production and energy flow.* IBP Handbook no. 18, Blackwell Sci. Publ., Oxford.
Popovici, J. (1977). The influence of atrazine on soilfauna. *Pedobiologia*, **17**, 209–215.
Potts, G. R. (1970). Recent changes in farmland fauna with special reference to the decline of the grey partridge. *Bird Study*, **17**, 145–166.
Prasse, J. (1978). Die Struktur von Mikroarthropodenzönosen in Agro-Ökosystemen und ihre Beeinflussung durch Herbizide, *Pedobiologia*, **18**, 381–384.
Price, D. W. (1975). Passage of *Verticillium albo-atrum* propagules through the alimentary canal of the bulb mite. *Phytopathology*, **66**, 46–50.

Raw, F. (1959). Estimating earthworm populations using formaline. *Nature, London*, **184**, 1661–1662.

Rhee, J. A. van (1969). Inoculation of earthworms in a newly drained polder. *Pedobiologia*, **9**, 128–132.

Rhee, J. A. van (1972). Het effect van bestrijdingsmiddelen op regenwormen. In *Landbouw en Plantenziekten*. Publ. Landb. Voorl. Dienst, Verslag Februari 1972, 93–102.

Rhee, J. A. van (1977). A study of the effect of earthworms on orchard productivity. *Pedobiologia*, **17**, 107–114.

Rivière, J. L. (1976). Les effets secondaire des pesticides agricoles II. Action de quelques herbicides sur la croissance et la fécondité de *Blattella germanica* (L.). *Ann. Zool.-Ecol. anim.*, **8**, 543–550.

Rogaar, H. and Boerwinkel, J. W. (1978). Some soil morphological effects of earthworm activity; field data and X-ray radiography. *Neth. J. Agric. Sci.*, **26**, 145–160.

Sanocka-Woloszyn, E. and Woloszyn, B. W. (1970). The influence of herbicides on the mesofauna of the soil. *Meded. Rijksfac. Landbwet. Gent*, **35**, 731–738.

Sheals, J. G. (1955). The effects of DDT and BHC on soil Collembola and Acarina. In *Soil Zoology*, (D. K. McE. Kevan, Ed.). Proc. Univ. Nottingham Second Easter School in Agric. Sci., London, pp. 241–252.

Shew, H. D. and Beute, M. K. (1979). Evidence for the involvement of soil borne mites in Pythium pod rot of peanut. *Phytopathology*, **69**, 204–207.

Snider, R. J. (1971). Dietary influence on growth and fecundity of *Onychiurus justi* (Denis) (*Onychiuridae: Collembola*). Proc. IV. Coll. pedobiologiae Dijon 1970. *Ann. Zool. Ecol. anim.* 225–234.

Southley, J. F. (1970). Laboratory methods for work with plant and soil nematodes. Ministry of Agriculture, Fisheries and Food, London. Her Majesty's Stationery Office.

Stam, P. A., Clower, D. F., Graves, J. B. and Schilling, P. E. (1978). Effects of certain herbicides on some insects and spiders found in Louisiana cotton fields. *J. econ. Entom.*, **71**, 477–480.

Subagja, J. D. (1978). Effects of herbicides Paraquat and Atrazine upon Collembola, *Folsomia candida* (Willem) and *Tullbergia granulata* Mills. M.Sc. Thesis, Michigan State University.

Tomlin, A. D. (1977). Culture of soil animals for studying ecological effects of pesticides. In *Crop Protection Agents—their Biological Evaluation.* (N. R. McFarlane, Ed.), 541–555, Academic Press, London and New York.

Van Voris, P., O'Neill, R. V., Shugart, H. H. and Emanuel, W. R. (1978). *Functional Complexity and Ecosystem Stability: an Experimental Approach.* Environmental Science Division publ. no. 1123. Oak Ridge National Laboratory, Oak Ridge, Tennessee.

Verhoef, H. (1978). *An ecological study on water relations in Collembola.* Thesis, Free University, Amsterdam.

Wallwork, J. A. (1970). *Ecology of Soil Animals*, McGraw-Hill, London.

Wallwork J. A. (1976). *The distribution and Diversity of Soil Fauna.* Academic Press, London and New York.

Westeringh, W. van de (1972). Deterioration of soil structure in worm free orchard soils. *Pedobiologia*, **12**, 6–15.

Wiggins, E. A. and Curl, E. A. (1979). Interactions of Collembola and Microflora of Cotton Rhizosphere. *Phytopathology*, **69**, 244–247.

Wilcke, E. (1968). Unkrautbekämpfung im Obstbau-Nebenwirkung auf die Bodenfauna. *Zeitschr. Pfl. Krankh. Pfl. Schutz, Sonderh.*, **4**, 163–167.

Witkowski, T. (1973). Significance of chemical improvement of the environment for the soil nematodes (Nematoda), Part II. Observations upon survival of the soil nematodes in aqueous solutions of herbicide substances. Roczniki Gleboznawcze T. XXIV, z. 2: 367–387, Polish (Eng. summ.).

Witkowski, T. (1974). Observations of the effect of contact herbicides on two saprobiotic nematode species. *Zeszyty Problemowe Postpow Nauk Rolniczyck nr.*, **154**, 443–456. Polish (Eng. summ.).

Witkowski, T. (1975). Fruchtbarkeit von *Rhabditis elongata* und *Rhabditis longicaudata* in Wasserlösungen des Herbizids Agroxone-4, Contribution to OILB/WPRS Soil Pests Group Meeting Zeist, Holland 17–21 March.

Witkowski, T. and Losinski, J. (1978). The sequent effect of Alar on soil Nematodes and Collembola. *Studia Societatis Scientiarum Torunensis, Torun, Pologna*, **10**, 69–88.

Witt, J. M. and Gillett, J. W. (1979). *Terrestrial microcosms and environmental chemistry*. National Science Foundation NSF/RA 79-0026, Washington DC.

# Summary

This book attempts to review knowledge of the behaviour of herbicides in the soil in relation to agricultural practice, legislation and academic science. The specific information requirements of these 3 categories often vary, but they are based on a common conceptual background. A summary of the preceding chapters must therefore consider how far apparently accepted principles explain and predict observations and then outline the need for further research if they are to be improved.

In most aspects of the subject there seems to be a reasonable theoretical framework with which to devise logical experimental programmes. Indeed in a few instances the principles seem sufficiently well established that the need for further work at the moment should be seriously questioned. For example, the relationships between adsorption measurements, movement on soil thin layer plates and movement in laboratory leaching columns are now well established so that a properly reported measurement in one of these systems can be used to predict the performance of a herbicide in the others. In order to estimate the relative mobility of a new compound, therefore, only one of these measurements is needed even though the relationships between sorption and mobility are not fully resolved. In a more academic context the apoplastic movement of herbicides in plants seems to be reasonably well understood so far as currently available compounds are concerned.

At the other extreme, the work with perhaps the least satisfactory theoretical basis is that which deals with effects on soil fauna and micro-organisms. Essentially, the problem is that not enough is known of the natural variation in numbers and activities of such organisms or of their importance in maintaining soil fertility and environmental stability. It follows that until their roles can be better defined there is uncertainty as to the relevant measurements which should be made. In the meantime there are problems in developing suitable methods for making these basic observations. It may well transpire that crop growth and development is the most useful measure of the activity of the soil population. Until these issues can be resolved the scientific basis for framing registration requirements is inadequate. Therefore, it is reasonable to suggest that it would be unwise to devise any rigid legislation in this area until a proper scientific foundation has been laid.

These examples mark the range of our current levels of understanding but the subject as a whole is characterized by the existence of a substantial body of plausible concepts which cannot be tested properly, or applied to practical problems, because of inadequate experimental evidence. This is true even of the more physical aspects but is a particular problem where biological and climatic factors are involved.

Quite elaborate simulation models have been developed to describe the movement of compounds in the soil in both liquid and vapour phases but their potential cannot be exploited because of the dearth of suitable data. A basic need is for quantitative information about rates of adsorption and desorption and the related question of the reversibility of sorption processes. There have been few reliable laboratory studies of these problems and virtually no field measurements. There are in addition some remaining uncertainties about mechanisms of adsorption and interactions between water and herbicides at adsorption sites that may be important, particularly with respect to vapour phase movement. Although, quite rightly, often required for registration, measurements of so fundamental a property as vapour pressure are notoriously unreliable. This imposes a considerable restriction on our understanding of movement in the vapour phase. The other major constraint on our ability to predict herbicide redistribution in the soil stems from the difficulties involved in describing water balance and flow phenomena. Because of the enormous variability of soils it is unlikely that it will ever be possible to do this precisely but even stochastic models cannot be developed without large quantities of data. This can only be accumulated if all future mobility studies include careful measurements of hydraulic relationships, an aspect which has been frequently overlooked in the past.

The simulation model, which can include sub models for decomposition and uptake by plants as well as for transport, has considerable potential for streamlining registration procedures. Given simple information relating to the physical and chemical properties of the compound together with some data on persistence, it is relatively easy to predict approximately its distribution in future time and space. Hence, compounds that need more detailed experimental study can be identified quickly. It is probably in this situation that properly validated models will be most useful, at least in the short term.

Registration procedures can hope only to establish general behaviour patterns in order to evaluate risks. Prediction of detailed performance is not necessary. Following this line of argument there are possibilities for considerably reducing the data requirements for legislation in the future. Reliable figures for water solubility, vapour pressure, and octanol-water partition, and an assessment of stability based on the most labile group in the molecule, should allow order of magnitude predictions to be made of persistence, mobility and distribution in a wide range of soil and climatic conditions. Unfortunately it is uncertain that enough figures will ever be assembled to test this proposition adequately.

The other aspects of registration requirements are those that deal with transformations in the soil. Although the probable products can be predicted from knowledge of the chemistry of the compound we cannot yet be confident that a theoretical exercise of this sort is adequate. Hence experimental work is still necessary. There is a strong case for the information requirements of different countries to be standardized as far as possible, so moves towards harmonization are welcome. The only reservation is that legislation should allow the research worker the maximum flexibility in the techniques he may use to produce this information because no laboratory method is perfect and technical improvements and innovations constantly arise. A major problem is that suitable techniques with which to define the biological activity of a soil sample are still in the development phase. This is important because biological activity usually determines the rates of the various transformation processes and hence the relative amounts of products at any particular time. Such a method is also needed in studies both of the effects of herbicides on soil organisms and of the role of micro-organisms in transformation processes.

In so far as they have been studied, the transformation products obtained in laboratory experiments seem to be the same as those produced in the field. Since herbicides are relatively simple molecules, there are only limited possibilities for transformation, so we might expect such result but there is still some scope for confirming that this is so in the rhizospheres of different plants and in the presence of other agricultural chemicals.

There seems to be good evidence that so-called "bound" residues in the soil are of no biological significance and this is now recognized by most registration authorities. Definitions of such residues essentially mean "Unextractable $^{14}C$". In most cases this is only a small fraction of that originally applied as labelled herbicide and some of it is likely to be carbon incorporated in the structures of soil organic matter following degradation of the herbicide to smaller units, so the reduction in interest is justified.

Transformations in the soil can be brought about by chemical and biological processes. Although it is relatively easy to demonstrate for any compound the potential chemical and microbiologically mediated reactions it is more difficult to show their relative importance. Perhaps this distinction scarcely matters for practical (including registration) purposes. There is still much to be learnt of the ways in which micro-organisms act upon herbicides. In particular there is considerable scope for studies of population dynamics, interactions between organisms and metabolic processes, many of which will first require the development of suitable methods.

The other facet of degradation of interest is the rate at which it occurs. From the viewpoint of environmental pollution current herbicides and their metabolites do not seem to pose problems as a result of long persistence periods. We have some understanding of the effects of temperature and soil moisture but very little can be predicted quantitatively of the effect of soil type. Improvements in this respect could reduce the number of observations needed to characterize the persistence of the compound.

Although we consider that relatively simple information is adequate to assess the environmental risks arising from the use of a new compound this is not true of that needed to provide reliable agricultural performance. Because both fields and plants vary, more detailed analyses of the interactions are needed in this context and the lack of data to do this causes major difficulties.

The response of a plant to a herbicide in the soil we know to be affected by a large number of factors including its state of development, climatic conditions, soil moisture, concentration of herbicide, the position of the herbicide and the length of exposure time. What we do not know with any precision are the quantitative relationships between these factors for even one herbicide with one plant species. Hence, although in theory it should be possible to estimate the activity of a compound in a particular soil on the basis of adsorption behaviour, mobility, persistence, and its inherent phytotoxicity, we cannot do this because we do not know the requirements in terms of concentration, time and space for activity against a particular plant species.

To unravel these problems by careful studies in controlled conditions in which each of the possible variables are successively manipulated independently is clearly beyond present and foreseeable research capabilities. This, taken in conjunction with the fact that under field conditions some of the variables cannot be controlled anyway, leads to the conclusion that we will always have to accept a measure of unreliability in the performance of soil applied herbicides. However, there is still much useful and profitable research that can be done.

It seems that most compounds can enter almost any part of the plant. Entry is often easier or faster through the roots than other organs but it does not follow that the root is necessarily the major site of entry in the field because the herbicide may never reach the root region. More knowledge of site of uptake-concentration-response interactions is a prerequisite for improvements in the efficiency with which the herbicide can be delivered to its target. In addition, understanding of the mechanism of symplastic movement and of the part played by phytohormones in uptake and transport, although at first sight primarily of academic interest, could well lead to significant advances in the way herbicides and also other pesticides are used.

Uncertainty over the quantity of herbicide that must reach the plant to produce a response leads to uncertainty in interpreting other observations particularly those concerned with movement and persistence. This is compounded by deficient knowledge of herbicide concentrations in soil solution in the field which in turn raises uncertainty about whether adsorption *per se* is a controlling factor in herbicide activity or whether it is something else with which adsorption happens to be correlated.

If we take an optimistic view and assume that in future we will know more about where and when the herbicide needs to be if it is to kill weeds or damage crops, then more refined data will be needed so that predictive models can be applied more specifically. As already mentioned the degree to which this is possible is constrained by uncontrollable factors particularly those relating to the climate. Hence a realistic goal will be to use models to predict the likely outcome of a particular set of circumstances together with an estimate of the range of possible results. Given the size of the task this can only be achieved by the combined efforts of a large number of workers. Major collaborative programmes are not feasible but much progress could be made if reports of field experiments always included basic climatic and soil data even if some of it is not relevant to the particular experiment reported. A list of recommended measurements is included as an appendix to this chapter.

This list is conventional in that it includes the items generally considered to play a part in controlling pesticide behaviour in the soil. It is however very restricted compared with the information used to classify a soil. Marbut's (1951) classical characteristics are:

1. Geological origin.
2. Mode of formation.
3. Colour.
4. Topographic position.
5. Drainage.
6. Profile.
7. Chemical composition.

The components of our list deal only with items 5 and 7. Similarly the appendix does not include any consideration of soil structure. Batey (1975) lists 4 methods of visual assessment of structure so the methodology seems to be available. In addition, Eagle (1976) has described a texture classification based on the "feel" of soil moulded between finger and thumb which showed promise as a basis for adjusting herbicide dose in different soils. These examples, together with the review of technical classification systems by Bartelli (1978), indicate that there are many possibilities for broadening our view of the soil properties which may be important.

The question was asked in the Introduction whether we are able to use soil applied herbicides more efficiently now than in 1960. The answer must be a qualified "yes". Unquestionably we are now much better equipped to predict the consequences of the introduction of a new xenobiotic into the environment. The chances of significant unforeseen effects to non-target organisms are now small and many would argue that the price of safety is high in terms of the inhibitions it imposes on innovation. It is not so easy, but not impossible, to identify the ways in which the use of soil applied herbicides in practical agriculture has improved. Certainly application rates can now be related to soil characteristics more carefully, although there is a good deal of empiricism involved, and problems with residues can be predicted and explained with much more confidence. It is in reducing the variability of performance that less progress has been made. Nevertheless, the conceptual bases for studying this problem are much improved and the challenge now is to produce enough data to allow the most effective use to be made of the principles that have emerged.

## Appendix

Recommendations for Data to be either included in reports of field experiments or available on request:

Soil mechanical analysis.
Organic carbon content.
pH.
Available N, P, K.
Water holding capacity.
Maximum and minimum daily air temperatures.
Daily rainfall.
(Weather data from the nearest meteorological station will usually be adequate.)

Experiments reporting mobility studies should also include:

Bulk density.
Adsorption-desorption isotherms.
Soil moisture profiles at the sampling dates.

## References

Bartelli, L. J. (1978). Technical classification system for soil survey interpretation. *Advances in Agronomy*, **30**, 247–289.

Batey, T. (1975). Soil examination in the field. In *Soil Physical Conditions and Crop Production*. Ministry of Agriculture, Fisheries & Food Tech. Bull., **29**. Her Majesty's Stationery Office, pp. 207–217.

Eagle, D. J. (1976). Soil texture classification for the adjustment of herbicide dose. *Proc. 1976 British Crop Protection Conference*—Weeds. 981–988.

Marbut, C. F. (1951). *Soils: Their Genesis and Classification.* Soil Science Society of America, Madison.

## Résumé*

Ce livre est un essai de mise au point bibliographique sur le comportement des herbicides dans le sol en relation avec les pratiques agricoles, les problèmes réglementaires et les connaissances scientifiques générales. Les informations spécifiques à ces trois domaines sont de nature différente, mais elles reposent toutes sur des connaissances de base communes. Un résumé des chapitres précédents doit donc à la fois, essayer d'apprécier dans quelle mesure les données acceptées permettent d'expliquer et de prédire le comportement des herbicides, et tenter de définir la nature des futures recherches qui peuvent paraître nécessaires.

Un ensemble de connaissances théoriques suffisantes pour la plupart des aspects de ce sujet semble exister et peut permettre de mettre en place des programmes de recherches expérimentales. Dans certains cas, les données acquises paraissent suffisamment bien établies pour que de nouveaux travaux ne soient pas absolument nécessaires. Par exemple, les relations entre l'adsorption et les déplacements dans des couches et des colonnes de terre sont maintenant assez bien connues pour qu'un type de mesure fait sur un système suffise pour prédire qualitativement le comportement d'un herbicide. Il est donc ainsi possible d'estimer la mobilité relative d'un nouveau composé bien que les relations entre les phénomènes de sorption et la mobilité ne soient pas complètement décrites. Dans un autre domaine, les mouvements apoplastiques des herbicides dans les plantes semblent assez bien compris du moins en ce qui concerne les composés couramment utilisés.

Au contraire, les travaux sur la faune et la microflore du sol ne peuvent pas encore s'appuyer sur des bases théoriques suffisantes. Il y a en effet peu d'informations sur les variations naturelles du nombre et de l'activité de ces organismes et sur leur rôle effectif dans le maintien de la fertilité des sols et de l'équilibre des écosystèmes. Une incertitude existera toujours sur la nature des études à mettre en place tant que ce rôle ne sera pas mieux connu. Par ailleurs, un certain nombre de difficultés doivent encore être surmontées pour l'élaboration de méthodes, adaptées à ce type d'étude. La croissance et le développement des végétaux cultivés semblent être la mesure la plus utile de l'activité des organismes vivants du sol. Tant que ce problème ne sera pas résolu cependant, il n'existera pas de base scientifique convenable pour proposer des réglementations applicables au niveau d'une procédure d'homologation. Il est donc raisonnable de ne pas


* Traduit par R. Calvet.

faire de propositions d'ordre règlementaires dans ce domaine tant que des données scientifiques appropriées n'auront pas été solidement établies.

Ces exemples montrent que notre niveau de compréhension peut être très variable. D'une façon générale, ce sujet est toutefois caractérisé par l'existence d'un ensemble important d'informations qui ne peuvent pas être correctement évaluées et appliquées à des problèmes pratiques en raison de l'absence de vérifications expérimentales. Ceci est vrai même dans le cas des aspects physiques et devient un problème majeur quand des facteurs biologiques et climatiques sont concernés.

Des modèles de simulation parfois très élaborés ont été développés pour décrire les mouvements des composés à la fois dans la phase liquide et dans la phase gazeuse du sol, mais ils n'ont pas pu être exploités du fait du manque de données expérimentales. Il existe un besoin fondamental d'information concernant les vitesses d'adsorption et de désorption ainsi que sur la réversibilité de ces phénomènes. Les études de ces problèmes au laboratoire sont peu nombreuses et sont très rares sur le terrain. Il existe en plus quelques incertitudes concernant les mécanismes moléculaires de l'adsorption et les interactions entre les herbicides et l'eau au niveau des sites d'absorption, interactions qui peuvent être particulièrement importantes pour le transport en phase vapeur. Bien que souvent fort justement demandée pour l'homologation, la mesure d'une propriété comme la pression de vapeur est notoirement peu fiable. Ceci impose des limites importantes à notre compréhension des modalités des déplacements des herbicides à l'état gazeux. Par ailleurs, la possibilité de prédire la distribution des composés dans le sol est sérieusement limitée par les difficultés rencontrées au niveau de la description des transferts de l'eau. Une prédiction précise est en fait peu probable en raison de la très grande variabilité des sols, mais des modèles stochastiques ne peuvent cependant pas être développés sans un grand nombre de données. Celles-ci ne pourront être obtenues que si les futures études de transport sont accompagnées d'une détermination convenable des propriétés hydrodynamiques des milieux utilisés, ce qui n'était pas toujours le cas dans le passé.

Les modèles de simulation, pouvant inclure aussi bien des sous-modèles pour la dégradation et le prélèvement par les plantes que pour le transport, constituent un puissant moyen pour élaborer des procédures d'homologation. Partant d'informations élémentaires sur les propriétés physiques et chimiques des composés et sur leur persistance il est assez aisé de prédire approximativement leur distribution dans l'espace. Les composés nécessitant des études expérimentales plus détaillées peuvent alors être rapidement identifiés. C'est probablement de cette façon que des modèles convenablement vérifiés pourront être les plus utiles, au moins dans un proche avenir.

Les procédés d'homologation ne peuvent seulement espérer établir qu'un schéma général du comportement permettant d'évaluer des risques. Des prédictions détaillées ne sont pas nécessaires et dans ces conditions, il sera possible de réduire considérablement les informations utiles au législateur dans le futur. Des données chiffrées valables sur la solubilité dans l'eau, la pression de vapeur, le partage octanol/eau, et une évaluation de la stabilité basée sur les groupements moléculaires les plus fragiles devraient permettre de prédire l'ordre de grandeur de la persistance, de la mobilité et de la distribution pour un grand nombre de sols et de conditions climatiques. Il n'est malheureusement pas certain que des valeurs suffisamment nombreuses puissent être rassemblées pour apprécier convenablement l'intérêt de cette proposition.

D'autres aspects relatifs à l'homologation concernent les transformations dans le sol. Bien que la nature des métabolites puisse être prédite à partir des propriétés chimiques il n'est pas encore certain que l'approche théorique soit suffisante. Des travaux expérimentaux sont donc toujours nécessaires. C'est là, d'ailleurs, une situation montrant la nécessité de standardiser les informations demandées par les différents pays et toute tentative d'harmonisation sera la bienvenue. La seule réserve est que les dispositions réglementaires devraient laisser le maximum de liberté au chercheur pour obtenir ces informations, étant donné qu'il n'existe pas de méthode de laboratoire parfaite et que des améliorations techniques ou des innovations sont constamment proposées. Un problème majeur provient du fait que les techniques appropriées d'évaluation de l'activité biologique d'un échantillon de sol sont encore dans une phase de développement. Ceci est important car l'activité biologique détermine habituellement la vitesse de différents phénomènes de transformation et donc la variation dans le temps des quantités de produit présentes dans le sol. De telles méthodes sont nécessaires, à la fois, pour les études des effets herbicides sur les microorganismes et pour les études sur le rôle des microorganismes dans les phénomènes de transformation.

Dans la mesure où ils ont été étudiés, les métabolites obtenus au cours d'expériences de laboratoire semblent les mêmes que ceux formés en plein champ. Etant donné que les herbicides sont des molécules relativement simples, les possibilités de transformation sont assez et un tel resultat n'est pas surprenant. Il n'en reste pas moins qu'il existe encore un certain champ de recherche pour s'assurer qu'il en est bien ainsi dans la rhizosphère de différentes plantes et en présence d'autres produits chimiques utilisés en agriculture.

Il semble qu'il existe de bonnes indications pour considérer que les résidus dits "liés" n'ont pas d'importance biologique et cela est maintenant reconnu par la plupart des instances d'homologation. La définition de ces résidus est essentiellement basée sur l'existence de $^{14}C$ non extractible. Dans la majorité des cas, c'est seulement une petite fraction de l'herbicide appliqué sous forme de molécules marquées qui est probablement incorporée à la matière organique du sol à la suite de la dégradation. Ces considérations font que l'intérêt limité de ces résidus est justifié.

Les transformations dans le sol peuvent être dues à des phénomènes biologiques et non biologiques. Bien qu'il soit relativement aisé de montrer l'existence possible de l'un ou l'autre de ces phénomènes, il est plus difficile d'apprécier leur importance relative. Cette distinction n'est peut être pas importante pour des problèmes pratiques (homologation) mais par contre il reste encore beaucoup à apprendre sur les voies d'action des microorganismes sur les herbicides. Il existe en particulier un champ de recherche considérable pour les études concernant la dynamique des populations et les relations entre les microorganismes et le métabolisme des herbicides dans le sol. Il faut cependant noter que beaucoup de ces études nécessitent tout d'abord le développement de méthodes appropriées.

La vitesse de dégradation est un autre aspect intéressant. Du point de vue de l'environnement les herbicides couramment utilisés et leurs métabolites ne semblent pas poser de problèmes du fait de leur persistance. Si l'on a quelques informations sur les effets de la température et de l'humidité du sol, il n'est par contre, presque pas possible de prédire quantitativement l'effet du type de sol. Un accroissement des connaissances dans ce domaine devrait réduire le nombre d'observations nécessaires pour caractériser la persistance d'un composé.

Bien que du point de vue de l'environnement, l'on considère que des informations assez simples soient suffisantes pour évaluer les risques résultant de l'utilisation de nouveaux composés, on connaît mal ce qui est nécessaire à l'obtention de performances agricoles convenables. Il faut tenir compte dans ce contexte de la variabilité des plantes et des sols et des études détaillées sont indispensables quoique difficiles du fait de l'absence de données précises.

Il est connu que la réponse d'une plante à un herbicide apporté dans le sol dépend de nombreux facteurs: l'état de développement de la plante, les conditions climatiques, la teneur en eau du sol, la concentration en herbicide, la position de l'herbicide et le temps d'exposition de la plante. Les relations entre ces facteurs ne sont pas connues avec précision même pour un herbicide et une espèce végétable. En conséquence, il n'est pas possible d'estimer l'activité d'un composé dans un sol donné sur la base d'informations concernant l'adsorption, la mobilité, la persistance et sa phytotoxicité propre. Pour que cela soit réalisable, il faudrait avoir des données chiffrées sur la concentration et les variations de la distribution de l'herbicide dans l'espace et dans le temps en relation avec l'activité biologique.

Résoudre ces problèmes par des études minutieuses faites dans des conditions controlées permettant de faire varier successivement et indépendamment chaque variable possible est tout à fait hors des capacités prévisibles de la recherche. Ceci, ajouté au fait que certaines variables ne peuvent pas être controlées au champ conduit à la conclusion que l'on devra toujours accepter un certain degré d'incertitude dans les performances des herbicides appliqués aux sols. Cependant, beaucoup de recherches utiles et profitables peuvent encore être faites.

Il semble que la plupart des composés puisse pénétrer dans la plante par n'importe quelle partie. La pénétration est souvent plus facile ou plus rapide à travers les racines qu'à travers d'autres organes mais ceci ne veut pas dire qu'elles soient le site principal d'absorption car au champ, les herbicides peuvent ne jamais atteindre la région racinaire.

Une plus grande connaissance des relations entre les sites de prélèvement, la concentration et la réponse est une condition préalable à une amélioration des modalités d'apport des herbicides. Malgré son intérêt à priori très théorique, la compréhension des mouvements symplastiques et le rôle joué par les hormones végétales dans le prélèvement et le transport pourrait conduire à des progrès dans les techniques d'utilisation des herbicides et aussi d'autres pesticides.

L'incertitude sur les quantités d'herbicide qui doivent atteindre une plante pour provoquer un effet biologique conduit à une incertitude dans l'interprétation d'autres observations comme celles relatives aux mouvements et à la persistance. Ceci s'ajoute à la mauvaise connaissance au champ de la concentration en herbicide dans la solution du sol qui entraine à son tour une interrogation sur le caractère limitant de l'adsorption en tant que telle ou sur l'intervention d'un autre phénomène qui lui serait lié.

Dans une perspective optimiste et en supposant que dans le futur on saura où et quand l'herbicide doit se trouver pour détruire les mauvaises herbes ou endommager les cultures, on peut penser que des données plus précises seront nécessaires pour permettre une application plus spécifique des modèles prédictifs. Comme nous l'avons déjà signalé, la faisabilité de cette prédiction est limitée par des facteurs incontrôlables, notamment ceux du climat. Un objectif réaliste sera d'utiliser ces modèles pour prédire le résultat probable relatif à un ensemble de circonstances accompagné d'une estimation du domaine de confiance correspondant. Etant donné son importance, cette tâche ne pourra être achevée que par la conjonction des efforts d'un grand nombre de chercheurs. Des programmes importants menés en collaboration ne sont pas toujours réalisables mais beaucoup de progrès peuvent être faits si les rapports des expériences au champ contiennent systématiquement des données sur le climat et les sols même si elles sont sans rapport apparent avec l'expérience décrite. Une liste des mesures recommandées est donnée en appendice à la fin de ce chapitre.

Cette liste est conventionnelle en ce sens qu'elle contient des caractéristiques généralement considérées comme devant jouer un rôle dans le contrôle du comportement des pesticides dans les sols. Elle est cependant très restrictive comparée à celle qui est utilisée pour classer les sols. Les caractéristiques classiques selon Marbut (1951) sont:

1—L'origine géologique
2—Le mode de formation
3—La couleur
4—La position topographique
5—Le drainage
6—Le profil
7—La composition chimique

Notre liste ne se rapporte qu'aux points 5 et 7. Par ailleurs, l'appendice ne contient pas d'informations sur la structure des sols. Batey (1975) a donné une liste de 4 méthodes d'appréciation visuelle de la structure, de sorte qu'une méthodologie approchée semble disponible. De plus Eagle (1976) a décrit une classification texturale (basée sur l'examen manuel d'échantillons malaxés entre le pouce et l'index) qui montre certaines possibilités d'utilisation pour ajuster la dose d'herbicide dans différents sols. Ces exemples, joints à la revue des systèmes de classification technique par Bartelli (1978) indique qu'il existe plusieurs possibilités pour élargie nos vues sur les propriétés qui pourraient être importantes.

Nous nous étions interrogés dans l'introduction pour savoir si nous étions actuellement capables de mieux utiliser les herbicides appliqués au sol qu'en 1960. La réponse doit être un "oui". Nous sommes maintenant incontestablement mieux équipés pour prédire les conséquences de l'introduction de nouveaux composés étrangers dans l'environnement. La probabilité d'effets significatifs non prévus sur des organismes non visés est maintenant faible et beaucoup diront que le prix de la sécurité est élevé eu égard aux contraintes qui sont imposées à l'innovation. Il n'est pas facile, mais non impossible, de dire en quoi il y a eu une amélioration de l'utilisation en agriculture des herbicides appliqués au sol. En effet, les doses peuvent être reliées plus correctement aux caractéristiques des sols malgré un empirisme encore certain et les problèmes liés aux résidus peuvent être prévus et expliqués avec maintenant beaucoup plus de confiance. C'est dans la réduction de la variabilité des performances que le moins de progrès a été fait. Les bases conceptuelles pour étudier ce problème ont néanmoins été accrues et l'objectif est maintenant d'obtenir suffisamment de données pour permettre un véritable usage des principes qui ont été dégagés.

## Appendice

Liste recommandée de données devant figurer dans les rapports sur les expériences au champ ou pouvant être obtenues auprès des auteurs.

Analyse granulométrique
Teneur en carbone
pH
N, P, K assimilable
Capacité de rétention pour l'eau
Température maximum et minimum relevée quotidiennement
Pluviométrie quotidienne (Les données climatiques de la station météorologique la plus proche pourront généralement convenir).

Les expériences relatives aux études de mobilité devraient également être accompagnées des informations suivantes:

Densité apparente
Isotherms d'adsorption et de desorption.
Caracteristiques hydriques des sols (relation potential—teneur en eau volumiques)

## Références

Bartelli, L. J. (1978). Technical classification system for soil survey interpretation. *Advances in Agronomy*, **30**, 247–289.

Batey, T. (1975). Soil examination in the field. In *Soil Physical Conditions and Crop Production.* Ministry of Agriculture, Fisheries & Food Tech. Bull., **29**. Her Majesty's Stationery Office, pp. 207–217.

Eagle, D. J. (1976). Soil texture classification for the adjustment of herbicide dose. *Proc. 1976 British Crop Protection Conference*—Weeds. 981–988.

Marbut, C. F. (1951). *Soils: Their Genesis and Classification.* Soil Science Society of America, Madison.

## Zusammenfassung*

Mit diesem Buch wird der Versuch unternommen, unsere Kenntnis vom Verhalten der Herbizide im Boden zusammenfassend darzustellen und aus dem Blickwinkel der pflanzenbaulichen Praxis, der Gesetzgebung und der Wissenschaft zu beleuchten. Die spezifischen Informationsansprüche, die diese drei Bereiche stellen, sind oft unterschiedlich; sie beruhen jedoch auf einer gemeinsamen Grundlage. Eine Zusammenfassung der Kapitel dieses Buches muß deshalb darauf eingehen, inwieweit allgemein anerkannte Prinzipien in der Lage sind, irgendwelche Beobachtungen zu erklären oder vorherzusagen und dort, wo sie verbessert werden müssen, den Weg zu weiteren Untersuchungen zu weisen.

In vieler Hinsicht ist ein verständlicher theoretischer Rahmen geschaffen, in dem praktische Experimente folgerichtig angelegt werden können. Und in einigen wenigen Fällen scheinen die Prinzipien so ausreichend gesichert zu sein, daß eine weitere Bearbeitung nach dem jetzigen Stand der Kenntnisse nicht mehr notwendig ist. So sind zum Beispiel die Zusammenhänge zwischen Adsorption und Mobilität in Boden-Dünnschicht-Platten und in Bodensäulen so gut bekannt, daß bei einer neuen Substanz eine exakt durchgeführte Messung mit einer der genannten Methoden ausreicht, um deren relative Beweglichkeit beurteilen zu können, obwohl die Beziehung zwischen Sorption und Mobilität noch nicht bis zum letzten geklärt ist. Betrachtet man als ein weiteres Beispiel den apoplasmatischen Transport von Herbiziden in der Pflanze, so dürfte dessen wissenschaftliche Grundlage geklärt sein, soweit es sich um bekannte Verbindungen handelt.

Ganz anders sind die Arbeiten über den Einfluß von Herbiziden auf die Bodenfauna und -mikroorganismen zu beurteilen, deren theoretische Basis bislang noch am wenigsten befriedigt. Das Problem liegt im wesentlichen darin, daß noch zu wenig sowohl über die natürliche Variationsbreite der Populationsdichte und -aktivität dieser Bodenorganismen als auch über ihre Bedeutung für die Bodenfruchtbarkeit unter verschiedenen Umwelteinflüssen bekannt ist. Solange ihre Funktionen nicht besser definiert werden können, bleibt die Aussage von entsprechenden Messungen unsicher, vor allem auch deswegen, weil die Methoden, nach denen die grundlegenden Erkenntnisse gewonnen werden sollen, in den letzten Jahren immer noch Probleme zeigen.

Wachstum, Entwicklung und auch Erträge der Kulturpflanzen sind möglicherweise der nützlichste Maßstab für die Aktivität der Bodenor-

* übersetzt von W. Pestemer und Th. Eggers.

ganismen. Solange diese strittigen Fragen noch nicht gelöst sind, ist die wissenschaftliche Grundlage für Prüfungsrichtlinien für die Zulassung von Herbiziden unzulänglich. Deshalb wird vorgeschlagen, noch keine verbindlichen Richtlinien zu erlassen, bevor nicht ausreichende wissenschaftliche Erkenntnisse vorliegen.

Diese wenigen Beispiele charakterisieren unseren Kenntnisstand über das Verhalten von Herbiziden im Boden. Sie zeigen auf, daß brauchbare Konzepte vorhanden sind, die Übertragbarkeit von Modellexperimenten in die Praxis jedoch problematisch ist, weil die Versuche nicht aussagekräftig genug sind. Dies gilt nicht nur für die physikalischen Aspekte, sondern vor allem dann, wenn biologische und klimatische Faktoren beteiligt sind.

Es besteht bereits eine Reihe von Simulationsmodellen, um die Mobilität von Substanzen im Boden in der flüssigen und in der gasförmigen Phase zu erklären, aber ihr Potential kann wegen des Mangels an brauchbaren Daten nicht voll genutzt werden. Quantitative Adsorptions- und Desorptionsdaten und Informationen über die Umkehrbarkeit der Sorptionsprozesse fehlen weitgehend. Es gibt zwar einige zuverlässige Laborstudien zu diesem Problem, aber keine Messungen unter Feldbedingungen. Daneben sind immer noch Unsicherheiten über Sorptionsmechanismen und Wechselwirkungen zwischen Bodenwasser und Herbiziden an Sorptionsstellen, besonders im Hinblick auf die Bewegung in der Gasphase, festzustellen. Aus der Angabe der Dampfdruckwerte, die für die Zulassung von Herbiziden regelmäßig verlangt werden, kann man für die Praxis jedoch kaum richtige Zusammenhänge ableiten, was uns dazu zwingt, unser Verständnis über die Bewegung von Herbiziden im Boden in der Gasphase mit Vorbehalt zu betrachten. Die Vorhersage der Herbizidverteilung im Boden ist wegen der schwierigen Analyse des Wasser- und Lufthaushalts in den vielfältigen Böden erheblich eingeengt. Wegen der enormen Variabilität der Böden ist es unwahrscheinlich, daß es überhaupt je möglich sein wird, eine sichere Prognose zu treffen, denn selbst stochastische Modelle können nicht entwickelt werden ohne großes Datenmaterial. Letzteres kann erst dann gesammelt werden, wenn allen zukünftigen Mobilitätsstudien sorgfältige Messungen der hydraulischen Beziehungen im Boden zugrundeliegen, ein Aspekt, der in der Vergangenheit immer wieder übersehen wurde.

Ein umfassendes Simulationsmodell, welches auch Untermodelle für Abbau und Aufnahme sowie Translokation durch die Pflanze einschließen kann, könnte einen bedeutenden Beitrag für ein ausgewogenes Registrierungsverfahren leisten. Wenn Informationen über die physikalischen und chemischen Eigenschaften eines Herbizids in Verbindung mit Daten über die Persistenz vorliegen, ist es relativ einfach, annähernd deren zeitliche und räumliche Verteilung im Boden vorherzusagen. Dann können Wirkstoffe, die ausführlichere experimentelle Untersuchungen erfordern, schnell erkannt werden. Es ist denkbar, daß unter diesen Umständen anerkannte Modelle zumindest in der nächsten Zeit einen brauchbaren Beitrag zu leisten vermögen.

Im Zulassungsverfahren können nur generelle Richtlinien festgelegt werden, um Risiken abzuschätzen. Eine detaillierte Ausführung braucht dabei nicht vorgeschrieben zu werden. Wenn man dieser Argumentation folgt, kann der erforderliche Datenbedarf für eine zukünftige Richtlinie beträchtlich reduziert werden. Wasserlöslichkeit, Dampfdruck und Oktanol-Wasser-Verteilung sowie die Beurteilung der Stabilität einer Verbindung, basierend auf der labilsten Gruppe im Ausgangsmolekül, sollten ausreichen, Persistenz, Mobilität und Verteilung in den wichtigsten Böden und unter den häufigsten Klimabedingungen größenordnungsmäßig zu bestimmen. Ob je genug Daten zusammengetragen werden können, um diesen Vorschlag ausreichend zu testen, bleibt leider offen.

Ein anderer Aspekt im Zulassungsverfahren betrifft die Umsetzungsprozesse von Herbiziden im Boden. Obwohl die möglichen Metaboliten und Abbauprodukte aus der Kenntnis der Chemie des Wirkstoffs abgeleitet werden können, kann man jedoch nicht sicher sein, daß theoretische Überlegungen dieser Art adäquat sind. Daher ist eine weitere experimentelle Bearbeitung der Abbauvorgänge im Boden notwendig. Die Anforderungen an die Daten sollten in den verschiedenen Ländern soweit wie möglich vereinheitlicht werden. Die einzige Einschränkung ist, daß die Gesetzgebung dem Forscher eine maximale Flexibilität einräumen sollte in Hinsicht auf die Analysentechniken, weil keine Labormethode perfekt ist und die Techniken immer weiter ausgefeilt werden. Im Rahmen dieser Untersuchungen ist die biologische Aktivität einer Bodenprobe ein wesentlicher Faktor, dessen Bestimmung und Bewertung—wie oben bereits erwähnt—jedoch noch in der Entwicklung stehen. Dies ist wichtig, denn gewöhnlich bestimmt die biologische Aktivität die Rate der verschiedenen Transformationsvorgänge und damit auch die relativen Anteile der Umwandlungsprodukte zu einem bestimmten Zeitpunkt. Eine derartige Methode wird auch gebraucht, um Nebenwirkungen von Herbiziden auf Bodenmikroorganismen und die Rolle der Mikroorganismen in Umsetzungsprozessen zu erfassen.

Nach den bisherigen Untersuchungen scheinen die Metaboliten in den Laborversuchen dieselben zu sein wie im Feld. Da Herbizide relativ einfache Moleküle sind, gibt es nur eine begrenzte Zahl möglicher Umsetzungsprozesse im Bodenkomplex, aber ihr Ablauf in der Rhizosphäre verschiedener Pflanzen und unter dem Einfluß anderer Agrochemikalien muß noch weitergehend erforscht werden.

Daß sogenannte "bound residues"—definiert als unextrahierbares $^{14}C$ eines radioaktiv markierten Herbizids—im Boden offensichtlich nicht von biologischer Bedeutung sind, wird auch von den Zulassungsbehörden allgemein anerkannt. Meistens handelt es sich hierbei nur um einen kleinen Teil des ursprünglich applizierten Herbizids. Einiges davon ist wahrscheinlich in der organischen Substanz des Bodens festgelegter Kohlenstoff, so daß diese "bound residues" aufs ganze gesehen unbedeutend sind.

Umsetzungen im Boden werden durch chemische und biologische Prozesse verursacht. Obwohl es relativ einfach ist, für jeden Wirkstoff die chemisch und mikrobiologisch beeinflußten Reaktionen aufzuzeigen, ist es schwieriger, deren relative Bedeutung zu erfassen. Vielleicht ist diese Unterscheidung für praktische Bedürfnisse (einschließlich Zulassung) kaum von Bedeutung. Auf welche Weise die Mikroorganismen auf Herbizide einwirken, bleibt weiterhin zu untersuchen.

Für Studien der Populationsdynamik, Wechselwirkungen zwischen Organismen und metabolischen Prozesse ist ein breites Arbeitsfeld gegeben, mit allen bereits genannten methodischen Schwierigkeiten.

Ein anderer interessanter Aspekt des Abbauverhaltens betrifft die Abbaurate. Vom Standpunkt der Umweltbelastung verursachen die gängigen Herbizide und ihre Metaboliten offensichtlich keine Probleme durch zu lange Persistenz. Gerade der Einfluß von Temperatur und Bodenfeuchte ist recht gut untersucht, aber der Einfluß verschiedener Böden läßt sich nicht generell voraussagen, so daß Vervollkommnungen, die in diese Richtung gehen, die Anzahl der Untersuchungen, die die Persistenz einer Substanz charakterisieren, einschränken können.

Obwohl wir der Meinung sind, daß relativ einfache Informationen ausreichen, um Umweltrisiken abzuschätzen, die aus dem Gebrauch einer neuen Substanz entstehen, kann man von einer neuen Verbindung nicht von vornherein sagen, welche und wieviele Daten ausreichen, um sie in der landwirtschaftlichen Praxis zuverlässig einsetzen zu können. Dies liegt darin, daß nicht nur die Kulturpflanzen, sondern auch die Anbauflächen sehr unterschiedlich sind. Deshalb muß man die Wechselwirkungen detailliert analysieren, wobei der Mangel an Daten hierbei die größten Schwierigkeiten verursacht.

Die Reaktion einer Pflanze auf ein Herbizid im Boden wird durch eine große Zahl von Faktoren wie Entwicklungsstadium, klimatische Bedingungen, Bodenfeuchte, Herbizidkonzentration, Position des Herbizids und Dauer der Einwirkung beeinflußt. Wir kennen die quantitativen Beziehungen zwischen diesen Faktoren nicht genau, nicht einmal für ein Herbizid mit einer Pflanzenart. Obwohl es in der Theorie möglich sein sollte, die Aktivität einer Substanz in einem bestimmten Boden auf der Grundlage des Adsorptionsverhaltens, der Mobilität, der Persistenz und ihrer physiologischen Selektivität zu beurteilen, können wir das nicht, weil die Phytotoxizität ein globaler Ausdruck der zeitabhängigen Herbizidkonzentration und -aktivität in einer Pflanzenart ist.

Diese Probleme durch sorgfältige Studien unter kontrollierten Bedingungen, in denen jeder Faktor variiert wird, zu entwirren liegt außerhalb der gegenwärtigen und erreichbaren Forschungskapazitäten. Dies, zusammengenommen mit der Tatsache, daß unter Feldbedingungen einige der Faktoren sowieso nicht kontrollierbar sind, führt zu dem Schluß, daß Bodenherbizide immer einen gewissen Grad an Unzulänglichkeit haben, womit wir uns abfinden müssen. Trotzdem lässt sich hier noch eine Reihe nützlicher und lohnender Forschungsarbeiten durchführen.

Die meisten Wirkstoffe können beinahe durch alle Pflanzenteile aufgenommen werden. Die Aufnahme durch die Wurzel ist oft leichter oder schneller als durch andere Organe, aber daraus läßt sich nicht ableiten, daß die Wurzel notwendigerweise der Hauptaufnahmeort unter Praxisbedingungen ist, da das Herbizid oft nicht in die Wurzelzone gelangt. Bessere Kenntnisse über die Beziehungen zwischen Aufnahmeort, Konzentration und Wirkung sind die Voraussetzung für eine erfolgreiche Applikationstechnik. Das Verständnis der symplasmatischen Translokation und der Rolle der Phytohormone bei Aufnahme und Transport von Herbiziden scheint auf den ersten Blick lediglich von akademischen Interesse zu sein, könnte aber zu bedeutenden Fortschritten in der Anwendung von Herbiziden und anderen Pflanzenbehandlungsmitteln führen.

Unsicherheit über die für eine Wirkung tatsächlich notwendige Herbizidmenge, die die zu bekämpfende Pflanze erreichen muß, führt wiederum zu Unsicherheiten der Interpretation anderer Beobachtungen, besonders der Mobilität und Persistenz. Dies ist mit mangelhafter Kenntnis der Herbizid-Konzentrationen in Bodenlösungen im Freiland eng verbunden, was wiederum darüber Unsicherheit erzeugt, ob die Sorption per se der limitierende Faktor der Herbizidaktivität ist.

Gehen wir davon aus, daß es gelingt, in Zukunft mehr Kenntnisse über den Ort und Zeitpunkt der Wirkung von Herbiziden auf Unkräuter und in Kulturpflanzen zu sammeln, werden darüber noch subtilere Daten benötigt, um Prognose-Modelle spezifischer anwenden zu können. Hier müssen wieder die unkontrollierbaren klimatischen Bedingungen als begrenzender Faktor genannt werden. Es kann ein realistisches Ziel sein, Modelle zu entwickeln, um das wahrscheinliche Zusammenwirken bestimmter Faktoren-Komplexe unter richtiger Einschätzung des Bereichs möglicher Ergebnisse zu bestimmen. Bei der Größe der Aufgabe kann das nur durch die gemeinsamen Anstrengungen vieler Forscher erreicht werden. Umfangreiche gemeinschaftliche Programme lassen sich nur schwer durchführen, aber es wäre von großem Vorteil, wenn in den zahlreichen

Berichten von Freilandversuchen immer die gleichen grundlegenden Klima- und Bodendaten enthalten wären, auch wenn einige davon für den jeweiligen Versuch irrelevant sind. Eine Liste von Vorschlägen für diese Messungen ist als Anhang zu diesem zusammenfassenden Kapitel gegeben. Die in der Liste aufgeführten Parameter betreffen die allgemeinen Faktoren, die das Verhalten von Wirkstoffen im Boden beeinflussen.

Betrachtet man dazu die von Marbut (1951) für die Klassifizierung von Böden gebrauchten Kriterien

1. Geologische Formation
2. Pedogenese
3. Farbe
4. Topographische Lage
5. Wasserhaushalt
6. Bodenprofil und
7. Chemische Zusammensetzung

erkennt man, daß sich unsere Parameter nur auf die Punkte 5 und 7 beziehen. Die Bodenstruktur wird gleichermaßen weder von Marbut noch von unserer Liste erfaßt.

Batey (1975) beschreibt 4 Methoden zur visuellen Beurteilung der Struktur; Methoden sind also vorhanden, die auch für unsere weiteren Untersuchungen gelten können. Daneben hat Eagle (1976) eine Einstufung der Bodenstruktur aufgrund der Fingerprobe vorgenommen, nach der erfolgversprechend Herbiziddosierungen in verschiedenen Böden gewählt werden können. Zusammen mit dem Überblick über Klassifikationssysteme für Böden von Bartelli (1978) zeigen uns diese Beispiele, daß es viele Möglichkeiten gibt, wichtige Bodeneigenschaften zu beschreiben.

In der Einleitung dieses Buches wurde die Frage gestellt, ob wir heute in der Lage sind, Bodenherbizide effizienter einzusetzen als 1960. Die Antwort ist ein klares "Ja". Ohne Zweifel sind wir jetzt besser ausgestattet, die Folgen, die bei der Einführung eines neuen Fremdstoffes in die Umwelt entstehen, vorauszusagen. Die Wahrscheinlichkeit, daß gravierende Nebenwirkungen auftreten ist jetzt gering, und man kann sagen, daß der Preis für die Sicherheit angesichts der Restriktionen und Auflagen bei neuen Zulassungen hoch ist. Es ist nicht einfach, aber auch nicht unmöglich, darzustellen, wie die Anwendung von Bodenherbiziden in der praktischen Landwirtschaft verbessert wurde. Die Herbizid-Dosierungen können jetzt je nach den Bodeneigenschaften gewählt werden, obwohl sie großenteils auf praktischen Erfahrungen beruhen. Rückstandsprobleme können mit höherer Sicherheit vorausgesagt und erklärt werden.

Abschließend läßt sich festellen, daß die Grundlagen zur Erforschung des Problemkreises sehr viel besser erkannt sind und es nun an der Zeit ist, ausreichend Daten zu erarbeiten, um für die Belange der Praxis den größtmöglichen Nutzen zu ziehen.

## Literatur

Bartelli, L. J. (1978). Technical classification system for soil survey interpretation. *Advances in Agronomy*, **30**, 247–289.

Batey, T. (1975). Soil examination in the field. In *Soil Physical Conditions and Crop Production*. Ministry of Agriculture, Fisheries & Food Tech. Bull., **29**. Her Majesty's Stationery Office, pp. 207–217.

Eagle, D. J. (1976). Soil texture classification for the adjustment of herbicide dose. "Proc. 1976 British Crop Protection Conference—Weeds." pp. 981–988.

Marbut, C. F. (1951). "Soils: Their Genesis and Classification." Soil Science Society of America, Madison.

## Anhang

Empfehlungen für Versuchsdaten, die in allen Berichten über Freilandversuche wiedergegeben oder auf Nachfrage mitgeteilt werden sollten:

Korngrößenzusammensetzung
C-Gehalt (organisch)
pH-Wert
Verfügbares N, P und K
maximale Wasserkapazität
Luft-Temperatur (minimal/maximal), täglich
Niederschläge, täglich

(Meteorologische Meßdaten der nächstgelegenen Wetterstation sind gewöhnlich ausreichend.)

In Mobilitätsstudien sollten darüber hinaus folgende Daten erfaßt werden:

Volumengewicht
Adsorptions-Desorptions-Isothermen
Profil der Bodenfeuchtigkeit zur Zeit der Probenahme.

# Index

B

N

## Q



## R

## S